Weldability of ferritic steels

Published by Abington Publishing, Abington Hall, Abington, Cambridge
CB1 6AH, England

Published in North America by ASM International, The Materials Information
Society, Materials Park, Ohio 44073, USA.

First published 1994, Abington Publishing and ASM International

© 1994, Woodhead Publishing Ltd

Conditions of sale
All rights reserved. No part of this publication may be reproduced or transmitted
in any form or by any means, electronic or mechanical, including photocopy,
recording, or any information storage and retrieval system, without permission in
writing from the publisher.

The views expressed in this book do not necessarily represent those of TWI.

British Library Cataloguing in Publication Data
A catalogue record for this book is available from the British Library.

ISBN 1 85573 092 8

Designed by Geoff Green (text) and Chris Feely (jacket).
Typeset by BookEns Ltd, Baldock, Herts.
Printed by Galliard (Printers) Ltd, Gt Yarmouth, England.

Weldability of ferritic steels

NORMAN BAILEY
BMet, CEng, FIM, FWeldI

The Materials Information Society

ABINGTON PUBLISHING
Woodhead Publishing Ltd in association with The Welding Institute
Cambridge England

Contents

	Introduction	*viii*
	Acknowledgements	*xi*
1	**Factors influencing weldability**	**1**
	The welding process	1
	Steel types	3
	Weld cooling cycle	9
	The heat-affected zone	14
	The weld metal	18
	Cutting and gouging	28
	Residual stresses	29
	Post-weld heat treatment	32
	Weldability formulae	35
	References	43
2	**Potential welding problem areas**	**45**
	Cracking	45
	Welding faults	45
	Inspection for defects	46
	Joint integrity	46
	Problems associated with PWHT	46
	Service problems	47
	Repair	47
	Joining dissimilar steels	47
	Joining dissimilar metals	50
	References	53
3	**Solidification cracking**	**54**
	Description	54
	Role of composition	66

	Solidification cracking and the welding process	74
	Control and avoidance of cracking	80
	Liquation cracking	86
	Copper pick-up	91
	Ductility dip cracking	93
	Detection and identification	93
	References	101
4	**Lamellar tearing**	**102**
	Description	102
	Role of inclusions	104
	Susceptible joint types	109
	Other factors	115
	Control and avoidance of cracking	118
	Control tests	119
	Detection and identification	128
	References	132
5	**Hydrogen cracking**	**133**
	Description	133
	Hydrogen	134
	Susceptibility to embrittlement	140
	Stress to give cracking	144
	Temperature of embrittlement	145
	Techniques for avoiding cracking	146
	Weld metal hydrogen cracking	155
	Detection and identification	159
	Other aspects of hydrogen in ferritic steels	162
	References	168
6	**Reheat cracking**	**170**
	Description	170
	Role of composition	172
	Other influences	173
	Control techniques	174
	Detection and identification	178
	References	179
7	**Faults of welding**	**180**
	Faults of shape	180
	Inclusions	183
	Cavities	185
	References	186

8	**Inspection for defects**	**187**
	Inspection techniques	188
	Inspectability	188
	Consequences of metallurgical features	189
	References	190
9	**Joint integrity**	**191**
	Strength and ductility	191
	Toughness	193
	Hardness	204
	Creep resistance	204
	Corrosion resistance	205
	References	209
10	**Service problems**	**210**
	Fatigue and corrosion fatigue	210
	Corrosion	211
	Stress corrosion	211
	Loss of toughness in service	212
	High temperature service problems	218
	Graphitisation	220
	References	222
11	**Repair**	**224**
	Analysis prior to repair	224
	Limitation of preheat	228
	Need for post-weld heat treatment	228
	Special circumstances	230
	References	235
	Further reading	*236*
	Glossary	*238*
	Index	*271*

Introduction

The topic of weldability is fascinating, ranging as it does from the simple ability to tack two pieces of metal together without having them come apart, to meeting the very sophisticated needs of modern specifications for offshore platforms, high temperature components and pressure vessels for nuclear, petrochemical and other applications. An earlier booklet on the topic[1] was produced by Ken Richards at the British Welding Research Association (BWRA) – the precursor of The Welding Institute (TWI) – over 25 years ago. Although TWI has produced several specialised booklets[2-6] in the interim, as well as many papers on most aspects of weldability, no general book has been produced from that source, although the Institute has been a leading organisation in studying the topic in its broadest sense for many years and books on various aspects of weldability have appeared over the years from other sources.[7-16]

The present book is intended to complement the revised version of the hydrogen cracking book,[3] to update and expand the texts on lamellar tearing[2] and solidification cracking,[4] and to include the other topics relevant to the weldability of ferritic steels. It is beyond the author's competence to tackle the problems of the weldability of stainless steels, cast irons[5] and non-ferrous metals, nor do the numerous, and often confusing, methods of testing for weldability[6] fall within the book's scope. The book, being to some extent a reflection of the author's interests while at TWI, is mainly concerned with the conventional fusion welding processes and their problems.

As indicated in the opening paragraph, the definition of weldability can be elastic and, in effect, depends on the application. Weldability is the ability to weld a metal without introducing cracks or other harmful defects and, at the same time, to achieve the properties required by the application. In fact, a onetime colleague has claimed that people are not interested in weldability – only in unweldability!

In considering the weldability of a steel, several types of cracking need to be avoided, several welding processes may be involved, various weldment properties need to be achieved and the restraint and thickness of the metals under consideration may vary considerably. Hence, it is impossible to use a simple number to characterise weldability. As a further complication, the composition of metals supplied to a single specification may vary considerably and at least one type of cracking is influenced mainly by the processing route of the steel.

Although the major problem in welding ferritic steels (and many other metals) is to avoid cracking of one type or another, other types of faults (often described as 'welder-induced' or 'process-induced' as opposed to 'metallurgical') and how they interact with precautions taken to avoid cracking are briefly described. Such faults should only be referred to as defects if they are sufficient to make the weld unacceptable to the standard used for the inspection. Lesser faults may be referred to as 'imperfections', although the term 'discontinuities' is also used.

Unfortunately there is only one term for a crack – and by nearly all inspection standards, a crack is not acceptable, even though, over a quarter of a century ago, techniques of fracture mechanics were developed which allow cracks to be accepted if they have been analysed and shown to be harmless for the intended service conditions. Sometimes the terms 'microcrack' and 'tear' have been used to try to overcome this barrier, but they are not widely accepted and, in fact, 'tearing' should be strictly restricted to ductile fracture by microvoid coalescence, either in service or by such processes as lamellar tearing (Chapter 4).

One aim of this book is to try to unravel the confusion over the different names and types of cracking that are possible when welding steels, and reduce them to a few distinct types, thus making rectification and avoidance of future cracking easier.

The book will be of interest to students studying welding engineering and welding metallurgy, as well as to practical welding engineers, inspectors and metallurgists. In addition to those people concerned with actual fabrication and manufacture, it will also be of importance to those concerned with design and specification. It is always more satisfactory and economical to foresee any problems and deal with them at the design stage, rather than to go through the trauma of detecting and rectifying faults, coping with the resultant delays and disputing the rights and wrongs of the resultant problems. I can recall arguments in the past over whether cracks discovered *after* post-weld heat treatment were hydrogen cracks (and thus the fault of the fabricator) or reheat cracks and therefore – with the steel in question – an 'Act of God' (because reheat cracking was previously unknown with that steel).

References

1. Richards, K.G., *Weldability of Steel*, BWRA, Cambridge, 1967.
2. Farrar, J.C.M. and Dolby, R.E., *Lamellar Tearing in Welded Steel Fabrication*, TWI, Cambridge, 1972.
3. Bailey, N., Coe, F.R., Gooch, T.G., Hart, P.H.M., Jenkins, N. and Pargeter, R.J., *Welding Steels Without Hydrogen Cracking*, Abington Publishing, Cambridge, 1993.
4. Bailey, N. and Jones, S.B., *Solidification Cracking of Ferritic Steels during Submerged-arc Welding*, TWI, Cambridge, 1977.
5. Cottrell, C.L.M., *Welding Cast Irons*, TWI, Cambridge, 1985.
6. Pargeter, R.J. (ed.), *Quantifying Weldability*, Abington Publishing, Cambridge, 1988.
7. Graville, B.A., *The Principles of Cold Cracking Control in Welding*, Dominion Bridge, Montreal, 1975.
8. Easterling, K., *Introduction to the Physical Metallurgy of Welding*, Butterworth, London, 1983.
9. IIW, 'Guide to the welding and weldability of C–Mn and C–Mn microalloyed steels', IIW-382-71, International Institute of Welding, 1971.
10. IIW, 'Guide to the welding and weldability of quenched and tempered C–Mn and C–Mn microalloyed steels', IIW DOC IX-1036-77, International Institute of Welding, 1977.
11. IIW, 'Guide to the welding and weldability of pearlite-reduced and pearlite-free steels', IIW DOC IX-1305-83, International Institute of Welding, 1983.
12. IIW, 'Guide to the welding and weldability of cryogenic steels', IIW/IIS-844-87 TWI for IIW, 1987.
13. Lancaster, J.F., *Metallurgy of Welding*, 4th edn, George Allen and Unwin, London, 1987.
14. Lancaster, J.F., *Welding Structural and Corrosion Resistant Steels*, Abington Publishing, Cambridge, 1992.
15. Stout, R.D., *Weldability of Steels*, WRC, 1987.
16. Granjon, H., *Fundamentals of Welding Metallurgy*, Abington Publishing, Cambridge, 1991.

Acknowledgements

Although middle aged at the time, I came to the old BWRA as a 'new boy' in 1966, and owe a great debt to Bob Baker, Frank Watkinson and Brian Graville, who gave me such a good grounding and set me in pursuit of the elusive goal of weldability. Subsequently other colleagues at TWI, particularly Peter Hart, Richard Dolby, Trevor Gooch, Louise Davis, Frank Coe, Norman Jenkins, Graham Carter, Phil Threadgill, Richard Pargeter and David Abson, have tried to keep me on the straight and narrow, but my biggest debt goes to the technicians who actually carried out the welding and assessment of many experiments. I would particularly like to compliment Peter Kerr and others responsible for the high quality of the micrographs. Thanks are also due to those in industry whose problems have contributed to my understanding of the practical aspects of weldability. It is one thing to study various cracking problems in the laboratory, but it is far harder to restore a cracked component to working order with the minimum cost and delay to a multi-million pound project. Finally I must thank the Awards Committee of the American Welding Society, who boosted my confidence by awarding me their Lincoln Gold Medal in 1973 for reference 16 in Chapter 1. Thanks are due to TWI for permission to use the illustrations in the book.

1 Factors influencing weldability

The welding process

Achieving the required soundness and properties in a fusion weld requires the weld metal to be sound and to have a composition, properties and a microstructure that are adequate for the purpose. The heat-affected zone (HAZ) should also be uncracked and have the required properties; this requirement applies to both the visible HAZ and the region beyond it – the sub-critical HAZ, which can suffer loss of hardness and strength in a very strong quenched and tempered steel. A non-fusion weld obviously cannot contain defects associated with fusion (such as solidification cracks and entrapped slag), but defects may still occur, associated with included oxides at the bond line, incomplete bonding and pores in diffusion bonds, whilst care may be needed to achieve adequate properties in the bond area.

In both fusion and non-fusion welds the need to provide adequate heat for fusion must be balanced against the detrimental effects of heating. Heating a parent steel will inevitably coarsen its prior austenite grain size if the temperature much exceeds about 1100 °C, as is inevitable in the parts of the HAZ of any fusion weld near the fusion boundary. Sufficiently rapid cooling from maximum temperature will give a martensitic microstructure, which is likely to be very hard, with poor toughness and at risk of cracking if the steel has a medium to high carbon content, i.e. above about 0.17%. On the other hand, a very slow cooling rate can give a coarse transformed microstructure largely of bainite which, although softer than martensite, is often less tough.

The cooling rate, which is governed by a number of factors including weld bead size, also governs the weld metal properties of fusion welds. Martensitic and bainitic microstructures occur less often in weld metals than in HAZs; nevertheless a coarse microstructure, resulting from slow cooling, is less tough than a fine one of similar strength and hardness.

Effect of process on weldability

Texts on welding engineering should be consulted for details of the different welding processes. This section gives a brief idea of the effect of changing welding processes on likely metallurgical problems. The processes are listed roughly in the order of steel thickness which can be welded with one pass.

Liquation cracking, lamellar tearing and reheat cracking are little affected by the welding process, except that the consumable should always be chosen to minimise weld strength above that required to match the parent metal yield strength by a small margin. The toughnesses of the HAZ and (to some degree) the weld metal are more affected by heat input than welding process, although some processes can give higher heat inputs than others.

Manual metal arc (MMA) welding gives a relatively high risk of hydrogen cracking, unless basic electrodes are used correctly, i.e. properly stored, dried and issued. Risks of solidification cracking are low, and MMA welding with basic electrodes is a good way of coping with the sulphur in free-cutting steels. Toughness can be good, provided suitable electrodes and heat inputs are used.

Tungsten inert gas (TIG), laser and **plasma welding** all give very low risks of hydrogen cracking and, provided welding speeds are not too high, of solidification cracking. Plasma and TIG welding with filler are capable of giving high toughness, provided multipass welding is used to maximise grain refinement. For autogenous TIG and laser, weld metal toughness depends on the steel being welded, particularly for single pass welding.

CO_2 welding gives a low risk of hydrogen cracking; solidification cracking behaviour depends on welding speed; toughness is moderate, as the weld metal is of too high an oxygen content to give the best upper shelf toughness.

Metal active gas (MAG) welding (solid or cored wires) is similar to CO_2 welding, except that better weld metal toughness is possible with shielding gases of low oxygen and low CO_2 contents; cored wires can vary considerably in their hydrogen potential, giving hydrogen cracking risks ranging from low to moderate.

Self-shielded welding gives moderate resistance to both hydrogen cracking and toughness, provided a multirun technique to maximise weld metal refinement is used when reasonable toughness is required.

Submerged arc welding gives a moderate to slight risk of hydrogen cracking, including the weld metal type, depending particularly on the flux and its condition. Fast welding speeds give a risk of solidification cracking, and poor toughness is possible at high heat inputs and with less basic fluxes.

Electron beam welding gives a very low risk of hydrogen cracking, but some risk of solidification cracking. Toughness depends entirely on the steel being welded, as the process is autogenous and single pass.

Factors influencing weldability

Electrogas, electroslag and **consumable guide welding** give low risks of hydrogen cracking (although fine weld metal cracking is possible). Solidification cracking limits possible welding speeds. Toughness is usually poor in the HAZ, owing to very slow cooling, so that normalising may be required to achieve acceptable values.

Steel types

In describing the weldability of steels, it is essential to have some understanding of the types of steel which may have to be welded. Apart from different standards – international, European, national, industry-wide, military and private – steels are often referred to by their type. Steels can be classified by their compositional type (e.g. C:Mn, Cr:Mo), by their end use (tool steel, rail steel), by their characteristics (high temperature, stainless), by their method of manufacture (semi-killed, calcium-treated), by a descriptive or trade name (maraging, Corten), by their microstructural type (bainitic, martensitic) or by a combination of these (austenitic stainless, rare-earth-treated C:Mn).

High alloy and austenitic stainless steels, together with the cast[1] and wrought irons, are outside the scope of this book.

Although some guidance is given below, the welding engineer must often seek advice from organisations such as libraries, trade associations or TWI if encountering an unfamiliar designation, particularly if it is part of a once well-known, but now obsolete, specification. A typical example of this is the persistence of the old (pre-1970) designations for British engineering steels, e.g. En24, which were part of the old British Standard BS970: 1955 and which were many years ago superseded by a six-digit designation, e.g. 817M40, in the revised version of the same standard: a further complication is that now some of the British BS Standards have been replaced by European EN Standards: En24 becomes 34 CrNiMo 6 in EN 10083-1: 1990. Another common failing is not to be given the strength level of the steel; this is essential in order to be able to select a welding consumable to give the appropriate strength. For example, the 817M40 steel was available in eight strength levels, whose minimum yield strengths varied by a factor of nearly two, from 650 to 1235 N/mm^2. Matching the lower yield strength levels is not difficult; few fillers, if any, will be found to meet the upper strength levels in the as-welded condition.

Compositional types

Carbon steels are alloyed with anything up to about 1% C, and most contain about 0.6% Mn, unless they are of the free-cutting type, when up to about 0.25% S and 1.2% Mn may be present. When carbon steels have carbon levels below about 0.2%, they are referred to as **mild** steels. The carbon steels are by far the more common (and cheap) type, mild steel being the commonest and the easiest to weld. Weldability of carbon and other steel types becomes more difficult as the carbon content is increased.

Alloy steels contain additions which usually improve hardenability and, hence, strength in heavier sections. In addition, alloying confers other advantages in resisting service environments, such as corrosion, oxidation, high pressure hydrogen and high temperature creep. Alloy steels are divided into several types.

Carbon : manganese (C:Mn) steels, contain 1–2% Mn and are so widespread (Mn provides the cheapest method of improving hardenability after carbon) that they may be regarded as forming a major steel type. **Micro-alloyed** steels are usually of the C:Mn type with small additions (generally <0.2%) of elements such as Nb, V, Ti or B, usually to confer the advantages of alloying without seriously impairing weldability; many are a sub-category of the C:Mn type.

Low alloy steels contain a few per cent (typically between 1 and 7%) of elements such as Cr, Ni, Mo and V, with carbon contents within the same range as C steels. This category includes several sub-types, such as C:Mn (which may be micro-alloyed) and which forms the major category of high strength constructional steels, used, *inter alia*, for bridges and offshore structures. Other types are **chromium** steels, which are used for resisting high temperature oxidation, creep and hydrogen and contain up to 5% Cr with usually 0.5 or 1% Mo. Chromium steels continue into the high alloy category and the stainless steels with the 9% Cr and higher Cr steels. Steels of the **Ni:Cr:Mo** type are typical engineering steels, with between about 1 and 7% of these elements in total. With low carbon contents these provide the highest strength, commonly used weldable steels. The **nickel** steels are used at low temperatures; the low alloy types contain 3.5 and 5% Ni whilst 9% Ni steel is in the high alloy classification.

High alloy steels contain larger amounts of alloying elements, and merge into the stainless steels, which are outside the scope of this text. Apart from those previously mentioned, high alloy steels usually contain Cr and Mo with additions of such elements as V and W, and are used for resisting abrasion and for making tools. They usually have relatively high carbon contents and poor weldability. Some specialised, very high strength, low carbon weldable steels, such as the maraging and 9Ni:4Co steels, fall into the high alloy steel category.

Method of manufacture

The steelmaking method is not particularly useful in judging weldability, but if an **electric** steelmaking method is not used (e.g. **open hearth**), it suggests that the steel was made by an older technique and may be of uncertain quality, particularly with regard to non-metallic inclusions (Chapter 4).

Steels made using the older techniques may be of the **rimming, semi-killed** (or **balanced**), **Si-killed** or **Al-treated** types. In the order given, the steels are of decreasing oxygen content, within the approximate range 0.2–0.001% O.

Modern steels are usually electrically melted and continuously cast (**concast**) and for this process need aluminium to be added. Further additions at the molten steel stage include **Ca treatment** and treatment with rare earth metals (**RE-treated**). These treatments are generally known as **inclusion shape control** and are used for steel intended to be rolled to plate, which requires an inclusion population not likely to give rise to lamellar tearing (Chapter 4) when the plate is welded or 'hydrogen-induced cracking' (HIC) when used in water containing hydrogen sulphide (Chapters 9 and 10). Electric melting can be sub-divided into various types, ranging from the normal arc melting to **VIM–VAR** (vacuum induction melting–vacuum arc remelting) and **electroslag melting**, both of which are intended to give very low inclusion levels.

Cast steel may be subsequently processed by **hot rolling** to **plate** or **sections**, it may be **forged**, **extruded** or used as **castings**. Subsequent processing includes **cold rolling** to **sheet** or **strip**, **rod** as well as **wire** drawing, and deep drawing of sheet to forms. Hot rolling may be carried with or without strict temperature control (**controlled rolling** and **TMCP** (thermomechanically controlled processing) fall into this category). The TMCP steel relies on a strict rolling schedule for its mechanical properties and does not require further heat treatment such as is given to most other types of steel.

Descriptive steel types

Weldable steels have low carbon contents, i.e. less than about 0.2% C, generally decreasing as the amount of alloying elements is increased. They range from mild steels to very high strength high alloy steels.

Structural or **constructional steels** are mild steel types; within this category the term 'high strength' denotes the use of C:Mn steels (including microalloyed and weather-resistant steels), rather than any form of alloy steel. Such steels are used for general construction of buildings, bridges, ships, harbour works, etc.

Weather-resistant steels, sometimes known by the trade name Corten, are a version of structural steel containing small additions of elements such as Cr, Cu and P to improve corrosion resistance in normal atmospheric conditions (i.e. not so severe as marine atmospheres containing salt spray). Weldability is similar to the C:Mn steels.

Engineering steels, frequently in bar form, may be of any type from mild steel (e.g. bright mild steel) to high alloy types with high carbon contents and poor weldability. They are mostly covered in the UK by the steels in BS970 (including the old 'En steels') and its successor EN 10083-1 and –2: 1991 and in the USA by the AISI/SAE steels. They may be also of the free-cutting type. Engineering steels are usually quenched and tempered (QT or Q&T; the terms oil hardening and air hardening refer to the quenching medium required to achieve the required hardness) so that care is needed in selecting post-weld heat treatment (PWHT) temperatures to avoid over-tempering and softening the parent steel. In some cases preheat temperatures and heat inputs may need to be controlled to avoid softened HAZs.

Free-cutting (or free-machining) **steels** contain up to about 0.25% S with about 1.2% Mn, possibly with additional elements such as Pb or Se, and range from mild steels to the low and high alloy engineering steels. Such steels require special precautions during welding to minimise the risks of solidification and liquation cracking (see Chapter 3). For this reason it is important to check whether a steel is of the free-machining type, particularly when repairing engineering and similar components.

High strength steels in any of the above categories may range from structural C:Mn steels of 350 N/mm^2 yield strength to maraging steels with a yield or proof strength approaching 2000 N/mm^2. The Q&T weldable high strength steels are an important category, because of their relatively good weldability in relation to their strength. Such steels include the so-called HY steels (the number following 'HY' denotes the minimum yield strength in ksi*), and a range of trade names, including the British RQT steels, the American T1 series, the OX series, the AX steels and many others. Frequently, precautions are required to minimise loss of HAZ and weld metal properties by restricting preheat and interpass temperatures and heat inputs and guidance should be sought from the steelmaker for recommended welding parameters.

An important sub-category of the high strength steels is the high strength, low alloy (HSLA) types, which typically have minimum yield strengths of 550 or 690 N/mm^2. In fact, high strength low *carbon* would be a better description of the type, as several of these steels contain less than 0.05% C; i.e. often at a lower level than those of the weld metals needed to achieve matching strength. Such steels are very weldable as regards HAZ properties – the danger is that the *weld metals* will suffer hydrogen cracking if this risk is not taken into account when devising welding procedures.

Heat and **creep-resisting** steels are of the Cr:Mo type, generally with low carbon contents (<0.20% C), although some authorities restrict the term to high alloy steels with 9% or more Cr. Chromium additions confer good resistance to high temperature oxidation and creep; these steels also have a good resistance to hydrogen attack at high temperatures and pressures. The 0.5 Cr:0.5 Mo:0.25 V steel, the 0.5% Mo and several other steels also

---

* 1 ksi = 6.9 N/mm^2

fall into this category. These steels require care when being welded, because of their high hardenability.

Tool steels are usually of high carbon, low to high alloy content and are quenched and tempered. Weldability is likely to be poor in terms of likelihood of cracking during welding; precautions may also be needed to avoid general or local softening.

Rail steels are of two types. For normal use a carbon steel containing about 0.6% C is used, although a version of this containing about 1% Cr is also available, as well as a low carbon bainitic steel. For very heavy wear resistance, a 12% Mn austenitic steel (**Hadfield's Mn steel**) is used. The two types have such different weldabilities that joining them by normal fusion welding techniques is virtually impossible.

Armour steels are of the low to medium alloy type, generally with about 0.3% C. For thin-sectioned armour, low alloy–microalloy types have been developed. Weldability is usually not good, and control of welding procedures may be needed to minimise softening of HAZs.

Deep drawing steels are sheet steels, usually of very low carbon content and not alloyed. Weldability is good, although the HAZ may be softened if the strength of the steel has been enhanced by heavy cold working, particularly at bends of sharp radius.

Surface hardening steels are steels, usually of the low alloy type, with a fairly wide possible range of carbon contents, which are suitable for surface hardening by different techniques. Steels which are surface hardened by induction or laser hardening are not of any specific type; the surface hardness will depend on the carbon content of the steel. For carburising or nitriding, special compositions are normally used, most of which have fairly low carbon contents and are reasonably weldable, although consideration has to be given to the influence of the hardened layer. In fusion welding, this has a negligible effect on the composition and properties of the weld metal, but it will lead to a HAZ which is harder at the surface than expected.

Microstructural types of steel have a name indicative of their microstructure, which does not normally give much indication of their likely weldability. Such steels may be of the bainitic, martensitic or acicular ferrite types. One special version is the maraging steel category of very low carbon, high alloy type. These steels are of very high strength, relatively good weldability and derive their strength from an age hardening treatment within a martensitic microstructure produced by air hardening.

Heat treatment

The major heat treatments applied to ferritic steels are as follows. **Homogenising** is normally carried out above about 1000 °C to homogenise the as-cast microstructure, which is not uniform because of the segregation which took place during solidification. It gives a coarse prior austenite grain size, leading to poor toughness and is, therefore, followed by one of the high temperature austenitising treatments given below.

Normalising involves heating the steel within the austenite range between about 800 and 1000 °C and cooling in still air. The austenitising temperature should be controlled to give a fine grain size and reasonable toughness. Some rolling mills incorporate a schedule which allows the steel to be described as **normalised rolled**; unless the steel composition contains an adequate Ti content, normalised rolled steels may be susceptible to strain ageing.

Annealing is similar to normalising, except that cooling is in the furnace, so that the steel is as soft as possible.

Quenching and tempering is similar to normalising, except that the cooling is speeded up to ensure that the steel is hardened through the full section thickness. Quenching may be into water (cold or hot liquid or spray) or oil; however, air cooling may be adequate for steels described as air-hardening. The steel is then tempered (softened) by reheating within the temperature range 100–750 °C; the upper part of the range may be the same as for the post-weld heat treatment (PWHT) of welds. When the quenching is carried out directly at the rolling mill, a more efficient quench is possible and the steel is referred to as **roller quenched**; such steels should not be reheated within the austenitising range (e.g. for hot forming), as their properties cannot be recovered.

Steel types and welding problems

Lamellar tearing is possible with all types of steel plate (except those made with inclusion shape control), depending on inclusion content and joint type. For the same inclusion content, risks are greater for steels of higher strength. The actual risk of welding problems generally depends very much on the actual composition, as most specifications are so wide that a wide range of weldability behaviour is sometimes possible for compositions provided to the same specification.

Mild steels give few problems; hydrogen cracking is unlikely with thicknesses below about 50 mm, unless restraint is very high; solidification cracking is possible at high welding speeds or with high dilution.

Carbon:manganese steels are more prone to hydrogen cracking, but are otherwise similar to mild steels.

Microalloyed weldable steels exhibit a resistance to hydrogen cracking depending on carbon content; heat input may need to be restricted to maintain toughness; some compositions containing boron and molybdenum may be prone to reheat cracking.

Quenched and tempered weldable steels are similar to the previous group but with a greater risk of cracking due to their higher strength; with low carbon contents hydrogen cracking in the weld metal may be a problem.

Low alloy weldable (pressure vessel) **steels** will need preheating to avoid hydrogen cracking, and may be prone to reheat cracking if they contain vanadium with molybdenum.

Engineering steels (medium to high carbon, with or without alloying elements) have a high risk of hydrogen cracking; liquation cracking is likely, as is solidification cracking with automatic welding at relatively high speeds.

Tool steels have an extremely high risk of hydrogen cracking, which needs an understanding of the steel's metallurgy to overcome.

Free-cutting steels will contain liquation cracks and pose a high risk of solidification cracking unless low dilution welding procedures are used.

Weld cooling cycle

Heat input

The microstructures of the weld and the HAZ, as for any steel, are controlled by their compositions and heat treatment. In an as-deposited weld, the heat treatment is that given to each weld run by the heating and cooling applied when that run was deposited and modified by heating from any subsequent weld runs. The effects of this heat treatment may be subsequently modified by any PWHT given to the weld, as is discussed later in this chapter.

The main influences on the cooling of the weld are the heating produced by the welding process, the thickness of the steel being welded and the temperature of the steel as each weld run is deposited; ambient conditions usually have only a minor influence.

The heating produced by welding is normally described in terms of the heat input per unit length of weld run, usually in units of kJ/mm if the welding speed is measured in mm/min. However, kJ/cm or kJ/in are also in use if welding speeds are in cm/min or in/min. Heat input is derived from arc energy, which is calculated (in the same units as heat input) from the welding current, arc voltage and welding (travel) speed as follows:

$$\text{arc energy} = \frac{\text{arc volts} \times \text{welding current} \times 60}{1000 \times \text{welding speed (mm/min)}} \quad [1.1]$$

The value '60' is required because welding speed is conventionally measured in mm/min, whereas the welding power, amps × volts, is in watts/sec. The value of 1000 is to convert joules to kilojoules. Should multiarc welding be used to give a single weld pool, the individual arc energies for each arc are totalled. Where alternating current (a.c.) welding is used, the values should preferably be the root mean square values.

Because of inevitable heat losses, mainly due to radiation, in transferring

Table 1.1

Process	Relative arc efficiency
MMA	1.0
MAG	1.0
Cored wire	1.0
TIG	0.8
Submerged arc	1.25

the heat to the steel being welded, the value given by Eq [1.1] is factored by the arc efficiency to obtain the heat input to the steel being welded. However, in practice, measurements of arc efficiency are rarely made and the term 'arc energy' is not commonly used. Currently, the term 'heat input' is being used in current European standards concerned with welding and national standards in Europe will presumably follow, although it is not always clear whether arc energy or heat input is intended. When using controlled heat inputs, care should always be taken to deduce from the standard or specification being used whether the value given is unfactored, i.e. arc energy, or factored, i.e. heat input.

The arc efficiency of the different welding processes in transferring heat to the steel being welded varies; the values in Table 1.1, taken from the current British Standard,[2] give an approximate idea of arc efficiencies of other processes in comparison with MMA welding.

Steel thickness

For the HAZ, welding is a heat treatment process, but the effect of steel thickness on the cooling of the steel is the direct opposite of its effect during conventional heat treatment, where an increase in section thickness reduces the cooling rate. When welding, the cooling of the steel is not externally imposed, but results from cooling by conduction from nearby steel which has not been heated by welding. Hence, the rate of cooling of a weld made at a constant heat input is *increased* when the thickness being welded is increased. For single- and two-pass welds, the heat input needed to complete the joint is roughly proportional to the thickness being welded, so that the cooling of such welds is roughly independent of thickness.

Although the cooling rate can be calculated from a knowledge of the relevant welding parameters, the calculation is complicated by the onset

Factors influencing weldability

of three-dimensional cooling, which usually operates at thicknesses greater than a value between 20 and 50 mm. The actual value depends on a number of factors; the thicknesses given above apply to cooling through 300 °C, at which temperature transformation from austenite is usually complete. At very small thicknesses the cooling is controlled by steel in the same plane as the weld, i.e. in the x and y directions. In very thick steel, cooling occurs equally from above (and below) the plane of the weld run (the z direction). Hence, for constant heat input and preheat, the cooling rate at a given temperature increases as steel thickness is increased up to a maximum value, above which three-dimensional cooling operates and no further change occurs.

Only the weld cooling **rate** has been mentioned so far, as it was used for many years in the early days of welding. Current practice is to use the cooling **time** between two given temperatures, as it is a value which can be

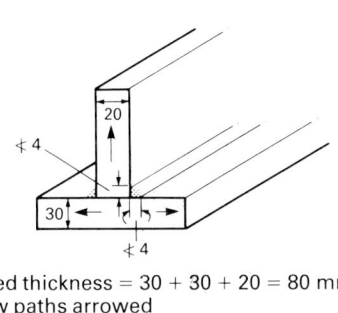

Combined thickness = 30 + 30 + 20 = 80 mm
Heat flow paths arrowed

(a)

Combined thickness = $t_1 + t_2 + t_3$

t_1 = average thickness over a length of 75 mm

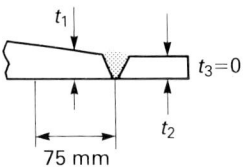

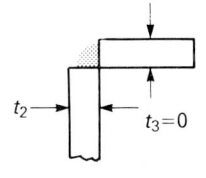

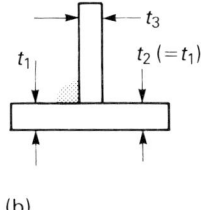

For directly opposed twin fillet welds combined thickness = $\frac{1}{2}(t_1 + t_2 + t_3)$

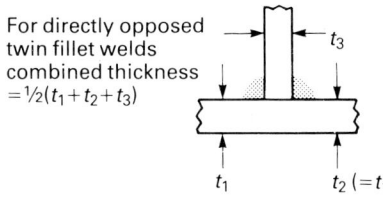

(b)

1.1 Heat flow paths and combined thickness: (a) typical example; (b) treatment of different joint configurations.

directly measured. The most common temperatures are from 800 to 500 °C, which is the temperature range within which most C and C:Mn steel HAZs transform on cooling from austenite. This cooling time, $T_{800-500}$ or $T_{8/5}$, is most commonly used when the transformation of the steel, particularly in the weld HAZ, is under consideration. For lower temperature cooling, i.e., when diffusion of hydrogen out of the welded area is the important factor, $T_{3/1}$ is sometimes used.

The cooling of a welded joint depends on the number of heat flow paths available. A fillet weld may give two or three paths, depending on its configuration, whereas a simple butt weld has two paths (Fig. 1.1). When the pieces being joined are of different or varying thickness, the thicknesses of those parts which provide the heat flow paths at the joint are added together to obtain the **'combined thickness'** value, as illustrated in Fig. 1.1. Varying thicknesses are conventionally averaged over a distance of 75 mm from the weld line. This convention can also be used if one of the pieces is shorter than 75 mm. For example, if a flange 15 mm thick and 50 mm wide is being joined along its length to a web (Fig. 1.2), its contribution to the combined thickness is $2 \times (50/75 \times 15) = 20$ mm.

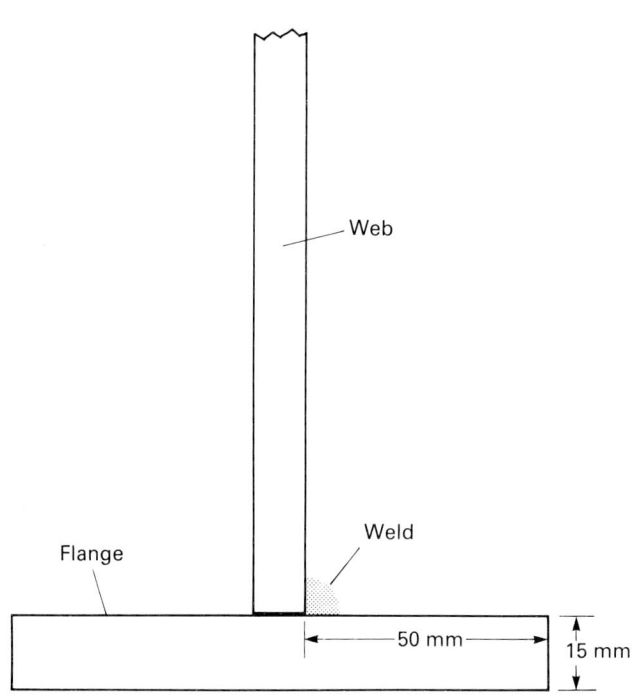

1.2 Combined thickness for flange of limited width.

Preheat and interpass temperature

The temperature of steel immediately a weld run is started is the third major factor affecting weld cooling; for the first run this is the preheat temperature, for subsequent runs the interpass temperature. To avoid hydrogen cracking (Chapter 5), steels are often preheated to temperatures up to 250 °C before welding. To control weldment properties (Chapter 9), maximum interpass temperatures (usually up to 300 °C) may also be specified. These temperatures should be measured close (~50 mm) to the weld line immediately before a weld run is deposited. If the preheating is applied from one side only, e.g. by a gas flame, temperatures should preferably be measured on the side that is not heated. If this is not possible, the flame should be removed shortly before the temperature is measured; a waiting time of 1 min/25 mm of individual steel thickness is recommended.[2]

Preheat and interpass temperatures within the ranges given above have less effect on the cooling rates at high temperatures than at low. Consequently, preheating a simple C or C:Mn steel, which transforms at a relatively high temperature, will not have much effect on the cooling rate through the transformation range, and hence on the resultant microstructure and hardness. However, it will considerably slow up the cooling at lower temperatures when hydrogen diffusion from the steel will be a major consideration in avoiding hydrogen cracking.

On the other hand, low alloy steels, transforming at lower temperatures, can undergo significant changes to their HAZ microstructures and properties (not always favourably) by preheating to a higher temperature. Adverse effects on HAZ microstructure and properties may require maximum controls on preheat and interpass temperatures; the latter may lead to slower joint completion rates than are economically desirable and can give problems, for example when automatically welding circumferential seams in small diameter, thick-walled pipes.

External influences

Although the cooling of a joint by cold metal in the vicinity of a joint, but beyond the arbitrary limit of 75 mm given in the previous section, will not affect HAZ and weld metal transformation behaviour and properties, it may well hasten cooling at low temperatures, at which hydrogen diffusion is important. Although there is, as yet, no formalised way of taking this into account, care is needed in such situations. This effect may be enhanced where water is lying on the steel (or the steel is immersed in water) a few centimetres away from the weld line, even though the site of the weld itself has been adequately dried off.

For completeness, two further cases should be borne in mind. One is when welding on to pipelines containing liquids or gas under pressure (a practice known as **hot tapping**, Chapter 11). Obviously, special precautions are needed to avoid the risk of burning through, and guidance is available in this area.

The other example is in the specialised area of underwater welding (Chapter 11). Welding in water (wet welding) gives extremely fast cooling, and special precautions are needed to avoid cracking. Underwater welding in a dry atmosphere (hyperbaric welding) also gives an increase in weld cooling because of increased thermal conductivity of the (compressed) hyperbaric atmosphere. The resultant increase in cooling rate is much less than with wet welding, but should be taken into account, particularly when welding thin-sectioned steel.

The heat-affected zone

Description

The HAZ is the crucial area in welding because, once the steel has been selected, the HAZ and its properties have to be accepted, whereas a weld metal can be changed if necessary. In welding steels, the HAZ is important as an area where cracking can occur as well as a region whose properties can be reduced by welding. The parent steel may have undergone a complex cycle of thermomechanical processing and heat treatments to achieve its required properties. The HAZ must retain sufficient of those properties after a rapid heating to temperatures up to its melting temperature followed by fairly rapid cooling and, if the weld is multipass, by a series of reheatings to successively lower temperatures. Despite these reheatings, some of the HAZ will be in the as-welded condition (i.e. not reheated by more than a few hundred degrees Celsius), unless a PWHT is given.

Although it is possible to heat treat welds, the usual steel heat treatment cycle of austenitising, cooling and tempering is generally impracticable. Distortion would be difficult and expensive to avoid and the weld metal would need to match the parent steel in composition (particularly in carbon content) as well as in properties, and this will normally increase the risk of cracking to an unacceptable degree. Post-weld heat treatment is the usual option which, although expensive, is often required but can still give problems, as discussed in a later section of this chapter.

The HAZ can be divided into several regions, as shown in Fig. 1.3. Slight compositional changes at the fusion boundary itself, i.e. at the boundary between the molten weld metal and the unmelted parent steel, are described towards the end of the section on the weld metal. These normally result in the HAZ within one grain diameter or so of the boundary being very slightly leaner in composition and softer than the grains slightly further from the boundary. However, if the weld metal is richer in composition than the parent steel (as is likely if stainless steel fillers are

Factors influencing weldability

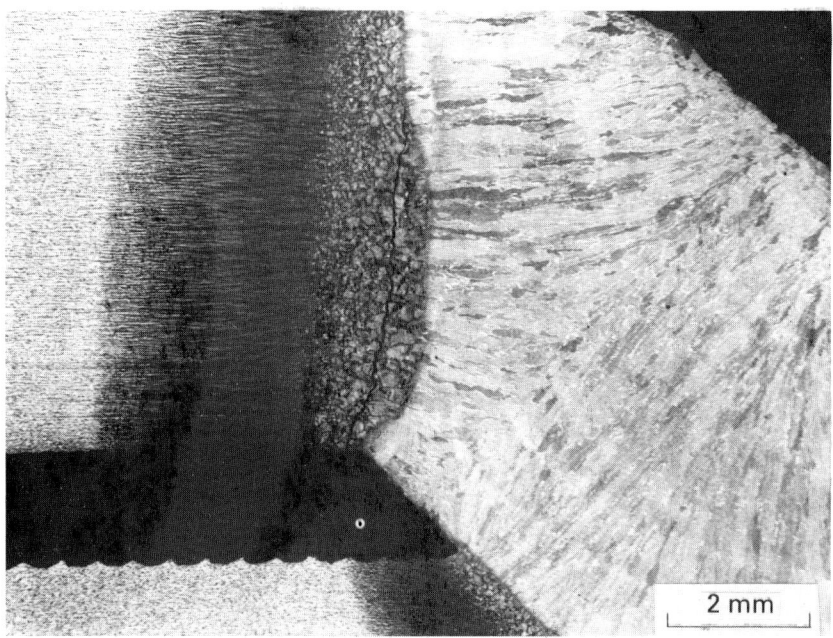

1.3 Macrosection of fillet weld showing coarse-grained HAZ (with hydrogen cracking), dark-etching fine-grained HAZ and lighter-etching intercritical region.

used), the reverse is likely to be the case. Near the fusion boundary the HAZ has been heated momentarily to its melting temperature, i.e. to about 1500 °C. This heating transforms the microstructure to the high temperature form, austenite, and coarsens its grain size to 100 μm or more, compared with 10–30 μm in a fine grained steel. On cooling, the austenite transforms to martensite, bainite, ferrite + pearlite, or mixtures of these constituents, depending on the cooling rate (Fig. 1.4). Unless heat inputs are unusually high (Fig. 1.4(c)), cooling is too rapid for pearlite to be formed, even in mild steel HAZs.

Because of the coarse prior austenite grain size, this region of the HAZ is more hardenable than would be expected from a conventional continuous cooling transformation (CCT) diagram of the steel, i.e. hard martensite forms more readily and soft ferrite less readily. Also the coarse prior austenite grain size results in coarse transformation products, which are inherently less tough than fine-grained ones, so that problems of poor HAZ toughness, and an increased liability to cracking, are readily understandable.

Further from the fusion boundary, the HAZ is heated to lower temperatures (1100 down to about 700 °C) and is therefore less coarse, so that it is less hardenable and the transformation products usually tougher. The

16 Weldability of ferritic steels

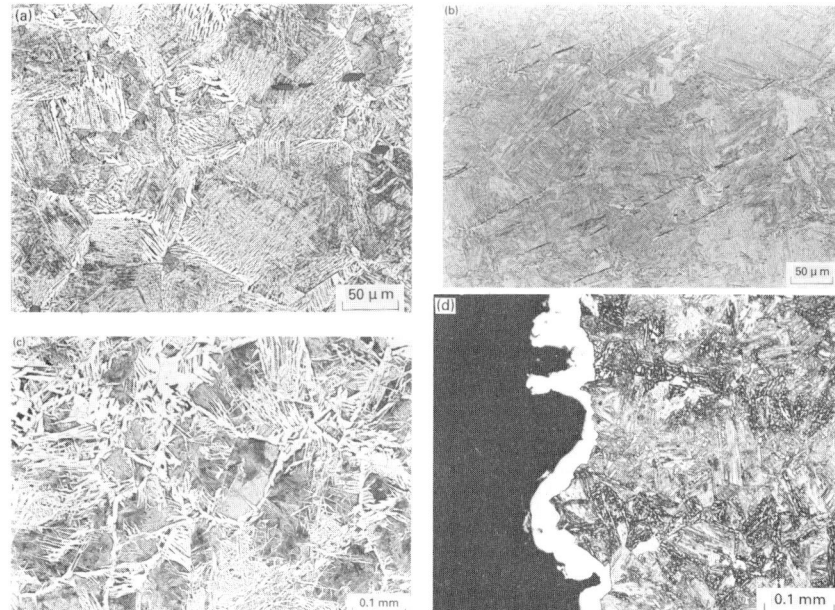

1.4 Typical HAZ microstructures: (a) C:Mn steel 25 mm thick welded at 2 kJ/mm showing mainly fine and coarse ferrite with aligned second phase and smaller amounts of interphase carbide, martensite and primary ferrite; (b) C:Mn:Nb steel 25 mm thick welded at 1 kJ/mm, predominantly martensite; (c) mild steel 25 mm thick welded at 7 kJ/mm, grain boundary ferrite, coarse ferrite with aligned second phase and finer ferrite with interphase carbide and some pearlite; (d) section through (nickel-plated) fracture surface in a low alloy Q&T steel HAZ showing dimpled ductile fracture through the lighter martensite regions and straighter cleavage fracture through the darker bainitic regions.

fine-grained HAZ, which is readily distinguishable on an etched macrosection of a weld (Fig. 1.3), is less of a problem area than the coarse HAZ. In fact, techniques have been developed to ensure that in multipass welds, all HAZ regions except those of the final run(s) are fine grained. These techniques are discussed in Chapters 6, 9 and 11.

The final region of the HAZ visible in an etched weld cross section is the intercritically heated region. Because the first two regions are heated above the transformation temperature of the steel, they are said to have been 'critically' heated. Outside the visible HAZ, none of the steel has been heated into the austenite region. In between these regions steel has been heated to a temperature where it is partly transformed to austenite (i.e. intercritically heated). In the intercritical region, microscopically small regions of higher carbon content transform on heating to austenite and on cooling transform to martensite or, less frequently, to pearlite.

Outside the visible HAZ, steel is heated to temperatures below about 700 °C. The upper end of this range is sufficient to give tempering if the steel is martensitic, but otherwise there is little obvious change in microstructure or hardness. However, in certain steels – those that are not fully killed with aluminium – heating to around 250 °C can give strain-ageing embrittlement, which can be very harmful if the tip of a sharp planar defect lies in or close to such a region. The steels in question, of the rimming, semi-killed, balanced and silicon-killed types, are not frequently made these days, because the current technique of continuous casting requires an aluminium addition to the steel. However, older fabrications may need welding for repair or modification, and weld metals (which do not normally contain sufficient aluminium to prevent strain ageing) are also prone to this problem.

A recent finding is that, to avoid the problem of strain-ageing, sufficient time is required at, and during cooling from, the normalising temperature to allow aluminium to combine with nitrogen (one of the elements responsible for strain-ageing). With a normalising heat treatment this condition is met, but if the steel is normalised rolled, and does not also contain an appropriate addition of the element titanium, strain-ageing is still possible.

Reheating

The above description of a HAZ is essentially that of a single-run weld. Reheating by subsequent passes modifies this description, because each HAZ overlaps that of the previous weld run and reheats it. Consequently some of the earlier HAZ is fully heated to give a new coarse-grained region, some is refined, some is reheated intercritically and some sub-critically. The refinement is normally beneficial and can be used to advantage. Intercritical reheating, particularly of original coarse HAZ regions, can, however, be harmful. This is because intercritical reheating does not give

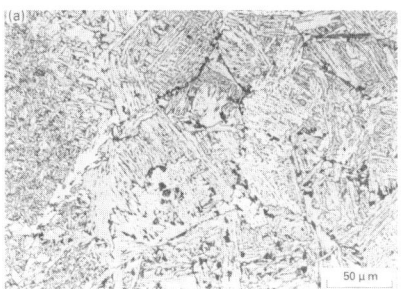
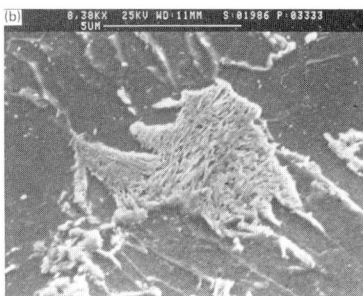

1.5 Microstructure of intercritically reheated coarse-grained HAZ in a weldable C:Mn:Nb steel: (a) showing small dark, re-transformed particles mainly at prior austenite boundaries; (b) detail of dark particle in SEM with a very fine pearlitic structure.

any refinement, so that the high carbon martensitic or pearlitic regions (Fig. 1.5) which result from the intercritical heating are larger and occur in coarser-grained microstructures than the intercritical region of a refined HAZ.

Problems of poor HAZ toughness encountered in this type of region are one cause of the so-called 'local brittle zones' (LBZs). With certain types of steel, sub-critical reheating can also give hard regions (and probably LBZs) by secondary hardening or precipitation hardening processes.

The weld metal

Composition

The composition of a weld metal depends not only on the composition of the filler used (if any) but also on the dilution (i.e. the incorporation of base metal into the weld pool) and the influence of the flux and/or shielding gas used. Unless the weldment is to be heat treated within the normalising temperature range, it is usual to aim for as low a carbon content as is compatible with the properties required. The reasons for this will become more apparent in later chapters, but they are essentially to reduce the risks of cracking and to improve toughness. Rimming steels of low carbon content were originally used for most welding wires and even today they are preferred for many applications; their relatively high oxygen contents are not a disadvantage and the normally low levels of impurities (particularly nitrogen) are advantageous. In fact, apart from oxygen and hydrogen, any impurity will strengthen the weld metal (excessively in many cases) and will often increase the risk of cracking and/or reduce toughness. The most harmful impurity is hydrogen, which is fully discussed in Chapter 5, as well as being mentioned in several others.

Apart from low carbon and low impurity levels, most welding consumables contain manganese and silicon as deoxidants. Manganese is also important in reducing solidification cracking (Chapter 3) as well as in conferring strength and toughness in the weld metal, with additions usually below 2%. Further strengthening additions are molybdenum and nickel, which can also improve toughness; chromium is added to confer resistance to high temperature oxidation and creep, as well as resistance to attack by hydrogen at high temperature and pressure (Chapter 10). In some cases boron is added to improve toughness at the same time as avoiding excessive hardness.

When no filler is added (as in autogenous welding) the parent steel provides all the weld metal and, with the usual autogenous welding processes (TIG, plasma, electron beam and laser), the composition is scarcely changed by the welding process. If two different steels are being joined, the resultant composition will depend on the proportions of each that are melted.

Even with added filler, a proportion of parent metal is always incorporated into the weld pool. This varies considerably with the welding process and the actual welding parameters. Manual metal arc welding will usually give a dilution of about 30% of parent steel into the weld pool; submerged arc welding may incorporate as much as 75%. With the TIG process, dilution can vary from about 20% to as much as 100% (i.e. if no filler is used). In multipass welding, some of the material diluted into the second and subsequent runs is weld metal deposited earlier. It is very difficult in a fusion welding process to maintain low dilution levels, even when this is required, as in surfacing with a material whose properties are impaired by iron, because of the difficulty of avoiding defects of welding. Non-fusion processes, such as friction welding, diffusion bonding, and also brazing and soldering, give no dilution.

The amount of dilution can be altered by varying the welding parameters. Welding current density has the greatest influence; increasing current or decreasing wire diameter increases both penetration and dilution. A large increase in the preheat or interpass temperature, as well as an increase in hydrogen input, will also increase dilution. In the case of submerged arc welding, the addition of metal or iron powder reduces penetration and dilution to a small degree. In most welding processes, dilution and penetration decrease when the current type is changed from d.c. positive to a.c. to d.c. negative.

All fluxes, even the so-called neutral submerged arc fluxes, produce some changes in composition between the added metal and the weld metal. The chemical reactions involved are complex; ref. 3 provides a useful and simple guide to flux types for fusion welding processes.

Fluxes and electrode coverings help in shaping the weld bead, as well as in providing shielding from the atmosphere and influencing weld metal composition. They are usually mixtures of minerals such as rutile, silica, silicates, bauxite, magnesia, fluorite, chalk and clays. The more 'acid' of these (silica, rutile and bauxite) result in relatively high weld metal oxygen contents and loss of Mn, Si, C and other readily oxidised elements from the weld metal. The more 'basic' minerals, such as fluorite, magnesia and chalk, result in smaller losses of these elements, lower weld oxygen contents and some removal of sulphur. Some minerals add appreciable amounts of alloying elements to the weld pool; for example, manganese is added from manganese silicate. Other elements may be added to the weld metal as such, or as ferro-hardeners, from the electrode covering or the flux.

Normally, because of the good mixing within the weld pool, the composition of each weld bead is uniform on a macroscopic scale over most of its volume, even though segregation on a microscopic scale is evident, as in all cast structures. However, the composition of adjacent weld beads may differ from each other because of differences in dilution and possibly

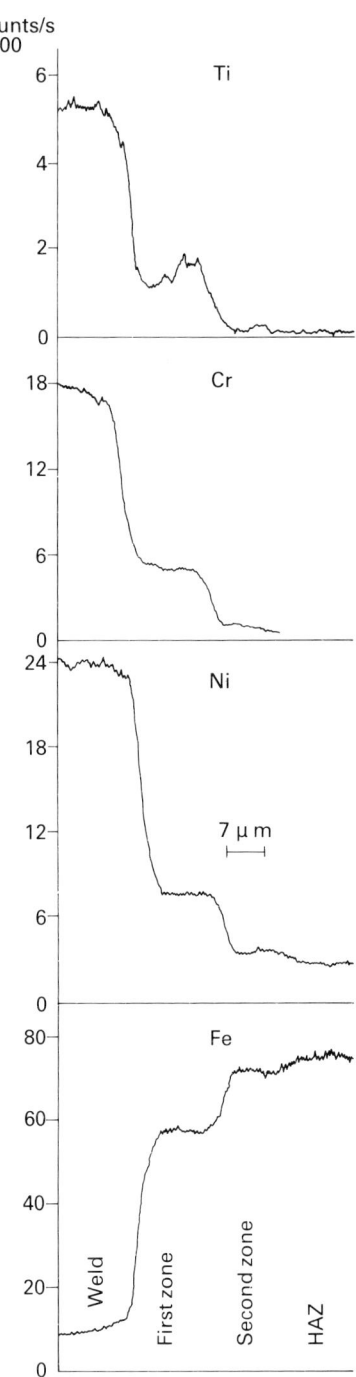

1.6 Steps in Ti, Cr, Ni and Fe contents across the fusion boundary between a nickel alloy weld metal and 9% Ni steel showing two regions between parent steel and weld metal.

Factors influencing weldability

shielding, even though the wire and parent steel may be of uniform composition. Between the weld metal and the parent steel there appear to be two compositional steps. The first step is the result of the essentially laminar flow of weld metal across the surface of the parent steel preventing full mixing of weld metal with melted parent steel near the fusion boundary. It consists of melted parent steel into which weld metal has diffused fairly rapidly while both were still in the liquid state. The second step is of parent steel that has not melted but into which less weld metal has diffused more slowly (Fig. 1.6).

Apart from this stepwise gradation of composition, tongues of melted parent steel can project into the weld metal (Fig. 1.7), with insufficient time before solidification and cooling for diffusion to even out the composition. When austenitic stainless steel fillers are used to weld ferritic steels, such regions can be highly alloyed, with hard martensitic microstructures, and are therefore at risk of cracking (see Chapter 5).

Although mixing of alloying ingredients added from electrode coverings and fluxes is normally uniform, occasions are known where alloying ingredients have not had time to dissolve into the weld pool and give a homogeneous microstructure. An addition at particular risk is ferromolybdenum hardener, whose melting temperature is relatively high; Fig. 1.8 shows a particle of ferro-molybdenum which has started to dissolve in the weld pool. Such partially dissolved particles are usually relatively harmless.

In some cases, however, the alloying particle has dissolved, but insufficient time has been available for complete mixing and diffusion. After

1.7 Tongue of unmixed parent metal in a non-matching weld pool.

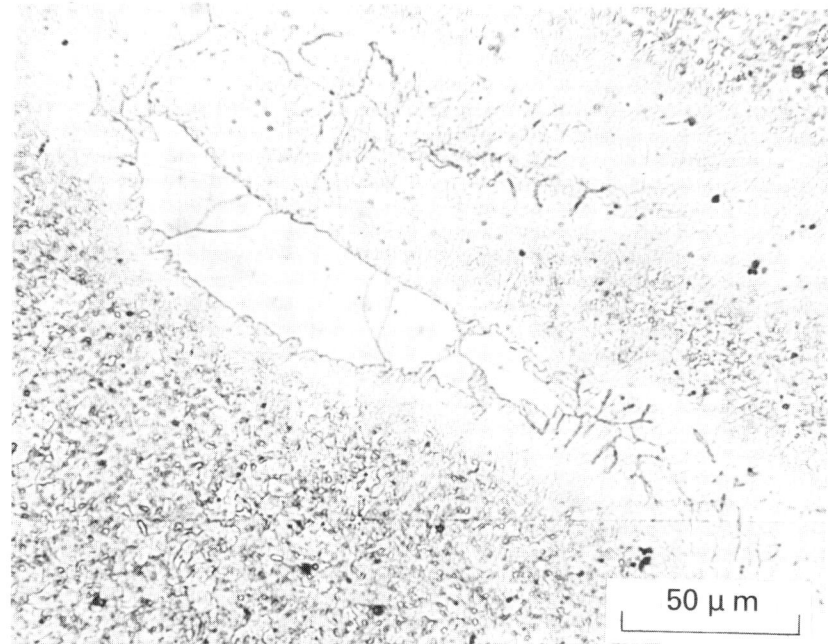

1.8 Partially dissolved Fe–Mo particle in MMA weld metal; note scalloped edges to particle, which has started to dissolve.

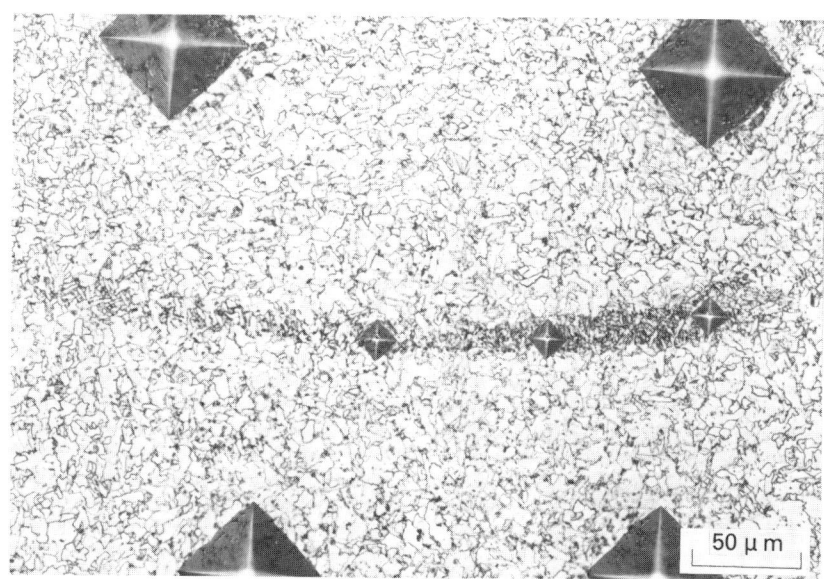

1.9 Site of partially dissolved Mn or Fe–Mn particle in weld metal marked by small microhardness indentations within square of full size hardness indents.

cooling, the sites of such particles become tiny regions more highly alloyed than the bulk of the weld metal (Fig. 1.9), often with hard martensitic microstructures which can give problems in service with stress corrosion cracking (Chapters 9 and 10). This type of 'hard spot' has given problems in submerged arc welds,[4] when manganese has been alloyed by the use of ferro-manganese added to the flux. Presumably in this case, some particles of Fe–Mn from the molten slag entered the weld pool just as it was starting to solidify. Occasionally, similar problems have occurred with MMA electrodes and cored wires, as well as the opposite problem – iron powder particles incompletely mixed; these tend to give soft spots which are less harmful.

Microstructure

Because the bulk of the weld metal solidifies from the HAZ, the widths of the solidifying grains are initially those of the coarse grains in the high temperature HAZ from which they have grown, as can be seen in Fig. 1.3, 5.7(a) and other illustrations in the text. However, their length is governed by the width of the weld and the direction of solidification. Except in rare cases (electroslag and similar welds of high heat input in which solidifying grains can be nucleated within the weld) most weld metal grains continue growing until they either meet grains growing from the opposite direction at the weld centre-line or until they reach the weld surface. Occasionally, in a weld metal of low impurity content and deposited with a high heat input (particularly by submerged arc and most types of electroslag welding) the weld grains can grow as in Fig. 1.10. This type of coarse grain size is not in itself detrimental to toughness, which depends on the grain size of the transformation product. If the latter is predominantly of acicular ferrite, then a coarse prior austenite grain size as shown in Fig. 1.10 will have no effect on toughness.

With the exception of the formation of acicular ferrite, weld metal

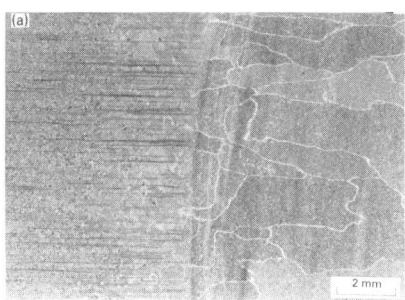
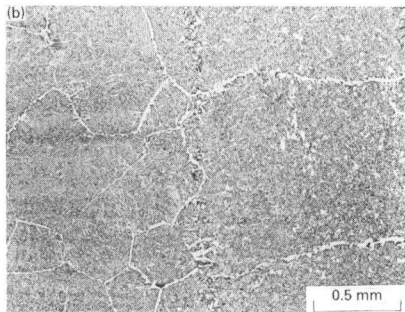

1.10 Grain growth in submerged arc weld metal: (a) general view, weld metal on right; (b) detail.

microstructures[5] resemble those of the HAZ. With a low content of carbon and alloying elements, the microstructure is predominantly of soft primary ferrite and a constituent called ferrite with aligned second phase (sometimes known earlier as 'ferrite sideplates' or 'ferrite with aligned MAC' – meaning martensite, austenite and carbides), Fig. 1.11(a). Although the 'second phase' is mainly the iron carbide cementite, other constituents, such as martensite and retained austenite (these two are sometimes known collectively as 'retained phases') are usually present in small quantities; Fig. 1.11(b).

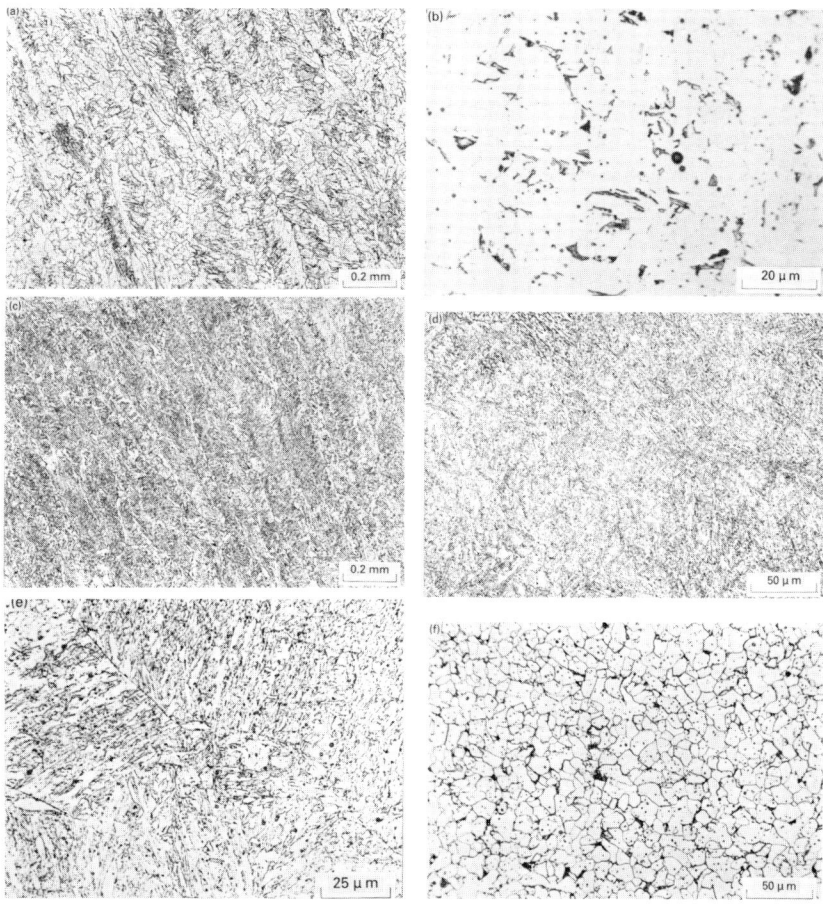

1.11 Selection of as-welded weld metal microstructures: (a) low carbon, low Mn submerged arc; primary ferrite with some ferrite with some aligned second phase; (b) submerged arc, so-called retained phases (SASPA–NANSA etch); (c) predominantly ferrite with aligned second phase in submerged arc weld; (d) bainite in 2.25%Cr:1Mo MMA weld metal; (e) martensite in high strength, low alloy MMA weld metal; (f) refined submerged arc weld metal.

Factors influencing weldability

Admixture of alloying elements increases the proportion of ferrite with aligned second phase at the expense of primary ferrite (Fig. 1.11(c)) until lower temperature transformation products appear. The first of these is bainite, typified by the microstructure of 2.25% Cr:1% Mo weld metal (Fig. 1.11(d)), and the second is martensite, common in very high strength low alloy weld metals (Fig. 1.11(e)). All of the constituents so far mentioned are preferentially nucleated at prior austenite boundaries, or stop their growth at such boundaries. Hence the coarser the prior austenite grain size, the coarser and less tough is the resultant microstructure.

Within the region where ferrite with aligned second phase occurs, it is possible for it to be replaced by acicular ferrite, which nucleates within the grains and – because so many more grains can be nucleated within a large grain than can be nucleated at its boundaries – the grain size of the acicular ferrite is fine in comparison with the other constituents. Its toughness, therefore, is very much better than that of any other constituent except perhaps low carbon martensite containing nickel or very soft ferrite.

The prior conditions for the formation of acicular ferrite are complex, as they appear to depend on the structure of non-metallic inclusions in the microstructure (Fig. 1.12). These inclusions, consisting mainly of deoxidation products (manganese silicates and alumina) and MnS are spherical and typically <10 μm in diameter. When coated (at least in part) with an oxide of titanium they appear to nucleate easy transformation of ferrite from austenite *within* the prior austenite grains. These so-called acicular ferrite grains cannot grow far before their growth is stopped by the presence of other growing grains. This blocking of grain growth by closely adjacent grains is the reason for the fine grain size of acicular ferrite

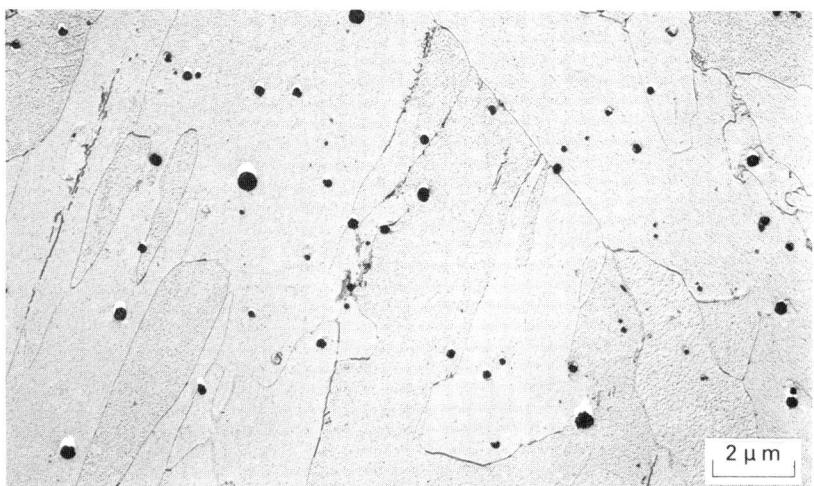

1.12 Typical inclusions shown in TEM replica of section, which has been prepared to emphasise the inclusions by shadowing.

and is in contrast to the behaviour of primary ferrite grains, usually growing from the prior austenite boundaries, as well as bainite and martensite.

Titanium is an essential constituent of inclusions that nucleate acicular ferrite, as it has been shown that if pure (i.e. titanium-free) chemicals are used to make a flux or electrode covering, the formation of acicular ferrite is prevented in otherwise favourable circumstances.[6] Other experiments have shown that additions of titanium compounds to a flux, giving as little as 0.003% Ti in an otherwise unfavourable weld metal, can induce the formation of acicular ferrite.[7] The only other element known to favour the formation of acicular ferrite is vanadium, an element not usually added to weld metals (partly because of its tendency to promote reheat cracking), and one whose effect is less efficient than that of titanium. However, the presence of vanadium can lead to the formation of some acicular ferrite in HAZs.

The titanium needed to promote acicular ferrite must be in the form of an oxide, because the presence of too much aluminium – an element which oxidises preferentially to titanium – is known to inhibit the formation of acicular ferrite,[8] as shown in Fig. 1.13. This inhibition occurs more easily if the weld oxygen content is low (Fig. 1.13(b)), particularly if it is below about 0.020% O. An excess of aluminium in this context means that suffi-

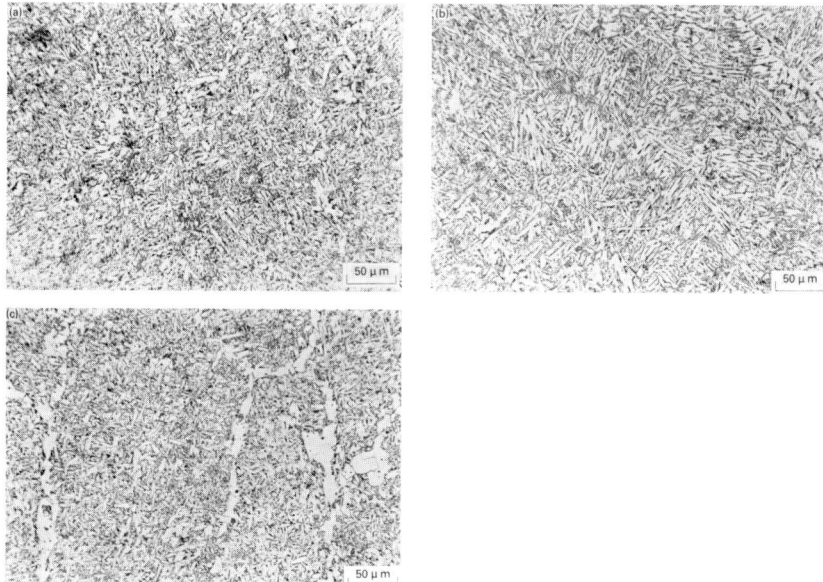

1.13 Effect of aluminium and oxygen on acicular ferrite content of similar C:Mn weld metals: (a) 0.008% Al, 0.020% O, 58% acicular ferrite, 35 J Charpy at −62 °C; (b) 0.029% Al, 0.017% O, 15% acicular ferrite, 35 J Charpy at +28 °C; (c) 0.028% Al, 0.047% O, 49% acicular ferrite, 35 J at −50 °C.

cient aluminium is present for it to combine with all of the oxygen available in the weld metal to form Al_2O_3, thus leaving no oxygen to form titanium oxide. If an accurate weld metal analysis is known, then the aluminium content should not exceed 9/8ths of the oxygen content by more than about 0.002%[7] to allow acicular ferrite to be nucleated.

If the weld metal oxygen content exceeds about 0.055% O, the presence of acicular ferrite is unusual because conditions will have been so oxidising that most titanium will have been oxidised out of the weld metal into the slag, unless special consumables have been used that deliberately add titanium.[7] In marginal cases the nucleation of acicular ferrite can be uncertain because titanium may or may not be present as impurities in welding wire, parent steel or flux but, if present, may find its way into the weld metal. Sufficient titanium will always be present where minerals such as rutile (TiO_2) are constituents of the flux or electrode covering. With mild and C:Mn steel weld metals, acicular ferrite is most likely in those made from basic MMA electrodes, basic and alumina-basic submerged arc fluxes and in MAG welds made using low oxidising shielding gases. Although conditions are just about suitable for acicular ferrite nucleation with rutile and cellulosic electrodes, the degree of alloying is normally insufficient to permit extensive formation of acicular ferrite. It should be noted that excessive amounts of titanium (i.e. >0.1%) can give weld metals having bainitic microstructures with very poor toughness.

Aluminium to inhibit acicular ferrite nucleation can originate from the same three sources as titanium, although it is rarely present in MMA electrode coverings and not at all in rimming steel wire which was once used exclusively for all MMA electrodes. However, aluminium metal is added as a major constituent to many self-shielded flux cored wires and is transferred to the weld metal in quantities sufficient to prevent the nucleation of acicular ferrite. Such weld metals achieve their toughness by the use of multipass refining techniques resulting in a fine ferrite microstructure.

Although alumina–rutile and alumina–basic submerged arc fluxes transfer aluminium to the weld metal, they do not give sufficiently low oxygen contents to prevent acicular ferrite nucleation. However, aluminium transfers readily from parent steel to the weld pool, so that high dilution welding processes, such as submerged arc when used with very basic fluxes (i.e. those that give weld metal oxygen contents well below 0.025% O), can make acicular ferrite nucleation difficult – especially in any weld runs that are more highly diluted with parent steel than usual. Such high dilution may occur in root runs and in any misplaced runs that dig into the parent steel at the side of the weld preparation.

Low oxygen welding processes such as TIG and plasma do not require much aluminium to be diluted from the parent steel to inhibit the formation of acicular ferrite, because of the very low oxygen contents associated with these processes. However, because the TIG welding process usually gives a high proportion of refined weld metal in multipass welds, the problem of uncertain acicular ferrite formation is less serious than might at first appear. This technique of adjusting the welding process to deposit a highly refined microstructure has been adapted for other welding processes where acicular ferrite is not usually produced, notably self-shielded cored wire welding.

As with the HAZ, the weld metal is reheated and transformed by subsequent weld runs. At the highest temperatures, the coarse columnar prior austenite grains are replaced by more or less equiaxed coarse grains with similar transformation products to the original as-welded grains. At lower temperatures, the grain size is considerably refined and the microstructure is usually of equiaxed fine ferrite with carbides at the ferrite grain boundaries – generally a tough microstructure. At still lower temperatures of reheating, the original as-deposited macrostructure persists, with little obvious change in the microstructure.

Cutting and gouging

Cutting and gouging are necessary adjuncts to the welding process, and should be treated with respect when used on many steels. Even mechanical methods – particularly grinding and chipping – can severely cold work, tear and leave residual stresses adjacent to the cut surface, which can be damaging in subsequent service. Thermal cutting and gouging leave thin heated and hardened surface layers, which have martensitic microstructures, whose hardness depends solely on the carbon content of the steel. If mishandled, both processes – but particularly carbon arc gouging – can also result in carbon pick-up, which exacerbates this problem.

However, if a cut surface is to be directly welded on, the damaged layer will be melted into the weld pool and will not thereafter be harmful. It is only where the cut or gouged surface is left on an unwelded area of the steel (or worse, on a region that will become a HAZ) are serious problems likely. However, it is possible for a badly mis-handled carbon arc gouging electrode to give sufficient carbon pick-up to harmfully raise the carbon content of any weld metal deposited on it.

Thermally cut and gouged surfaces should, therefore, be ground off to a sufficient depth to remove any damaged material and hardened layers unless they are going to be fully incorporated into a weld. With sensitive steels, preheat (less than that used for welding) is recommended in order to reduce hardness and avoid hydrogen cracking directly after thermal cutting – the hydrogen originating mainly from the flame or arc atmosphere.

Factors influencing weldability

Residual stresses

Residual stresses build up in welds because, on cooling, the weld metal and adjacent HAZ contract against the restraint of cooler, and hence more rigid, steel which has not been heated to high temperatures during welding. Because the weldment is being restrained from contracting, the residual stresses in a single-run weld are predominantly tensile and are balanced by compressive stresses remote from the weld, as shown in Fig. 1.14.

Residual stresses quickly build up to the level of yield strength (which is initially very low at the highest temperatures); strains above the yield strain are relieved by plastic deformation. As the yield strength of the

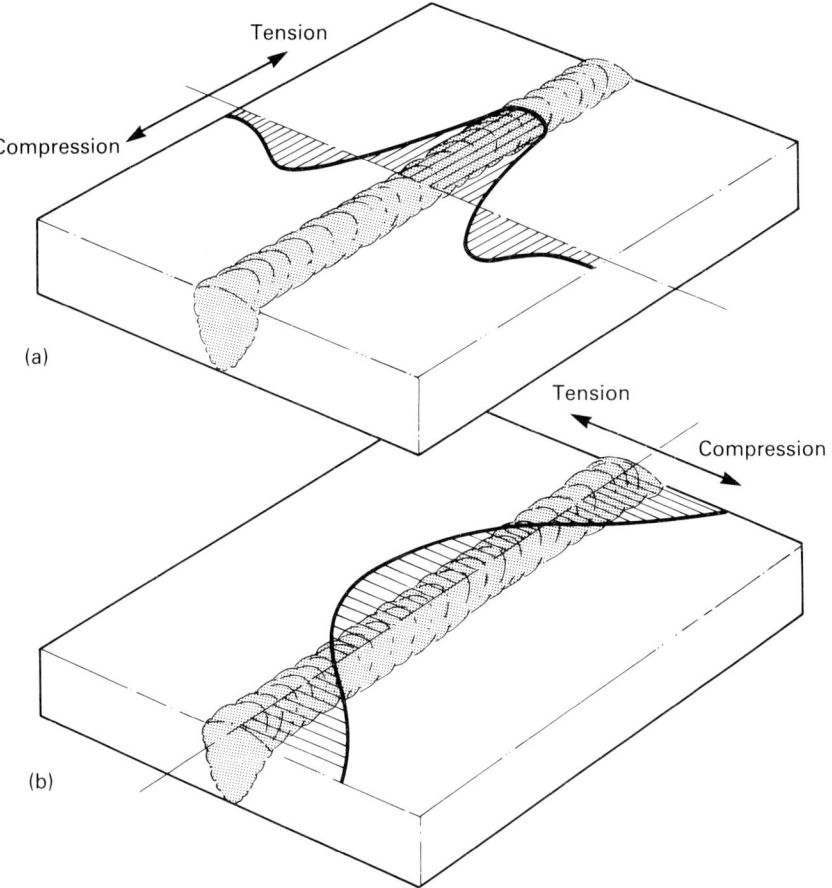

1.14 Sketch illustrating residual stress distributions: (a) longitudinal stress in single-pass weld; (b) transverse stress in single-pass weld.

metal builds up on cooling, residual stresses also build up. In a single-pass butt weld, where the heated zone and resultant contraction strain is usually wider at the face of the weld than at the root, this can result in the plates deforming upwards or 'winging' unless the weld is heavily restrained. If the weld is restrained, the deformation is accommodated by elastic and plastic deformation within the heated area. Contraction strains are greatest along the axis of a weld and least in the transverse directions.

In a simple metal, like aluminium or austenitic stainless steel, residual stresses continue to build up as the temperature falls and can be considered to be at the elastic limit (i.e. close to the proof stress) at ambient temperature. Any excess strain over the yield strain is accommodated by plastic deformation.

The behaviour of a ferritic steel is complicated by its transformation on cooling. Transformation from high temperature austenite to lower temperature transformation products (martensite, bainite and ferrite) is accompanied by an expansion, which works in the opposite direction to the contraction resulting from cooling. This means that the build up of the ambient temperature residual stresses in a ferritic steel essentially starts from the transformation temperature, as shown in Fig. 1.15(a).

Although most steels have sufficiently high transformation temperatures for the yield stress to be reached on cooling from the transformation temperature, some highly alloyed steels have such low transformation temperatures that residual stresses do not reach yield stress level (Fig. 1.15(b)). Hence steels such as the 9% Cr steel[9] may have a lower risk of cracking and a higher defect tolerance than steels of similar strength and toughness, but of lower alloy content and higher transformation temperature.

An important metallurgical result of the considerable plastic deformation undergone by a weld metal on cooling is that this plastic deformation leads to a very dense network of dislocations. This dislocation network is responsible for the yield strengths of ferritic steel weld metals being much higher than would be expected from their compositions and grain sizes.[10] This network, together with transformation hardening, accounts for the ease with which ferritic steel weld metals overmatch parent steels in strength, in contrast to most non-ferrous systems.

The dislocation network is little changed by tempering during PWHT but is destroyed when the weld metal is subjected to high temperature heat treatments within the austenite range, such as normalising or hardening and quenching. Because of this, considerable care in the selection of consumables is needed if a weld metal has to survive high temperature heat treatment (or hot working within the austenite temperature range) and retain strength properties to match the parent steel.

As a multipass weld is built up, the residual stresses in the first runs are altered and may be replaced by compressive stresses as the depth of the

Factors influencing weldability

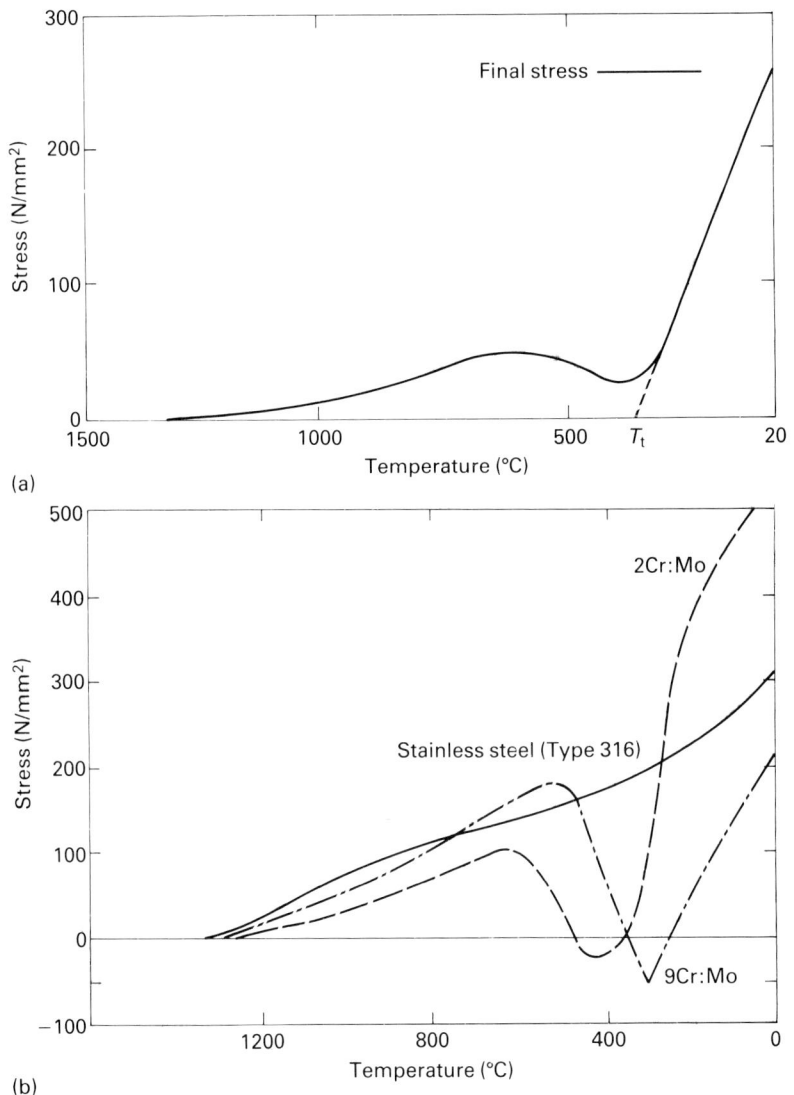

1.15 Build up of residual stresses in a ferritic steel showing decrease in stress as steel transforms below about 600 °C: (a) 3.5% Ni weld metal; (b) selected steels; note absence of contraction from stainless steel, which does not transform.

weld is increased (Fig. 1.16). This has an important effect for subsequent non-destructive examination (NDE) because the faces of any cracks formed in the first runs to be deposited are forced together later and are thus more difficult to detect than they would have been had they been held apart by the initial tensile residual stresses.

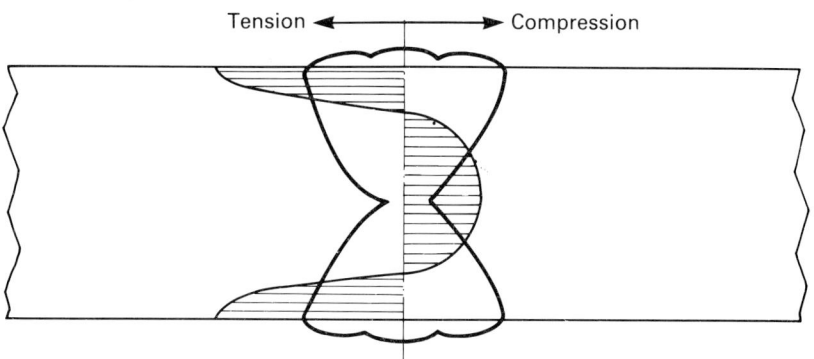

1.16 Final residual stress distribution in multipass weld.

A knowledge of residual stresses is essential when carrying out fracture mechanics analyses to establish the fitness for purpose of welded joints. For such analysis, the most important single parameter is the maximum value of tensile residual stress, as this, when related to the fracture toughness of the appropriate region, determines the maximum tolerable defect size that will not lead to failure when subjected to the maximum likely stress.

Although it might seem that the residual stress cannot exceed that of the parent steel (which is nearly always lower than that of the weld metal or HAZ), detailed measurements using the hole drilling technique[11] have shown that residual stresses near the surface in weld metal can approach the yield stress of the weld metal itself, even when this is higher than that of the parent steel. In weld metal, the residual stress varies over quite short distances, being higher within a weld bead than near the junction of two weld beads.[11]

Unless existing knowledge or future research shows otherwise, it is important to use the weld metal yield stress in estimating the maximum residual stress levels for fracture mechanics calculations involving weldments. Consequently, there is benefit in using weld metals with the lowest possible yield strength consistent with just overmatching that of the parent steel in situations where the best toughness is required. The reduction of residual stresses by PWHT is covered in the next section.

Post-weld heat treatment

Post-weld heat treatment is frequently needed for weldments in alloy steels and in C and C:Mn steels in section thicknesses typically in excess of about 40 mm. Generally, PWHT reduces residual stress levels and tempers, i.e. softens and reduces the hardness of any hard regions, particularly in the HAZ. After PWHT, a weldment should be tougher and also should resist such service hazards as stress corrosion cracking (SCC) and in-service hydrogen cracking more readily than in the as-welded condition.

The selection of PWHT temperatures is often aided or limited by the appropriate application standard, and is usually within the temperature range 550–750 °C. It should be remembered that PWHT at a lower temperature will not undo any excessive softening which may have resulted from the use of too high a temperature. However, if a weldment has been under-tempered by PWHT, a further heat treatment at a higher temperature is always possible.

Carbon, C:Mn and Ni steels are usually treated in the range 550–650 °C, more usually at about 600 °C. However, the lower end of this range is now known to give inadequate relief of residual stresses and temperatures around 600 °C are preferred. The reduction in residual stress of a simple C:Mn weldment after PWHT at 600 °C is approximately 70%, and a reduction of this extent gives a very useful improvement in the toughness of the joint in terms of fitness for purpose, i.e. its ability to tolerate cracks and other faults. However C:Mn steels microalloyed with Nb, V and, most particularly Nb + V, are known to give a slower tempering than a simple C or C:Mn steel. In fact, as a result of secondary hardening, the HAZs of C:Mn:V and C:Mn:V:Nb steels can exhibit increases in hardness and reductions in Charpy toughness when subject to PWHT at about 600 °C.

Low alloy steels containing chromium generally require PWHT temperatures at or above 650 °C, the temperature depending on the required properties and the application. The lower temperatures are used when the highest strength is required; optimum creep resistance results from an intermediate temperature, whilst the highest temperature within the range is used to obtain maximum softening and stress relaxation.

With other alloy steels, two important points should be noted. Firstly, the selected temperature should always be below the A_{c1} temperature of the steel, that is the temperature above which the steel begins to transform to austenite. Here it should be noted that the addition of chromium progressively increases the A_{c1} temperature and the addition of nickel progressively lowers it.

Secondly, when welding quenched and tempered steels which have been tempered to particular strength or hardness levels, it is important not to exceed the tempering temperature which was originally used, otherwise the whole component will be softened and weakened below its required level. It is normal to set the maximum of the PWHT temperature some 20 °C below the actual tempering temperature.

Unfortunately it is not always easy to discover the previous heat treatment history of a piece of steel, although it should be on the mill sheet. Some steelmakers give such a wide range for the tempering temperature of a particular steel as to be useless, for example 530–680 °C. In such cases, the supplier or steelmaker should be carefully questioned. If this fails to provide the actual tempering temperature, the only action is to remove a small piece of steel from an unimportant area and subject it to careful hardness measurements after laboratory heat treatments at progressively higher temperatures in order to estimate the original tempering temperature. These tests should start at the bottom of the steelmakers' range and finish either where the steel starts softening, or at a temperature such as 650 °C where a high degree of stress relief will be obtained.

During PWHT, care should be taken to ensure uniform heating of the components, which may be of a very complex structure. Most heat treatment specifications give guidance on heating and cooling rates and temperature differentials. The penalties for exceeding these limits include distortion, cracking and also the re-imposition of residual stresses (albeit not in exactly the same configuration, but possibly up to the same maximum level) which the PWHT is intended to reduce.

Although PWHT is preferably applied to completed fabrications, there are times when this is not possible. For example, available furnaces may be too small, the weld may be a repair in a structure too large to heat treat or there may be a component in the neighbourhood which would be damaged by the heating. Local PWHT is acceptable in such cases, always provided precautions are taken to maintain low temperature gradients between the heated and unheated regions and to avoid distortion or build up of high local stresses on cooling. Again, guidance is usually given in the appropriate application standard.

Other heat treatments

Heating of welds in ferritic steels is often carried out within the range 150–300 °C. This is not usually a tempering heat treatment (although it could be so for a high strength Q&T steel originally tempered at a low temperature), but is used to allow the diffusion of hydrogen out of the joint in order to avoid the risk of hydrogen cracking. As such, this topic is covered in Chapter 5.

Sometimes there is a requirement to give a welded joint a full heat treatment such as normalising or Q&T. Such heat treatment is common for electroslag, electrogas and consumable guide welds in order to achieve acceptable toughness in the HAZ. Except when welding certain low carbon steels (for example, some of the Cr:Mo types used for high temperature service in which weld metal and parent steel are of similar carbon content), consumables for other types of welding should be selected to ensure that the heat treated weld metal is of adequate strength and toughness. This is because the application of an austenitising heat treatment to a weld metal gives a drastic reduction of the as-welded weld metal strength to levels well below the as-welded or PWHT values.

Preheating is *not* a heat treatment.

Weldability formulae

Over the years, attempts have been made to provide single numbers to characterise the weldability of steels, either in general terms, or to cover specific types of cracking or other important aspects of welding behaviour. The earliest, and most important of these, cover HAZ hardenability and HAZ hydrogen cracking. In fact, these two problems are so closely related that users are not always clear which aspect some formulae cover.

Other formulae cover solidification cracking, reheat cracking, maximum martensite hardness and temper embrittlement, as well as weld metal properties, such as yield strength, tensile strength, hardness and microstructural constituents. Further formulae relate hardness to yield strength for HAZ and weld metal. The most significant of these will be summarised in this section. Unless otherwise stated, the use of the symbol for a chemical element in a formula denotes its content in mass percentage in the steel or weld metal.

Hardenability and hydrogen cracking

The first of the many weldability formulae was developed by Dearden and O'Neill over 50 years ago,[12] and the whole topic was recently reviewed in a seminar held to commemorate the golden jubilee of that paper.[13] Dearden and O'Neill developed formulae, based on steel chemical compositions, for yield and tensile strength, as well as for hardness; all were complicated by some of the divisors depending on the levels of elements present. Over 25 years later, the most useful of their formulae, that for hardenability, was simplified by a sub-committee of the International Institute of welding (IIW) into the generally accepted version – often known as the IIW formula – giving a 'carbon equivalent' (CE or CE_{IIW}):

$$CE_{IIW} = C + \frac{Mn}{6} + \frac{Cr + Mo + V}{5} + \frac{Ni + Cu}{15} \qquad [1.2]$$

This formula has been adopted into several standards and codes, e.g. ref. 2 and 14.

Other formulae were developed in the 1950s, 1960s and 1970s but the only ones which have gained wide acceptance are P_{cm} due to Ito and Bessho in Japan (quoted by Yurioka in ref. 13) and that of Düren in Germany,[13] which have been specially developed for low carbon steels for which the CE_{IIW} formula is not entirely suitable. Both these authorities consider that HAZ behaviour should be divided into three distinct types:

1. The fully hardened HAZ, i.e. with a fully martensitic microstructure.
2. An intermediate microstructure.
3. An unhardened HAZ, generally with a bainitic, or even a ferrite + pearlite microstructure.

The formulae for modern steels, particularly those used for pipeline manufacture, are intended for carbon contents no more than ~0.11%. Of these, the most well-known hardenability formulae is:

$$P_{cm} = C + \frac{Si}{30} + \frac{Mn+Cu+Cr}{20} + \frac{Ni}{60} + \frac{Mo}{15} + \frac{V}{10} + 5B \quad [1.3]$$

Düren's formula for this range is similar:

$$CEq = C + \frac{Si}{25} + \frac{Mn+Cu}{16} + \frac{Ni}{40} + \frac{Cr}{10} + \frac{Mo}{15} + \frac{V}{10} \quad [1.4]$$

Although CE_{IIW} was originally developed as a hardenability formula, it has come to be used as a hydrogen cracking formula, and is the basis of the nomograms to avoid cracking in ref. 2 and 15. The principal reason for this is that the original Dearden and O'Neill formula was developed for semi-killed, balanced and rimming steels with very low silicon contents. Subsequent research showed that, although silicon had an effect on hardenability similar to that of manganese,[16] its effect on weldability in terms of cracking behaviour was negligible,[15] so that its presence could be ignored when considering weldability.

It is important to understand this distinction, because if the CE_{IIW} is used to estimate the HAZ hardness of a single-pass weld in a steel containing silicon, a low result will be obtained. A more accurate estimate will be obtained if the factor of Si/6 is added to the CE_{IIW} formula,[16] i.e.:

$$CE_H = C + \frac{Mn+Si}{6} + \frac{Cr+Mo+V}{5} + \frac{Ni+Cu}{15} \quad [1.5]$$

In recent years, complicated formulae (e.g. ref. 13 and 17) have been developed, which include welding parameters, and the effects of inclusions, as well as chemical composition, on hardenability and/or weldability. The general view is that such formulae, although of scientific interest, are not practical for normal industrial use where a welding engineer needs to look at a mill sheet and quickly decide whether the steel is suitable for the welding required. However, this view may well change as the use of personal computers extends into industry.

Other formulae relevant in this area are those developed for calculating the hardness of martensite (HV_M), bainite (HV_B) and the cooling times to give approximately 100% of each in the HAZ of a steel when it is hardened. One series due to Düren[13] is:

$$HV_M = 802C + 305 \qquad [1.6]$$

$$HV_B = 350(C + \frac{Si}{11} + \frac{Mn}{8} + \frac{Cu}{9} + \frac{Cr}{5} + \frac{Ni}{17} + \frac{Mo}{6} + \frac{V}{3}) + 101 \qquad [1.7]$$

Calculations to determine the extent of, and the hardness within, the intermediate range are complicated and reference should be made to the original papers.[13,18]

Yurioka's formula[13] for a fully martensitic microstructure is:

$$HV_M = 884C(1-0.3C^2) + 294 \qquad [1.8]$$

Again the formulae for bainitic and intermediate microstructures are too complex[13] for the present text, particularly as the hardness depends to some extent on the type of steel and its inclusion content, as well as on the cooling rate.

Mechanical properties

For weld metals, approximate relationships exist between the CE_{IIW} of the weld metal itself (not that of the parent steel) and its strength, particularly for submerged arc welds.[19] Slightly different relationships have been developed for single, two-pass and multipass weld metal; the most useful of these are for multipass weld metal, as-welded:

$$\sigma_{yam} = 670CE_{IIW} + 230 \qquad [1.9]$$

for multipass weld metal after PWHT:

$$\sigma_{ysm} = 670CE_{IIW} + 200 \qquad [1.10]$$

for two-pass weld metal, as-welded:

$$\sigma_{ya2} = 720CE_{IIW} + 182 \qquad [1.11]$$

Other formulae for the yield and tensile strengths of submerged arc weld metal are given in ref. 19.

These equations should *not* be used for weld metal which has been normalised or quenched and tempered, because of the *loss of strength* of weld metals after heat treatments involving heating within the austenite range, i.e. generally above about 750 °C. This is a result of the destruction of the fine, dense network of dislocations which confer strength on weld metals to a degree greater than would be expected from their compositions and grain sizes.[10] It is important to be aware of this loss of strength, as it requires a special selection of welding consumables, generally of similar carbon content to that of the parent steel. Situations where this is liable to occur include the foundry repair of castings before they are heat treated, the hot working of partly fabricated (welded) components within the austenite range, and any unexpected or unauthorised heating (e.g. in fires) to excessively high temperatures.

Formulae have also been developed for estimating the yield strength of weld metals and HAZs from their hardness.[20,21] These yield strength formulae are particularly important as they allow fracture mechanics calculations of defect tolerance to be carried out on structures where insufficient material can be spared for a tensile test to be carried out.

For weld metal:

$$\sigma_{ywm} = 3.15HV - 168 \qquad [1.12]$$

For the HAZ:

$$\sigma_{yHAZ} = 3.25HV - 349 \qquad [1.13]$$

Although Eq. [1.12] gives reasonably accurate results, there is appreciable scatter in the values obtained from Eq. [1.13] because of the paucity of the data on which it was based.

Estimation of toughness from carbon equivalent formulae is not simple because of the complications of the different weld metal microstructures and microstructural refinement by reheating in multipass welds, and how these impinge on what would otherwise be a simple inverse linear relationship between toughness and strength (or hardness). For example, in the absence of conditions favouring acicular ferrite formation, manganese is harmful because it increases strength whilst reducing toughness. However, if acicular ferrite can form, then manganese is beneficial up to about 1.8% (in the absence of other alloying elements), because it replaces coarse primary ferrite by fine acicular ferrite and thus increases both toughness and strength. However, a direct inverse relationship has been found between Charpy upper shelf energy (U) (and also the slope of the transition part of the Charpy curve) and the weld metal inclusion content (I) of submerged arc welds[19] which, in turn, can be calculated from the weld metal composition:

$$U_{am} = 158 - 83I \qquad [1.14]$$

$$U_{sm} = 173 - 83I \qquad [1.15]$$

$$U_{a2} = 138 - 83I \qquad [1.16]$$

where the same notation is used as in Eq. [1.9–1.11]. Additional relationships are given in ref. 19.

The inclusion content, I as a volume fraction, is estimated by assuming that the oxygen content of the weld metal first combines with Al to form

Al_2O_3, then with Ti to form TiO_2 and finally with Mn and Si to produce $2MnO.SiO_2$; S forms MnS. Factors for wt% element to vol% inclusions are 4.09 for Al, 3.67 for Ti, 5.3 for Mn to MnS, 6.2 for remaining O to $2MnO.SiO_2$; 1% oxygen is equivalent to 0.89% Al and 0.67% Ti. These relationships appear to be valid for all types of fusion weld.

Microstructure

The proportion of primary or grain boundary ferrite in submerged arc weld metals can be estimated from the weld metal CE_{IIW} as a volume percentage:[19]

$$GBF_{mT} = 81 - 173CE_T \qquad [1.17]$$

$$GBF_{mR} = 96 - 197CE_R \qquad [1.18]$$

where the subscripts R and T refer to the root and cap runs, respectively. Unless microalloying elements such as Nb are absent from the weld metal, these equations are limited to ferrite contents up to about 60%; additional relationships are in ref. 19. From these relationships, zero grain boundary ferrite would be expected at a CE_{IIW} of about 0.48. Reference 19 indicates that (in the absence of Nb, Mo, etc.) weld metal microstructures tend to become fully martensitic when CE_{IIW} exceeds about 0.6–0.7.

Solidification cracking

These formulae inevitably relate to the composition of the weld metal under consideration, rather than to the parent steel. This is not a problem when considering well-shielded, autogenous processes such as TIG, laser and electron beam welding, where the composition of the parent steel approximates to that of the weld metal. A further step, however, is required for flux and active gas-shielded processes, particularly when they are used to deposit multipass welds, and the cracking is in the root region. This is because of the influence of the flux or shielding gas on the weld composition, because of the dilution of parent steel into the weld metal and because analyses of multipass welds are usually carried out on samples from the cap regions of low dilution.

In using any solidification cracking formula, care is needed to ensure that all the welds used to develop the formula are as nearly as possible identical in shape. In the type of test normally used to establish compositional effects, altering the welding parameters also alters the test conditions.

In all the formulae, the elements S, C and P promote cracking, whilst Mn is beneficial. One of the early attempts to develop a formula for TIG welding[22] led to a crack susceptibility factor (CSF):

$$\text{CSF} = 501\text{C} + 1560\,(\text{Mn}\cdot\text{S}) - 82.8\text{Mn} - 47.9 \qquad [1.19]$$

for steels having carbon contents between 0.14 and 0.31%.

Later work along similar lines showed that oxygen (the term (O) represents the weld metal oxygen content) had a beneficial effect on cracking at the low levels present in TIG welds, and a CSF for autogenous TIG welds in medium carbon engineering steels[23] was:

$$\text{CSF} = 42\text{C} + 847\text{S} + 265\text{P} - 10\text{Mo} - 3042(\text{O}) + 19 \qquad [1.20]$$

Submerged arc welding is probably the process most prone to give solidification cracking and a formula was developed,[24] using the British Transvarestraint test, in units of crack susceptibility (UCS) using standardised welding conditions; this formula has been adopted in the relevant British Standard:[2]

$$\text{UCS} = 230\text{C} + 190\text{S} + 75\text{P} + 45\text{Nb} - 12.3\text{Si} - 4.5\text{Mn} - 1 \qquad [1.21]$$

In this formula, values of C<0.08% are taken as being equal to 0.08%C; the reasons for this, together with the compositional limits within which Eq. [1.21] can be used, are discussed in Chapter 3 and ref. 24.

Unlike solidification cracking, there has been little effort to quantify compositional effects on liquation cracking in the HAZ, because it is normally a minor problem.

Reheat cracking

Reheat or stress relief cracking is limited to a few steels and no attempts appear to have been made to develop universal formulae for this type of cracking.

However some formulae have been developed for parameters (ΔG and P_{SR}) showing the susceptibility to cracking of ferritic steels when clad with austenitic stainless steels and then subjected to PWHT. For the Cr:Mo:V types used for such pressure vessels, the formulae include:

$$\Delta G = \text{Cr} + 3.3\text{Mo} + 8.1\text{V} - 2 \qquad [1.22]$$

The steel is not susceptible if $\Delta G<0$.[25] A modification of this formula, to take account of the carbon content of the steel, and where $\Delta G<2$ to avoid cracking, is:[26]

$$\Delta G1 = \text{Cr} + 3.3\text{Mo} + 8.1\text{V} + 10\text{C} - 2 \qquad [1.23]$$

A different formula, due to Ito and Nakanishi,[27] is:

$$P_{SR} = \text{Cr} + \text{Cu} + 2\text{Mo} + 10\text{V} + 7\text{Nb} + 5\text{Ti} - 2 \qquad [1.24]$$

The steel is susceptible if $P_{SR}>0$.

Factors influencing weldability

For the other major type of cracking, lamellar tearing, no formulae have been developed because, as explained in Chapter 4, the problem is one of inclusion population, rather than steel composition.

Embrittlement and other formulae

Types of embrittlement can also be characterised to some degree by chemical formulae. In addition, standard metallurgical formulae, given in appropriate textbooks and compilations of transformation data, can be used to estimate such parameters as transformation start and finish temperatures which may be required when determining weldability, welding conditions and/or heat treatments.

Strain ageing is caused by any free nitrogen and carbon dissolved in solid solution in the ferrite lattice interfering with the movement of dislocations (which allow plastic deformation) and thus embrittling the steel. Most attention has been concentrated on free nitrogen, the presence of which is likely in as-rolled and possible in normalised rolled steels (but not in steels that have been normalised and tempered) and in weld metals if the 'free' Al content is less than twice the nitrogen content, i.e. if:

$$Al_{free} < 2 \cdot N_{total} \qquad [1.25]$$

If a 'free' Al figure is not available, the following approximation should be used (but not for weld metals) to avoid the risk of strain-ageing:

$$(Al_{total} + 0.010) > 2 \cdot N_{total} \qquad [1.26]$$

Attempts have been made to modify Eq. [1.25] and [1.26] to allow for the beneficial effects of Ti and B as nitride formers. As these depend on the elements being in the 'free' state, they are not included here because they are not relevant to weld metals.

To avoid the risk of **temper embrittlement**, the so-called *J*-factor[28] has been used, particularly for the 2¼Cr:1Mo steel, which is particularly susceptible to this problem:

$$J = (Si + Mn) \times (P + Sn) \times 10^4 \qquad [1.27]$$

Values of *J* should be kept below a predetermined value (e.g. 180).

However, this formula is a simplified one, as other impurities such as As and Sb are also important. A more general parameter P_E, has been proposed for weld metals:[29]

$$P_E = C + Mn + Mo + Cr/3 + Si/4 + 3.5 (10P + 5Sb + 4Sn + As) \qquad [1.28]$$

The maximum value to avoid serious embrittlement depends to some degree on the welding process,[29] but it is apparent that values should be below about 2.8–3.0 where coarse grained weld metal is present.

Bruscato[30] has developed several slightly different formulae for different steel types. These all relate the (Mn + Si) value (in wt %) to different impurity parameters ($\overline{X}$, $\overline{Y}$ or $\overline{Z}$; all with impurity contents in ppm). For the 2¼Cr : 1Mo steels:

$$\overline{X} = (10P + 5Sb + 4Sn + As)/100 \qquad [1.29a]$$

For 3.5Ni : 1.75Cr : 0.5Mo : 0.1V steels:

$$\overline{Y} = (10Sb + 5Sn + 2P + As)/100 \qquad [1.29b]$$

For 3.5% Ni steel:

$$\overline{Z} = (4.5Sn + 2.5Sb + 2P + As)/100 \qquad [1.29c]$$

Low values of each parameter give low degrees of embrittlement on prolonged exposure to temper embrittling temperatures as described in Chapter 10.

Heat treatment and creep

The generally accepted method for comparing the possible effects of two different heat treatments, which differ not only in time but also in temperature, is to use a parameter that combines the two, such as the Holloman–Jaffe or Larson–Miller parameter:

$$H_P = T(20 + \log t) \times 10^{-3} \qquad [1.30]$$

where T is the temperature, in K (=°C+273), and t the time in hours.

Such parameters are useful to compare heat treatments, and also creep service and testing data. However, they should be used with some caution whenever there is the possibility of the onset of a metallurgical reaction at a temperature between those being compared.

Slags and fluxes

The chemistry of welding fluxes and slags is a complex subject, and reference should be made to ref. 3 for a simple introduction. There some of the more commonly used chemical compounds or mineral ingredients have been listed in order of increasing basicity. That list is reproduced in Table 1.2.

In much the same way as a carbon equivalent formula was developed to characterise the weldability of a steel in a single number, the basicity index was developed[31] as a simple way of characterising the chemical behaviour of welding fluxes for ferritic steels, and their slags. The basicity index (BI), which is in most widespread current use is:

$$BI = \frac{CaO + CaF_2 + MgO + K_2O + Na_2O + 0.5(MnO + FeO)}{SiO_2 + 0.5(Al_2O_3 + TiO_2 + ZrO_2)} \qquad [1.3]$$

Table 1.2 Compounds in order of increasing basicity

SiO_2	Silica
Al_2O_3	Alumina
TiO_2	Rutile
ZrO_2	Zirconia (often Zr silicate)
FeO	Iron oxide (common impurity)
MgO	Magnesia (usually in $MgCO_3$ or (Mg, Ca) CO_3
MnO	Manganese oxide (usually in Mn silicate)
CaO	Lime (usually $CaCO_3$ or other compounds)
Na_2O	Sodium oxide (in binder as silicate)
K_2O	Potassium oxide (in binder as silicate)

Although this formula has some disadvantages, particularly with the alumina type of submerged arc fluxes (in which the Al_2O_3 is less acidic than indicated by the BI formula), it has advantages of fairly widespread acceptance and simplicity, in that the compounds are in weight percentages, which are easier to obtain than the molar fractions which would be used on purely chemical grounds.

The major 'competitor' to the basicity index is the use of oxidation potentials, which are useful to welding chemists, studying chemical reactions during welding, but less easy for the welding engineer to use.

References

1. Cottrell, C.L.M., *Welding Cast Irons*, TWI, Cambridge, 1985.
2. British Standard BS 5135: 1984, 'Arc welding of C and C-Mn steels', BSI, London, 1984.
3. Davis, M.L.E., *Introduction to Welding Fluxes*, TWI, Cambridge, 1981.
4. Kotecki, D.J. and Howden, D.G., 'Submerged-arc weld hardness and cracking in wet sulfide service', *WRC Bull*. No. 184, June 1973.
5. Anon., *Compendium of Weld Metal Microstructures and Properties*, published by TWI for IIW, Cambridge, 1985.
6. Evans, G.M., 'The effect of Ti in SMA [MMA] C-Mn steel multipass deposits', *Weld. J.*, 1992, **71**, 447s-454s.
7. Bailey, N., 'Ferritic steel weld metal microstructures and toughness', in *Proceedings of the Conference on 'Perspectives in Metallurgical Developments'*, Sheffield University 1984, Metals Society, London, 1984.
8. Terashima, H. and Hart, P.H.M., 'Effect of aluminum in C-Mn-Nb steel submerged arc welds', *Weld. J.*, 1984, **63**, 173s-183s.
9. Jones, W.K.C. and Alberry, P.J., 'A model for stress accumulation in steels during welding', in TWI Conference on *Residual Stresses and their Effects in Welded Construction*, London, Nov. 1977, TWI, Cambridge, pp. 15-26.
10. Wheatley, J.M. and Baker, R.G., 'Factors governing the yield strength of a mild steel weld metal deposited by the metal arc process', *Brit. Weld. J.*, 1963, **10**, 23-28.
11. Leggatt, R., 'Residual stress measurements at repair welds in pressure vessel steels in the as-welded condition', TWI Res. Report 315/1986, Oct. 1986.

44 Weldability of ferritic steels

12 Dearden, J. and O'Neill, H. 'A guide to the selection and welding of low alloy structural steel', *Trans. Inst. Welding*, 1940, **3**, 203–214.
13 Bailey, N. (ed.), *Hardenability of Steels, a Select Conference*, TWI, Cambridge, 1990.
14 AWS, *Structural Welding Code – Steel*, ANSI/AWS D1.1-92, AWS, Miami, Fl., 1992.
15 Bailey, N., Coe, F.R., Gooch, T.G., Hart, P.H.M., Jenkins, N. and Pargeter, R.J., *Welding Steels Without Hydrogen Cracking*, Abington Publishing, Cambridge, 1993.
16 Bailey, N., 'The establishment of safe welding procedures for four low alloy steels', *Weld. J.*, 1972, **51**, 169s–177s.
17 Cottrell, C.L.M., 'Hardness equivalent may lead to a more critical measure of weldability', *Metal Construction*, 1984, **16**, 740–744.
18 Düren, C., 'Equations for the prediction of cold cracking resistance on field-welding large diameter pipes', *3R Intl*, 1985, **24**, 434–439; also IIW Doc. IX-1458-87.
19 Bailey, N. and Pargeter, R.J., 'The influence of flux type on the strength and toughness of submerged arc weld metal', TWI Report Series, Cambridge, 1988.
20 Pargeter, R.J., 'Yield strength from hardness – a reappraisal', *TWI Confidential Res. Bull.*, 1978, **19**, 325–326.
21 Hart, P.H.M., 'Yield strength from hardness data', *TWI Confidential Res. Bull.*, 1975, **16**, 176.
22 Huxley, H.V., 'The influence of composition on weld solidification cracking in C–Mn steel', *Metallurgia*, 1970, **82**, 167–174.
23 Morgan-Warren, E.J. and Jordan, M.F., 'A quantitative study of the effect of composition on weld solidification cracking in low alloy steels', *Metals Tech.*, 1974, **1**, 271–278.
24 Bailey, N. and Jones, S.B., 'The solidification cracking of ferritic steel during submerged arc welding', *Weld. J.*, 1978, **57**, 217s–231s.
25 Nakamura, H. et al., *Proceedings of the 1st International Conference on Fracture*, 2, Sendai, Japan, 1965, p. 863.
26 Haure, J. and Bocquet, P., 'Underclad cracking of pressure vessel components', Convention CECA 6210-75-3/303, Creusot-Loire, final report, Sept. 1975.
27 Ito, Y. and Nakanishi,M., IIW Doc. X-668-72, IIW, 1972.
28 Ishiguo, T., Murakami, Y., Ohnishi, K. and Watanabe, J., '2.25%Cr–1%Mo pressure vessel steel with improved creep rupture strength', in *Proceedings of the Symposium on Applications of 2.25%Cr–1%Mo steel for thick-wall pressure vessels*, Denver, Col., 1980, ASTM STP 755, ASTM, Philadelphia, PA, pp. 129–147, 1982.
29 Sugiyama, T., Hatori, N., Yamamoto, S., Yoshino, F. and Kiuchi, A., 'Temper embrittlement of Cr–Mo weld metals', IIW Doc. XII-E-6-81, IIW, 1981.
30 Bruscato, R.M., 'Embrittlement factors for estimating temper embrittlement in 2¼Cr:1Mo, 3.5Ni–1.75Cr–0.5Mo–0.1V and 3.5Ni steels', ASTM Conference, Miami, Florida, Nov. 1987.
31 Tuliani, S.S., Boniszewski, T. and Eaton, N.F., 'Notch toughness of commercial submerged arc weld metals', *Weld. Metal Fabrication*, 1969, **37**, 327–329.

2 Potential welding problem areas

Six potential problem areas which need to be considered in welding are summarised below and discussed in more detail in the chapters mentioned.

Cracking (Chapters 3–6)

Although cracking comes under many names, there are basically only four types to be considered, of which two are restricted to limited types of steels. However, because welding itself is a complicated process, their understanding and control is not always straightforward.

The four types are:

1 Cracking occurring at the late stages of solidification of the weld – solidification cracking of weld metal and liquation cracking, usually restricted to the HAZ in ferritic steels.
2 Lamellar tearing – restricted to plate steels.
3 Hydrogen cracking in both HAZ and weld metal.
4 Cracking during PWHT – restricted to a limited range of steels and weld metals and only when subjected to PWHT.

Other types of cracking have not yet been firmly distinguished from other more likely types (e.g. the term 'strain ageing cracking' is likely to be a variant of hydrogen cracking in which strain ageing is unusually significant), or are insignificant in the context of welding ferritic steels (i.e. ductility dip cracking, which is discussed in Chapter 3).

Welding faults (Chapter 7)

Although the control of welding faults is a matter of welding engineering, i.e. selection, control and maintenance of correct operating parameters, there are metallurgical aspects in the formation, presence and subsequent identification of welding imperfections and defects. These are discussed, as well as how measures taken to deal with metallurgical problems may influence welding faults.

Inspection for defects (Chapter 8)

The correct selection and operation of NDE is strictly outside the scope of this book. Nevertheless, certain aspects are discussed because some details of inspection techniques are very much related to the types of defect likely to be present, whilst the requirements of the inspection imposed may impinge on the selection of welding methods. Also, a good metallurgical understanding of the types of defects likely to be present influences the type, scale and time of NDE required, as well as the interpretation of its results.

Joint integrity (Chapter 9)

Joint integrity is concerned with the selection of consumables and welding parameters to meet the requirements imposed on weldments. The properties of most importance are weldment strength and toughness. The principles behind the selection of these features are discussed in the chapter. Other properties, such as resistance to creep, corrosion, oxidation and hydrogen content are more especially related to service conditions.

Problems associated with PWHT (Chapters 6, 9 and 10)

The most common problem in heat treatment is that of distortion. Careful control of heating and cooling, with jigging in difficult cases, is important. Reheat or stress relief cracking is a problem restricted to a limited number of steels when they are subjected to conventional PWHT, and is discussed in Chapter 6.

Temper embrittlement is a progressive embrittlement of steels when they are subjected to (often prolonged) heating within the temperature range 350–600 °C. The problem is confined to low alloy steels which contain relatively high levels of Mn, Si and the impurities P, Sn, Sb and As. Temper embrittlement is discussed as a service problem in Chapter 10, but it should be noted that steels with very susceptible compositions, and which are in thick sections, may cool so slowly through the sensitive temperature range after PWHT that they become embrittled.

Strain ageing is a problem of as-welded joints and its effects are completely removed by PWHT.

Potential welding problem areas

Service problems (Chapter 10)

In selecting steels, welding consumables and welding parameters, a fairly detailed knowledge of service conditions is needed, particularly where corrosion, high temperatures and/or the presence of high pressure hydrogen are involved.

Repair (Chapter 11)

Much welding is carried out to modify or repair components that have been in service, and that need to be welded under conditions which are less favourable to achieving an ideal weld than the shop conditions where the component was originally welded. The possibility of such repair welding should be borne in mind at the original design stage.

Difficulties may take the form of the need for positional welding, the inability to deploy the ideal welding process because of space limitations, restrictions on heating (either preheating temperatures or the inability to apply PWHT), because of nearby heat-sensitive material which cannot be moved or the location of the repair. In some cases there may be a requirement to weld pipework containing fluids (hot tapping) or to weld under water or in vacuum (e.g. in space).

Joining dissimilar steels

The selection of welding consumables and procedures when joining dissimilar steels is normally a relatively simple matter. However, the selection of PWHT temperatures can pose problems. This section is concerned solely with joining of two ferritic steels; joints involving austenitic stainless steels or cast irons are discussed in the next section, along with joining ferritic steels to other metals.

When joining two ferritic steels of different strength levels and different toughness requirements, simple considerations will show that only the strength of the weaker steel and the toughness of the less tough steel will be required at the joint. For example, when joining a 700 N/mm^2 yield strength steel, requiring Charpy toughness at -50 °C, to a mild steel with toughness requirements at ambient temperature, the properties required at the joint itself should be those of the mild steel. Similarly, if a steel selected for its oxidation and creep resistance has to be joined to a simple C or C:Mn steel, the consumable does not require any alloying additions, other than those necessary to achieve the strength of the weaker steel.

The selection of welding procedures, particularly preheat and interpass temperature and heat input, will normally depend on the requirements for the more highly alloyed of the two steels, bearing in mind the importance of carbon as an alloying element. Thus, in the previous example, preheat requirements and so forth should be in accordance with the needs of the high strength, alloyed steel.

If, however, the high strength steel (assuming it to be of low carbon content, e.g. <0.15%) is being joined to a simple medium carbon steel, then the requirements of the latter for preheat to avoid hydrogen cracking need to be taken into consideration, although some degree of compromise may be required. It will usually be helpful to use consumables giving ultra-low hydrogen levels. It may also be necessary to allow higher preheat and interpass temperatures than are normally acceptable for the tougher steel, on the grounds that full HAZ toughness of the higher strength steel will not be required at the dissimilar joint. In cases where a compromise is not possible, consideration should be given to the use of austenitic stainless steel or nickel alloy fillers, as discussed in the next section and in Chapter 5, because such fillers allow the use of low preheat temperatures (which are likely to be needed for the strong tough steels) for higher carbon steels.

The PWHT of dissimilar joints requires some care. Firstly, there is a need to avoid over-tempering and weakening any quenched and tempered or similar steel. Secondly, there is a need to avoid impairing the HAZ toughness (and other properties) of either steel. Some guidance in these matters is given in the British pressure vessel standard BS5500.[1] The basic principle is that one of the two steels must be the more important and the PWHT temperature range should be selected to ensure that its correct properties are achieved. Then the temperature is selected within that range to be as close as possible to what is required for the secondary steel. For example if a C:Mn steel (PWHT temperature 570–630 °C) is being welded to a major low alloy steel component (PWHT 620–680 °C) the aim PWHT temperature will be 630 °C, although particular care will be needed to achieve better than normal temperature control.

If, on the other hand, a Cr:Mo steel (requiring PWHT at 700–750 °C) is welded to the C:Mn steel, then an alternative strategy is needed. This involves buttering the weld preparation with two layers of the weld metal selected for the joint (usually a C:Mn weld metal) and applying PWHT at 725 °C. The joint to the C:Mn steel is then completed and the whole is given a PWHT at 600 °C.

The higher temperature will temper the Cr:Mo steel HAZ, and will not unduly harm the two layers of weld metal, which will have picked up some Cr and Mo by dilution from the parent steel and will also be partly melted out when the joint is completed. The subsequent 600 °C treatment will not harm the Cr:Mo steel and will be suitable for both the C:Mn steel and the weld metal. During this procedure, great care should be taken when the joint is being completed not to weld too close to the Cr:Mo steel after it has been buttered and given its higher temperature heat treatment. Low current, low dilution techniques should be used for the first few runs on to the Cr:Mo steel side of the complete joint, and the weld toes should approach it no closer than 2–3 mm. Such dissimilar steel joints must always be made *after* all of the joints requiring the higher temperature PWHT have been welded and heat treated.

Free-cutting steels

Although the use of such steels does not normally increase the risk of hydrogen cracking, their high sulphur contents give a serious risk of weld metal solidification cracking and of minute liquation cracks (also known as hot tears – Chapter 3) in the HAZ of the free cutting steel. Little can be done about the latter, but the risk of the potentially more serious solidification cracking can be considerably reduced by welding with a low dilution (i.e. low current) technique on to the free cutting steel and by using basic covered manual electrodes. The latter help to reduce the risk of cracking by removing some of the sulphur (an element which confers good machinability to steels) from the weld metal into which it has been diluted from the free-cutting steel.

Where large welds need to be made, it may be necessary to use **MMA** welding as a buttering technique on to the free-cutting steel before completing the joint with more efficient welding processes, such as **MIG** or submerged arc. When this approach is used, care should be taken to ensure that the more efficient welding process does not burn through the layer of buttering and pick up sulphur from the free-cutting steel.

Plated and coated steels

When welding plated and coated steels, the amounts of coating which enter the weld metal are sufficiently small to be ignored. Nevertheless, when welding these materials there can be problems, which tend to be more serious as the strength and degree of alloying of the steel are increased.

When welding painted steels, the main problem is that of hydrogen cracking in the HAZ, resulting from hydrogen picked up when the paint is heated and decomposed. Safe practice is to remove the paint within at least 3 mm of the weld preparation, although weld-through priming paints are available, which give little significant pick-up of hydrogen so that they can be safely welded over (except perhaps when welding high strength, medium to high carbon steels).

With two exceptions, steels coated with zinc can be welded without metallurgical problems, because the zinc usually boils off and is not picked up by the hot HAZ. However, welding such steels increases the amount of welding fume and may also affect the welding characteristics, for example by reducing penetration and increasing undercut and spatter. Problems may be expected if the zinc is trapped, e.g. between the two fillet welds of a T-joint, or if the steel is of high strength, medium to high carbon content and/or alloyed. In both these cases, it is possible for the zinc to penetrate the HAZ (and weld metal) by a mechanism of liquid metal penetration. Such penetration is not acceptable because of likely reductions in strength and toughness, as well as cracking occurring directly or later. In such cases the coating within about 3 mm of the weld preparation should be removed before welding. Alternatively, in the case of fillet welds, a deliberate root gap is helpful in reducing the likelihood of zinc penetration.

It has been claimed that weld metal of low Si content is particularly resistant to liquid metal penetration, although this effect in the original tests could alternatively have been due to a reduction in weld metal strength and, hence, residual stress level.

Similar precautions are advisable in the case of coatings of other low melting point metals, unless trials and previous experience have shown that removal is not necessary. In any event, a change to a higher strength steel should always require that the procedure should be checked.

Steels which have been case-hardened, either by carburising or nitriding, will not lead to injurious pick-up into the weld metal. Nevertheless, excessive hardening of the HAZ surface may occur, and this should be taken into account when devising welding procedures (including the need for PWHT).

Joining dissimilar metals

The joining of dissimilar metals is a large topic, because of the vast numbers of possible materials, and the large number of available joining techniques, which include brazing and soldering, bolting, riveting and adhesives, as well as welding. Most of the subject area is really the subject of a text on non-ferrous joining metallurgy, but some guidance is given here in the more likely areas where ferritic steels are to be welded to other metals. For further guidance, reference should be made to standard textbooks, and also the proceedings of a TWI seminar on the subject.[2]

Using fusion welding techniques, materials that have melting temperatures and general characteristics similar to ferritic steels can usually be welded to them. Such metals include austenitic stainless steels, weldable cast irons, many nickel alloys and some copper alloys. For other metals, non-fusion processes such as friction welding and diffusion bonding, as well as non-welding processes should be considered. In some cases, it is possible to prefabricate transition pieces by these means which can then

Potential welding problem areas

be fusion welded, on site if necessary, into the rest of the structure. One example of this technique is the use of friction welded transition pieces between aluminium and mild steel in order to join aluminium bus bars to mild steel anode holders.

Major considerations in dissimilar metal welded joints are corrosion and mechanical properties (including creep and toughness) at and near the joint line, where unexpected deterioration in properties may occur as a result of diffusion of elements (particularly carbon) across the interface. Such behaviour is particularly likely as a result of PWHT or other heat treatments, or during high temperature service. These effects can give rise to low toughness at lower temperatures and low creep ductility at high temperatures.

In corrosion service, accelerated corrosion of the more anodic metal should be considered where service conditions make this likely, particularly where the exposed area of the anodic material is smaller than the cathode.

Austenitic stainless steels

Joints to stainless steel are usually welded using fillers of the 29Cr:9Ni:3Mo, or similar type, designed for welding 'difficult' steels without preheat, although a filler appropriate to the stainless steel being welded may be a better option. Risks are posed by:

1 Hydrogen cracking of the ferritic steel. This is considered in Chapter 5, p 152–155.
2 The formation of the embrittling sigma phase during any PWHT required for the ferritic steel.
3 The consequences of the differences in coefficients of thermal expansion between the two types of steel, particularly during PWHT and in high temperature service and especially where thermal fatigue is likely. These effects can sometimes be minimised by the selection of a nickel alloy filler having an expansion coefficient intermediate between the two steels.
4 Deterioration of properties near the fusion boundary or joint line as a result of the diffusion of carbon and other elements during PWHT or high temperature service.

Cast irons

Not all cast irons can be welded, and most require careful, low current welding techniques, designed to minimise cracking and excessive hardening in the HAZ.[3] Frequently, pure nickel or its alloys are used as fillers, and these are normally used when welding cast irons to ferritic steels. Where low heat input techniques are necessary for the cast iron, such nickel-based fillers will ensure that preheating will not be required for the steel. On the other hand, some cast irons require such high preheat temperatures (up to 330 °C with MMA and MAG and 600 °C with gas fusion welding – ref. 3) that adequate HAZ properties may not always be possible in the steel.

Because of the high hardenability of cast irons, very hard HAZs can be expected, although these can be reduced by PWHT around 620 °C followed by furnace cooling to about 300 °C. However some cast irons will develop a white iron structure, which requires heating to 900–950 °C; such temperatures may irretrievably damage the properties of some steels and weld metals.

Wrought iron

Wrought iron is normally of very low carbon and alloy contents but contains much rolled-in slag. This slag can give rise to defects in the weld metal if too much parent metal is incorporated in the joint. To minimise dilution, a low current technique should be used on the wrought iron itself. As the carbon content is low enough to make hydrogen cracking unlikely, MMA welding with rutile electrodes is the preferred option, provided this does not conflict with the requirements for the steel being welded to the wrought iron. In such cases, basic iron powder electrodes of the lowest available strength should be used.

Nickel and its alloys

Most weldable nickel alloys can be welded to ferritic steels, using nickel alloy fillers, with little risk of hydrogen cracking, even though preheat may not be used. Solidification cracking is a major risk if the ferritic steel contains a significant sulphur content, and it may be necessary to select a filler with a sufficiently high niobium content to cope with this, provided it is compatible with the nickel alloy being welded, and also use low dilution welding techniques. Although many nickel alloys have thermal expansion characteristics similar to ferritic steels, diffusion across the interface may pose problems during heat treatment and in high temperature service.

Copper and its alloys

The major risk in fusion welding copper alloys to steels, using copper alloy fillers, is that, because of its low freezing temperature, the copper alloy can penetrate an appreciable way into the HAZ of the steel (Chapter 3). Although this may not lead to cracking directly, it weakens the joint, and such fusion welds are only suitable for lightly stressed service.

Light alloys

Aluminium, magnesium, beryllium and their alloys cannot be fusion welded to ferritic steels, nor are brazing and soldering advisable. Friction welding and, in some cases, diffusion bonding are possible, but riveting and the use of adhesives probably provide more general solutions. Special precautions are also necessary when handling beryllium, because of the poisonous nature of many of its compounds.

Clad steels

Steels clad with austenitic stainless steels or nickel alloys are usually welded so that the bulk of the ferritic steel (which is normally the thicker material) is welded first and the final runs (where the cladding is incorporated into the joint) are carried out using appropriate stainless steel or nickel alloy fillers.

Single-sided closing welds, and full thickness repairs which can only be made from the unclad side, provide a problem, in that the alloy filler must be used for the whole weld. This is because welding with a mild or low alloy steel filler on to the stainless steel or nickel alloy weld metal used to weld the cladding will, because of dilution, produce a highly alloyed layer of weld metal. This layer has poor properties and is at serious risk of cracking.

Other materials

Few rules can be given here, as the potential range of materials, joining processes and steels is too great. Major problems are particularly likely where one of the materials is of poor toughness, where properties (particularly expansion coefficients and elastic moduli) differ appreciably, and where relatively high temperatures are needed for the joining process.

References

1 British Standard BS5500: 1991, 'Unfired fusion welded pressure vessels', BSI, London, 1991.
2 Bailey, N. (ed.). 'Welding dissimilar metals', TWI Seminar, TWI, Cambridge, 1986.
3 Cottrell, C.L.M., *Welding Cast Irons*, TWI, Cambridge, 1985.

3 Solidification cracking

In addition to solidification cracking, which occurs exclusively in the weld metal, this chapter contains brief descriptions of liquation cracking – a HAZ phenomenon in ferritic steels – and ductility dip cracking. The last named is a type of high temperature cracking which is little more than a theoretical possibility in ferritic steels, although it has been identified in certain stainless steels.

Description

When the liquid weld metal solidifies, it contracts. The earliest stages of contraction are taken up by hotter liquid metal from near the arc moving against the direction of welding into the solidifying weld metal. As more metal solidifies, this flow is impeded and further contraction occurs by the partially solid weld metal settling down as its volume is reduced. At the end of a weld run – the weld crater – contraction takes place without any feed of hotter metal from elsewhere unless special crater filling techniques are used at this stage. The crater sinks appreciably below the normally fed weld metal elsewhere, as can be seen in Fig. 3.1, and usually contains shrinkage, porosity and/or solidification cracking (Fig. 3.1(c)), as the sinking of the surface is not usually able to compensate fully for the solidification contraction at the end of the weld.

Away from the weld crater, cracking is less likely, except when the weld metal has a long freezing range, particularly when most of the weld metal solidifies at high temperatures within this long freezing range. Such cracking is usually known as solidification cracking, although the term 'hot cracking' is also widely used.

When cracking does occur, it is because a fairly rigid structure has formed late in solidification, with only narrow channels between the solid grains through which any liquid metal can flow and counteract (or feed) the contraction which takes place on solidification and subsequent

Solidification cracking

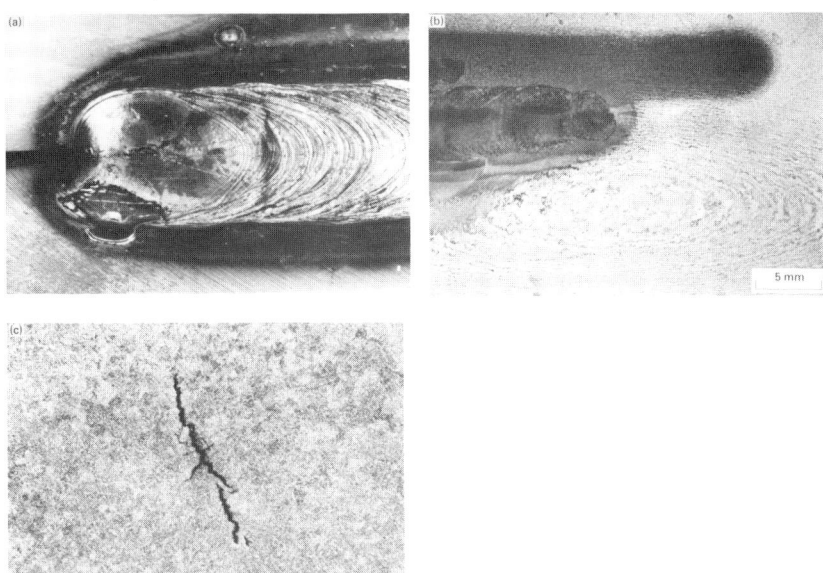

3.1 Weld crater: (a) surface showing open crack; (b) plan section; (c) detail of cracking.

cooling. The last stage of solidification is for the liquid in the intergranular channels to freeze on to the existing solid grains. When the weld metal freezing range is long, the liquid metal has a long distance to flow and there is a risk that voids, i.e. solidification cracks, will be left at the grain boundaries. Very rarely, these voids are short, isolated voids along the grain boundaries, as shown in Fig. 3.2; usually they form continuous solidification cracks, Fig. 3.3.

By analogy with the behaviour of castings (which suffer from 'hot tears', equivalent to weld metal solidification cracks), it is likely that gas dissolved in the weld metal is necessary to nucleate cracking. However, in ferritic steels some gas is always available, both as dissolved hydrogen and as carbon monoxide (a product of the reaction between oxygen and carbon in the weld metal).

The compositional factors giving rise to the freezing behaviour likely to result in solidification cracking are discussed in detail later, but the main cause is the presence of elements – particularly sulphur and phosphorus – giving rise to a small amount of compounds which give eutectics of low melting point with iron, thus extending the freezing range.

The solidification of a weld metal is progressive from the fusion boundary towards the weld centre-line and angled towards the welding

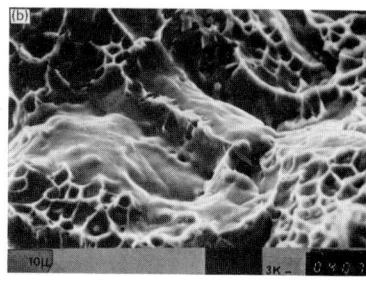
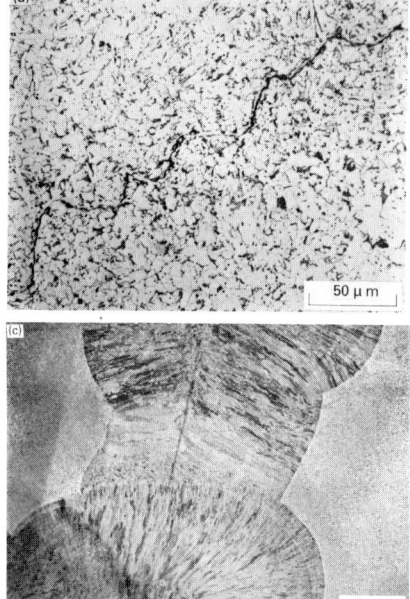

3.2 Discontinuous solidification crack: (a) microsection of part of crack; (b) SEM fractograph showing mixture of smooth solidification cracking and dimpled ductile tearing (microvoid coalescence) which occurred when the crack was broken open at ambient temperature; (c) macrosection of complete crack.

direction and to the face of the weld; any deficiency in feeding liquid metal to compensate for the contraction on solidification is likely to lead to a solidification crack at the weld centre-line. Provided the welding speed is not too fast, the contraction on solidification is made up by liquid metal fed from the hotter parts of the weld pool towards the solidifying weld metal.

If the welding speed is too fast, the growing grains, and the channels between them, become longer and narrower, thus making feeding more difficult. Similarly, if the weld pool is deep and narrow (i.e. it has a high depth/width ratio – see Fig. 3.4), or is very wide and shallow, feeding also becomes more difficult, and cracking is more likely than if the depth/width ratio is near the optimum. In addition, the solidification of a weld with a high depth/width ratio tends to be inwards from the sides, and this increases the severity of segregation of low melting material to the weld centre-line. With a less deep pool, the solidification tends to be upwards as well as inwards, thus reducing the severity of centre-line segregation

Solidification cracking 57

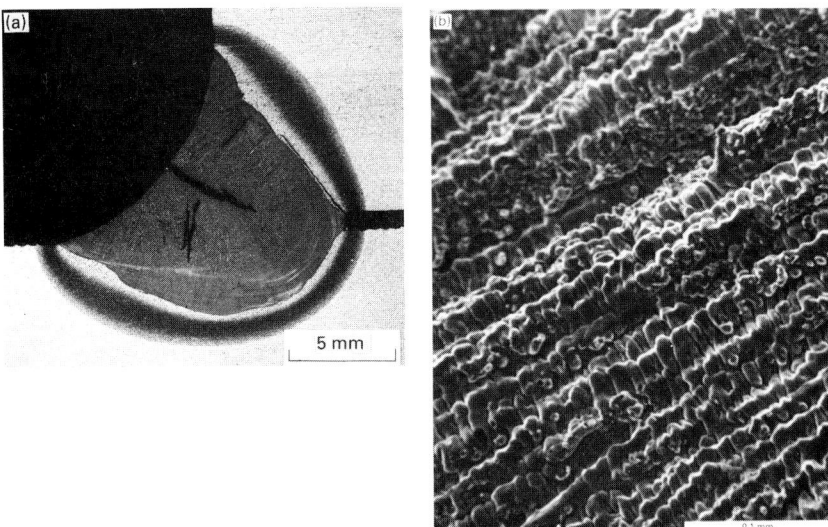

3.3 Typical continuous solidification cracks: (a) section through open centre-line and sub-surface cracks (also HAZ hydrogen crack) in fillet weld with root gap; (b) SEM fractograph of well-developed sub-surface crack.

and allowing more contraction to be taken up by sinking of the weld surface. Although a complete survey of the ideal values for the depth/width ratio does not appear to have been made, a figure of about 0.5 is likely to be somewhere near the optimum.

As the weld solidifies towards its centre-line, anything which locally delays solidification there is likely to encourage cracking, because this delay in solidification will hold back the solidification front at the weld centre from reaching the free surface, where the contraction is free. The most likely cause of delayed centre-line solidification is a root gap (Fig. 3.3(a), although severe root defects (Fig. 3.5) can have similar effects. A root gap is a very common cause of cracking and, unlike hydrogen cracking (Chapter 5), the effect becomes greater if the width of the root gap is increased. Other delays to solidification at the weld centre can be caused by lack of root fusion (Fig. 3.6) and, to a lesser degree, by the use of ceramic or very thin steel backing strips in square edged butt welds. These types of backing cause the weld to solidify directly inwards from the two sides, instead of inwards and upwards when efficient heat extraction from a thick steel backing is used.

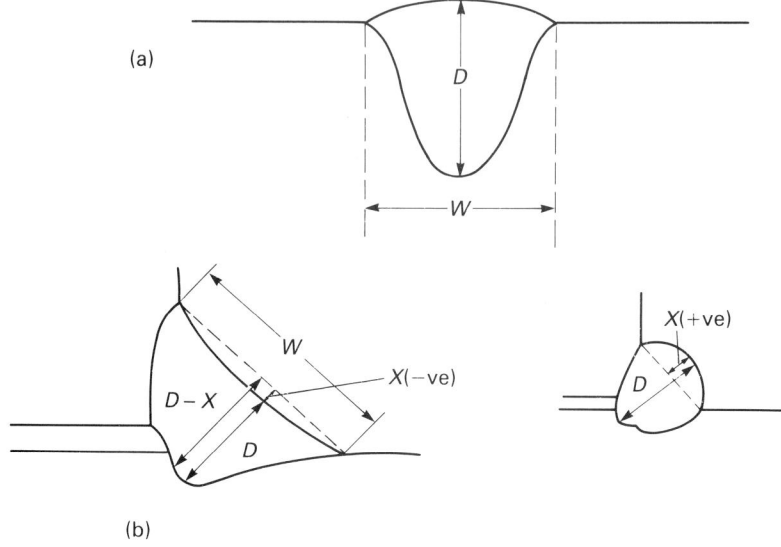

3.4 Depth to width (D/W) ratios: (a) conventional butt weld; (b) modified ratio, $(D-X)/W$, to take account of convexity or concavity of fillet weld shape.

Welds having long, narrow weld pools (associated with the use of fast welding travel speeds) are particularly at risk of cracking, and the influence of welding conditions on weld pool shape and cracking is discussed later.

Because welds normally solidify inwards from the parent metals being welded, the most likely location for solidification cracking is the weld centre-line, as shown in Fig. 3.3 and 3.5. The centre-line is that of heat extraction, rather than the geometric centre-line. This can be seen in Fig. 3.7, where an attempt has been made to weld two plates of different thickness with a single pass. Heat extraction during solidification has been predominantly through the thicker plate. The heat needed to achieve adequate fusion to the thicker steel has melted back the thinner steel so that the root gap has widened from near zero to almost 6 mm. This has led to a solidification crack at the heat centre, which is near to the fusion boundary on the thin plate. Welding was successfully carried out in two passes, because the small root pass did not involve fusing on to such a large area of the thicker steel and, hence, the thinner plate was melted back less.

As cracks occur in the last stages of solidification, they are often open to the surface, unless the weld metal forms a thin, solid skin during solidi-

Solidification cracking 59

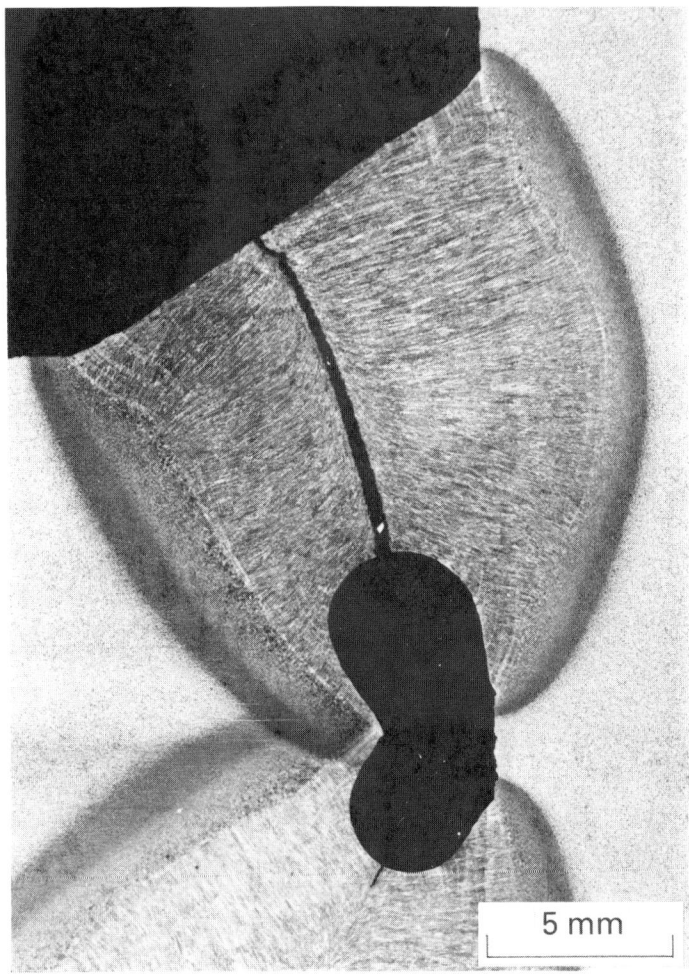

3.5 Macrosection of weld with root defect and crack.

fication (Fig. 3.8) or is so deep that the surface bridges over before the underlying weld metal solidifies (Fig. 3.9).

The resistance to cracking of welds made with convex surfaces is better than if made with sunken tops (compare Fig. 3.10 and 3.3(a) and Fig. 3.11(a) and (b)), even though the latter appear to have smaller depth/width ratios. In fact, for fillet welds, the depth/width ratio should be measured from a line joining the weld toes as shown in Fig. 3.4(b), rather than from the weld surface.

60 Weldability of ferritic steels

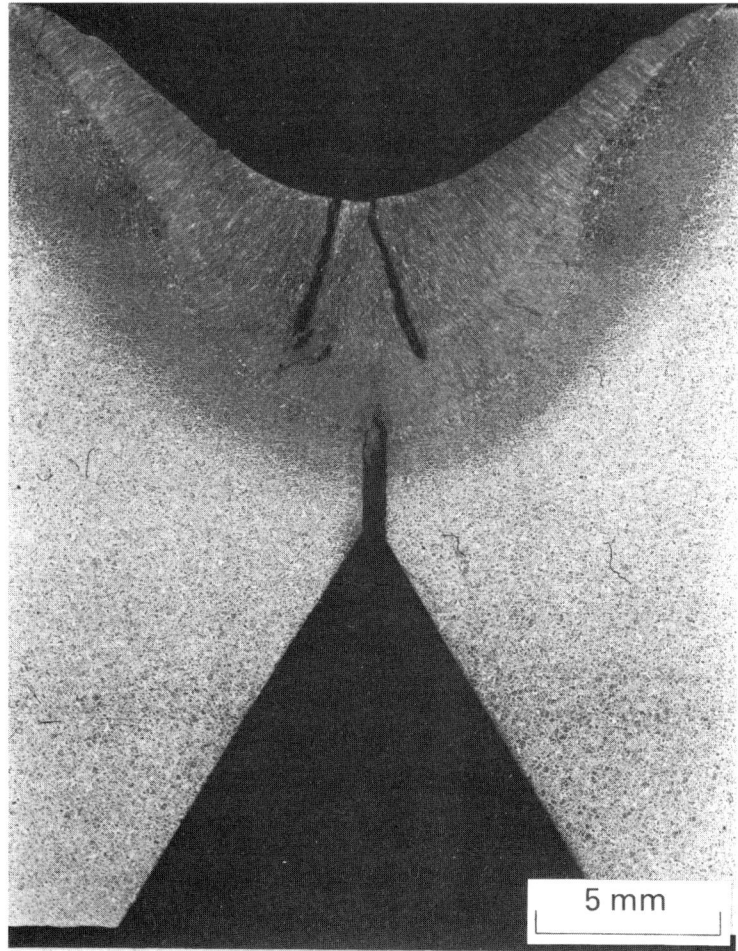

3.6 Solidification cracking in wide, shallow root run of butt weld.

Consequences of cracking

In ferritic steels, it is unusual for solidification cracks to extend by fracture of the solidified weld metal, as can be seen in the fractograph in Fig. 3.2(b) and by the macrograph, Fig. 3.8. Even where a second run of weld metal is deposited over a previous run containing a solidification crack, the first crack will not initiate a second, Fig. 3.12, unless those portions of the two runs that are the last to solidify are in line with each other. Because the crack tips are usually rounded on a microscopic scale (Fig. 3.13), they are less damaging than hydrogen cracks, fatigue cracks and other cracks which have occurred through solid metal.

Solidification cracking 61

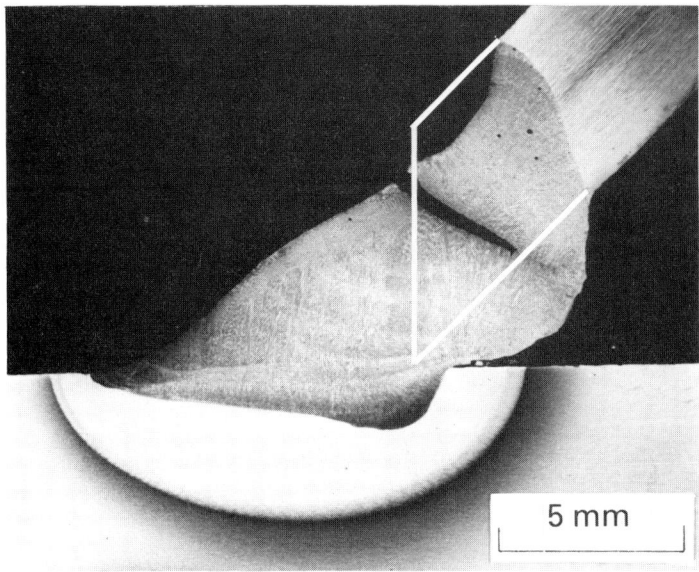

3.7 Macrosection of MAG weld between thin and thick sections in maraging steel, white line is original preparation.

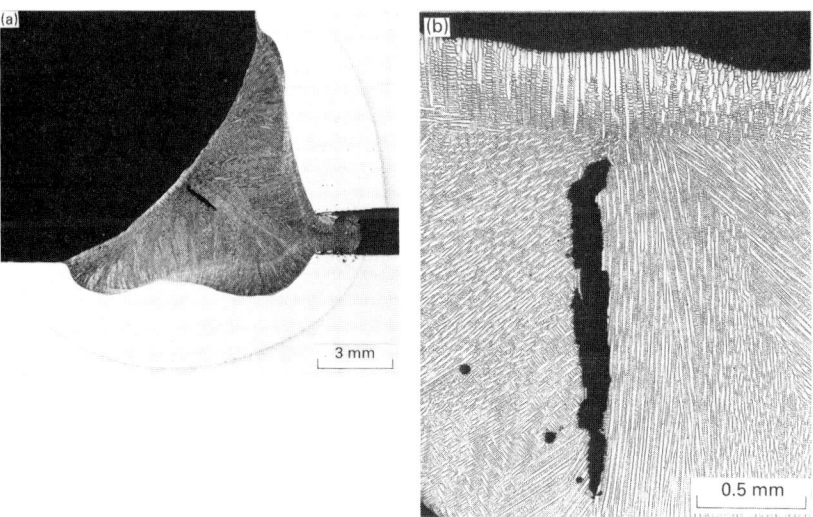

3.8 Sub-surface crack in maraging steel MAG fillet weld with root gap: (a) macrosection (note reduction of root gap near weld root due to contraction of weld); (b) detail of crack showing thin layer of weld metal which has solidified from the surface.

62 Weldability of ferritic steels

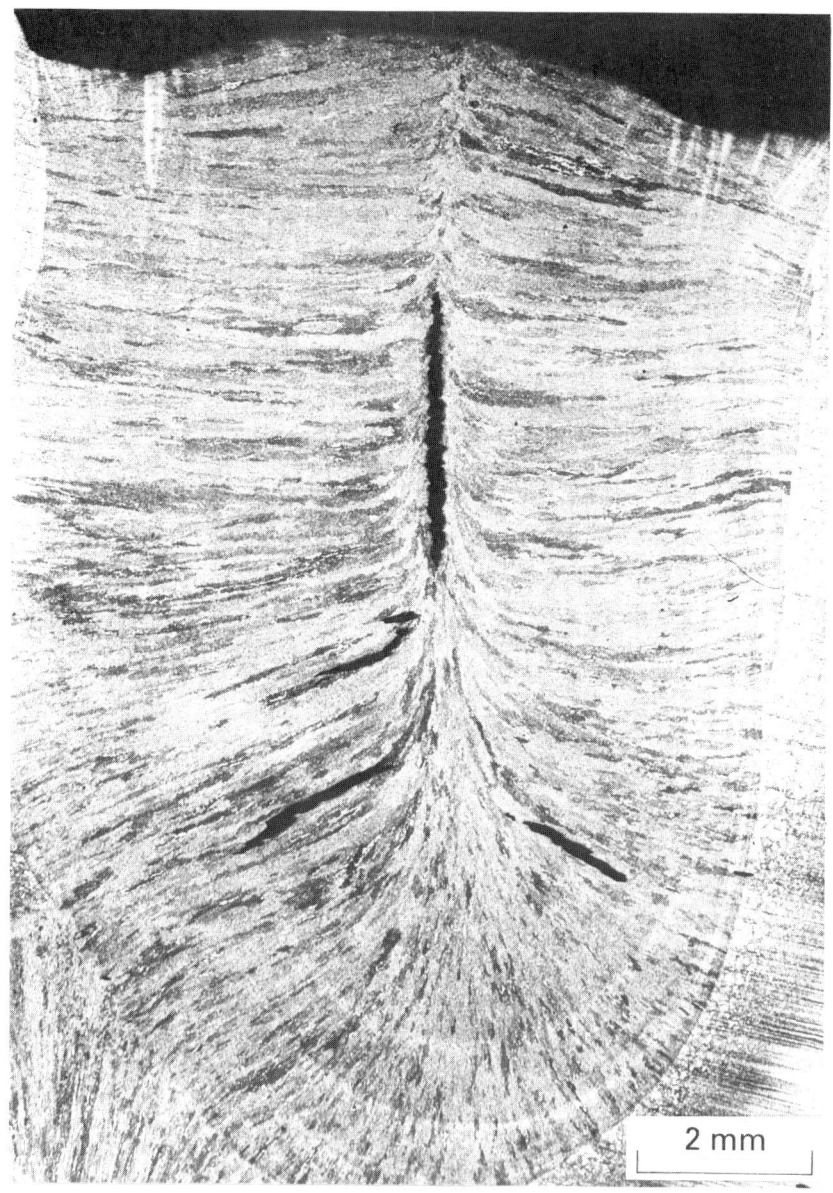

3.9 Centre-line and intercolumnar sub-surface solidification cracking in deep, narrow submerged arc weld.

Solidification cracking

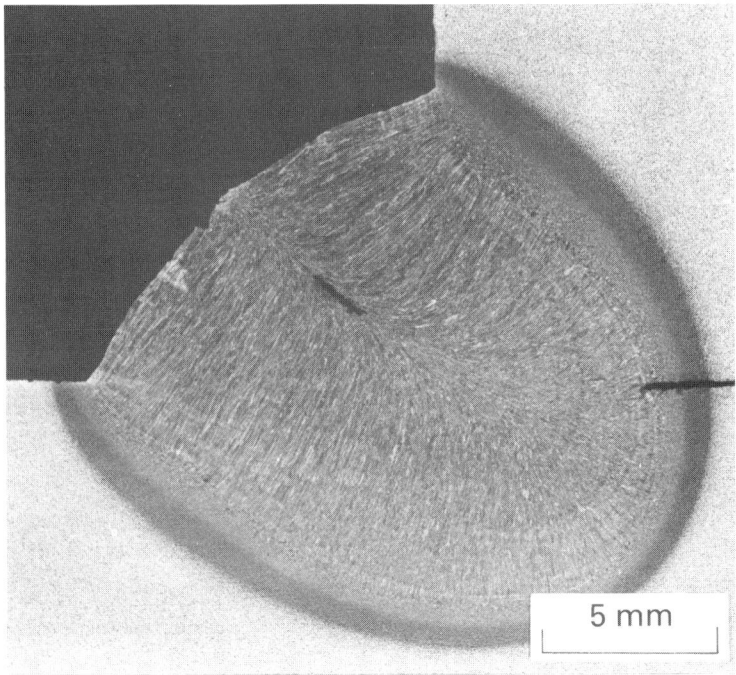

3.10 Convex-topped submerged arc fillet weld showing minor sub-surface solidification cracking.

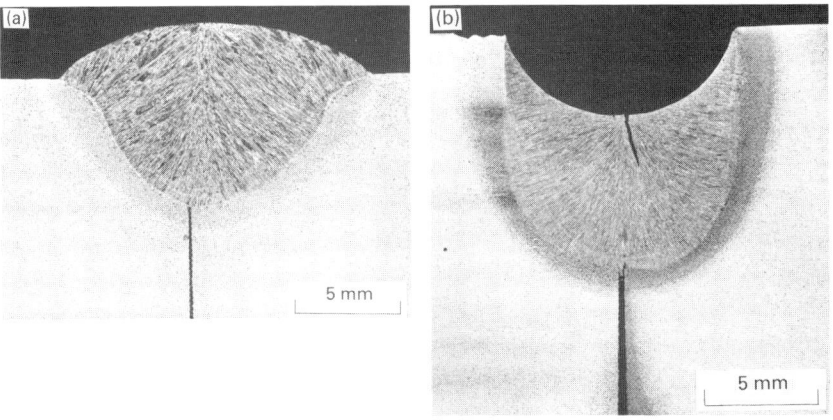

3.11 Similar submerged arc butt weld runs having: (a) convex top; (b) concave top with solidification crack.

64   Weldability of ferritic steels

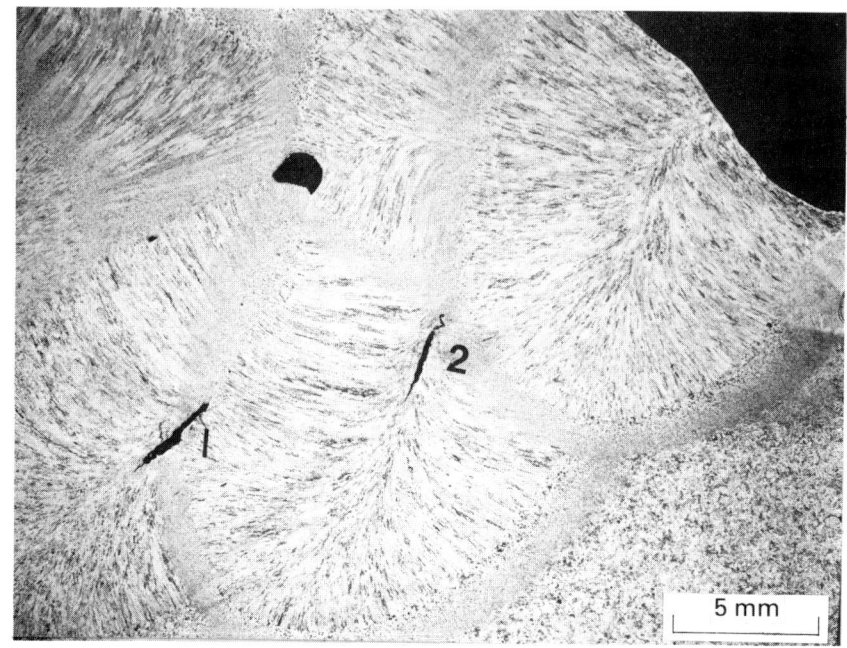

3.12 Section of part of a heavy submerged arc fillet weld showing that crack 2 was not initiated from the earlier crack 1.

Solidification cracking 65

3.13 Detail of cracks of Fig. 3.9 showing relatively blunt crack tips.

Role of composition

In any alloy system, the risk of solidification cracking is very low when welding a pure metal and using the same pure metal as filler. Any alloying element that lowers the melting temperature and increases the freezing range will increase the risk of cracking. If a eutectic system is formed, the risk of cracking will be reduced after the melting temperature range has reached a maximum, until the eutectic point is reached (Fig. 3.14(a)). Although the behaviour of steels is complicated by high temperature phase changes (i.e. the delta to gamma peritectic), this classical type of behaviour is shown by steels containing different amounts of boron[1] (Fig. 3.14(b)) or niobium. Additions of either element initially increase the risk of solidification cracking whilst further additions then reduce the risk. In general, elements that give a sharp drop in melting temperature to a low minimum (eutectic) value are the most dangerous. These elements are usually the ones with low solid solubilities, i.e. those that form a second phase, so that only a small addition is needed to give a long freezing range.

In steels, the presence of the peritectic (Fig. 3.15) complicates the situation, because most elements have different solubilities in the two solid phases, high temperature (delta) ferrite and austenite (gamma). This is particularly the case for sulphur (a very common impurity in steels) which is much less soluble in austenite than in ferrite. Hence, elements that promote solidification as austenite, rather than ferrite, generally increase the risk of solidification cracking. The most influential of these elements is carbon (Fig. 3.15); only 0.08% C is needed for solidification to be completed with some austenite, and carbon additions above 0.18% C also lead to an increase in the freezing range.

However, if the carbon content is so low that solidification is completely to delta ferrite, it has been claimed[2] that the subsequent transformation to austenite (with its volume contraction) adds to the existing contraction strains sufficiently to increase the risk of cracking with very low carbon weld metals (i.e. 0.09% C down to about 0.04% C).

Ferrite formers, such as Cr, Mo and V, reduce the risk of cracking. Nickel is the only significant austenite former generally to promote solidification cracking; although at very low carbon levels, the influence of nickel on preventing solidification as delta ferrite (with its subsequent contraction on transformation to austenite) appears to reverse this trend,[2] but only with weld metals of very low carbon content.

The other important austenite former is manganese and its behaviour is complicated by its interaction with sulphur. Without the presence of manganese, iron forms a eutectic with iron sulphide (FeS), which has a particularly low melting point – about 990 °C. Addition of sufficient

Solidification cracking

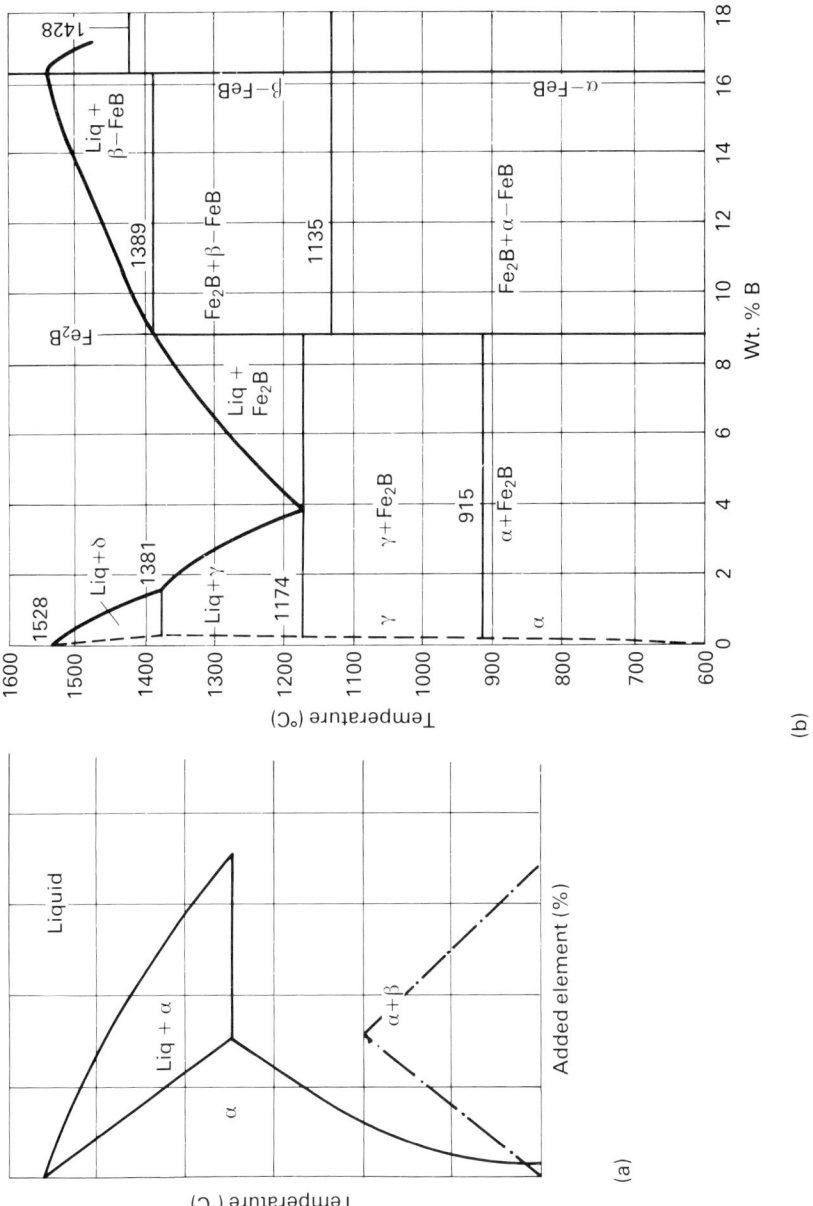

3.14 Parts of typical phase diagrams containing eutectics: (a) showing how the risk of cracking (chain-dotted line) increases as the freezing range increases; (b) Fe–B system showing low-melting eutectic, present at very low boron contents.

Weldability of ferritic steels

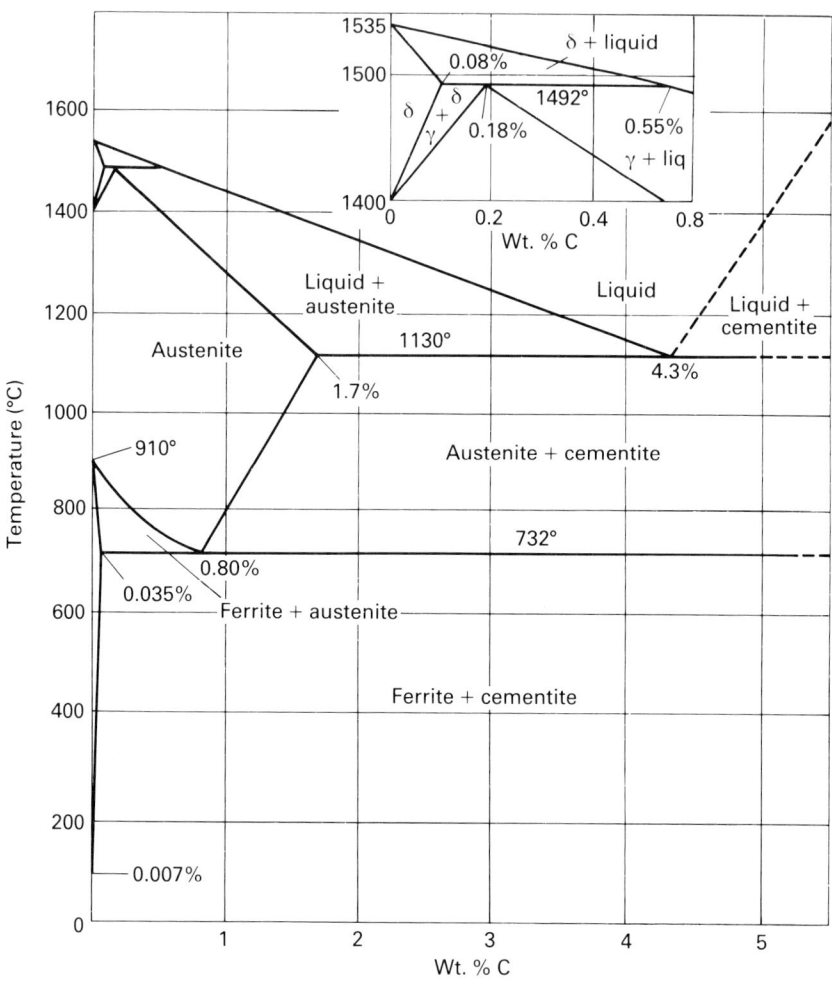

3.15 Part of phase diagram of Fe–C system.

manganese to the weld metal replaces FeS by a mixed sulphide, (Mn,Fe)S, and this beneficially raises the eutectic temperature if sufficient is present. It is claimed[3] that in TIG weld metal and in the absence of carbon, FeS occurs up to a Mn^3:S ratio of about 0.83:1. With ratios from about 0.83 to 3.0 the (Mn,Fe)S is film-like whilst above a ratio of about 3.0 solidification is completed with a high melting eutectic and cracking is much reduced. With about 0.12% C in the weld metal, the ratio for the appearance of (Mn,Fe)S is unchanged but the film-like appearance only disappears with Mn^3 : S ratios above 6.7. If 0.2% C or 2–7% Ni are present, as much as 2–3% Mn is needed to avoid cracking.[3]

Because of the formation of a peritectic and of the different solubilities of sulphur in ferrite and austenite, the behaviour of carbon is quite complex. Although carbon is generally regarded as promoting solidification cracking, several investigators (e.g. ref. 4) have noticed that reducing the carbon content below some level (values have been quoted between 0.08 and 0.14% C) does not further reduce the risk of cracking, and may, in fact, worsen it.[2] However, on approaching a zero carbon content the resistance to cracking probably improves again.

Another element which interacts with sulphides is oxygen. At very low levels (probably up to about 0.010%) of oxygen, the sulphide (Fe,Mn) S is in the type II form, which has a low surface tension and spreads easily along the grain boundaries, thus making solidification cracking easy. Higher oxygen contents promote the formation of the Type I sulphide with a higher surface tension, making penetration of the grain boundaries, and hence cracking, more difficult. This effect of oxygen is only important with the low oxygen welding processes, such as TIG,[5] plasma, electron beam and laser welding. No effect has been found with the flux and active gas welding processes MMA, MAG, submerged arc and electroslag.

Several attempts have been made to quantify the effects of different elements on solidification cracking of steels. These have usually been developed for specific welding processes, and are discussed below for submerged arc welding, which has given the most serious problems in practice, and for TIG, which is often used for the most 'difficult' steels.

Submerged arc welding

A systematic study of the influence of compositional variables on solidification cracking was made some years ago,[4] and the results have been incorporated into the British Standard for structural steel welding.[6] This work led to a formula which relates the risk of cracking, when carrying out Transvarestraint tests under carefully standardised conditions, to the composition of the weld metal. Because submerged arc welding involves considerable dilution of the parent steel into the weld pool, and also involves widely different fluxes, which modify the weld metal composition from that simply calculated (i.e. from a knowledge of the base metal and wire compositions and dilution), guidance was given[4] on how to allow for these factors.

The formula relates crack susceptibility (in terms of units of crack susceptibility, UCS) to the composition of the weld metal (mass %):

$$UCS = 230C + 190S + 75P + 45Nb - 12.3Si - 5.4Mn - 1 \quad [3.1]$$

The equation is valid for the following range of weld metal compositions:

0.08*–0.23% C; 0.010–0.050% S; 0.010–0.045% P; 0.15–0.65%† Si; 0.45–1.6% Mn; 0–0.07% Nb.

The presence of elements up to the following amounts do not significantly influence the risk of cracking: 1% Ni; 0.5% Cr; 0.4% Mo; 0.07% V; 0.3% Cu; 0.02% Ti; 0.03% Al; 0.002% B; 0.01% Pb; 0.03% Co.

The significance of the values calculated from the formula is that, with submerged arc welding, values up to 10 UCS represent a very low risk of cracking and values of 30 and greater, a high risk. In a standardised restrained 19 mm thick, single-pass, T-fillet weld test, 20 UCS was just above the threshold for cracking when the weld depth/width ratio was approximately 1.0. In butt welds, the results were affected by root gaps, as well as by weld shape, but cracking became relatively common when the UCS exceeded 23 for welds with a depth/width ratio of about 0.8.[4]

Because the accuracy of analysis for carbon is usually no better than ±0.01% C, and this is equivalent to 2.3 UCS, differences of <3 UCS are not significant when comparing the likely cracking behaviour of weld metals. For this reason, UCS values should be given to the nearest whole number.

Although the amounts of the elements listed above have no significant effect on cracking, nickel and boron in amounts greater than those given can certainly increase the risk of cracking. It is also likely that the ferrite stabilisers Cr, Mo and V can reduce the risk of cracking. To a first approximation, it can be stated that elements present in the quantities given above have an effect on solidification cracking no greater than 3 UCS, i.e. similar to the influence of about 0.01% C (2.3 UCS).

Some of the elements not included in Eq. [3.1] are discussed in more detail in the next section. Of these, aluminium should have a mildly beneficial effect, but its variation during normal submerged arc welding is not sufficient to have a detectable effect. Oxygen is known to be beneficial to solidification cracking resistance, but the amounts present in submerged arc deposits are all above the level at which benefit can be obtained by increasing oxygen levels. Boron is known to be harmful, but the amounts known to be detrimental – 0.01% B is known to give cracking in bead-on-plate deposits[7] – are well above the levels added to submerged arc weld metal.

Nickel is known to be harmful, and it is likely to give problems if high speed welding procedures for clean C:Mn steels are adopted unchanged for 3% and 5% Ni weld metals. However, with the high strength weld metals of the Ni:Mo, Ni:Cr:Mo and similar types, the amounts of Ni are smaller and are, to some extent, counterbalanced by those of the other, ferrite-

---

*Carbon contents <0.08% should be taken as equal to 0.08%.
†Silicon contents above 0.65% may *increase*, rather than reduce the risk of cracking.

Solidification cracking

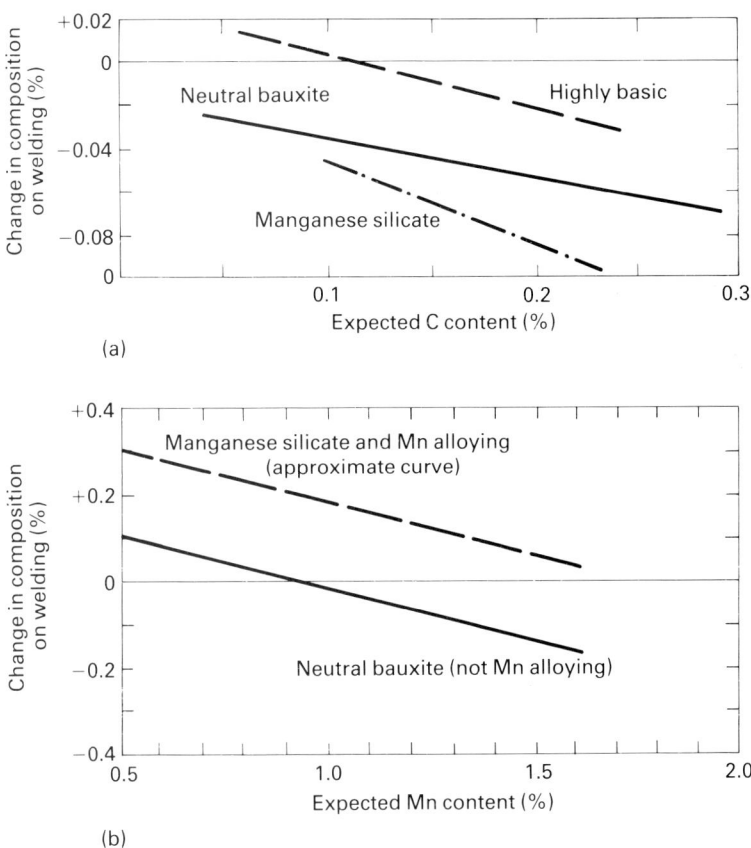

3.16 Effect of flux type on compositional changes during welding: (a) carbon (calcium silicate flux is intermediate between neutral bauxite and highly basic types); (b) manganese (for flux types not given in Table 3.1).

stabilising alloying elements present in such weld metals. The effect of nickel at very low weld metal carbon levels was discussed earlier in the chapter.

When the values that can be obtained from Eq. [3.1] are related to the normal compositional variations likely in submerged arc weld metal, it will be seen that the most harmful element is carbon. Whereas increasing sulphur from 0.010 to 0.050% (an increase that would be very difficult with present-day structural steels and consumables!) would give an increase of less than 8 UCS, this increase is less than occurs if the carbon content is increased by 0.04% – well within the likely variation in different runs of a multipass submerged weld.

Table 3.1 Effect of the flux type on weld metal compositional changes during submerged arc welding (changes, applicable to root runs and 1 and 2 pass welds, are approximate)

Element	Flux type	Approximate composition change from 'expected' level[a]
C	All	See Fig. 3.16(a)
S[b]	Basic	−0.004
	Neutral bauxite	+0.002
	Ca silicate	⩾+0.002
P	All	0 to +0.010
Nb	All	0
Si	Basic	Negligible
	Neutral bauxite	+0.05
	Ca silicate	+0.1
	Mn silicate	+0.2
Mn	Basic	−0.1, unless Mn-alloying (Fig. 3.16(b))
	Neutral bauxite	See Fig. 3.16(b)
	Ca silicate	0 to −0.15
	Mn silicate	See Fig. 3.16(b)
Ni	All	0
B	All	Loss

[a] Expected levels calculated from plate and wire compositions and dilution; changes are for single arc 1- and 2-pass welds at 500–900 A d.c. positive.
[b] At an expected level of 0.020% S (change may depend on flux batch).

Of the beneficial elements, manganese appears to be the more useful, because of the risk of reversing the beneficial effect of silicon if it is increased above about 0.65% Si, coupled with the risk of its impairing weld metal toughness at such levels. However, it is not easy to increase weld manganese levels significantly because of the high dilution inherent in the submerged arc welding process. With a dilution of 65% of parent steel in the weld metal, increasing the wire Mn content from 0.5% to 2.0% will increase the Mn addition to the weld by just over 0.5%, equivalent to 3 UCS, which has a barely perceptible effect on cracking. In practice, the influence of the flux on Mn transfer – which could change from a Mn gain to a Mn loss over the range selected (Fig. 3.16b) – will reduce the actual increase in weld Mn content still further. In practice, an improvement in cracking behaviour must be sought by other means, namely by altering the welding conditions, as discussed in the next section, or by changing the flux.

The influence of the flux was discussed in Chapter 1, but a brief guide to the influence of different types of flux on the changes in weld metal composition for the elements that influence solidification cracking is given in Table 3.1.

Other welding processes

Of the flux welding processes, submerged arc welding is the most likely to give solidification cracking problems, partly because of the high welding speeds which can be used and partly because of the high dilutions, which can give high weld carbon contents when welding normal structural steels. However, the latter problem is much less acute with many modern weldable steels – particularly of the line pipe type (which are manufactured by high speed submerged arc welding) and with some of the high strength low alloy (HSLA) steels – because their carbon contents are no higher than those of many undiluted weld metals.

The manual welding processes MMA and manual TIG with filler give least risk of cracking when welding comparable steels because welding speeds and dilution levels are both low. However, these are the processes that are most likely to be used for welding medium and high carbon steels, which are at particular risk of cracking because of their high carbon contents. Manual metal arc welding has certain advantages over TIG in welding medium-high carbon steels to avoid solidification cracking (but *not* other types of cracking) because weld metal carbon contents tend to be low (unless fillers of matching carbon content are used) and sulphur contents are reduced when basic electrodes are used.

Many of the earlier solidification cracking formulae were developed in this area, particularly for autogenous TIG welding, where the composition of the weld metal can be taken as being the same as that of the parent steel.

Huxley's crack susceptibility factor:[8]

$$CSF = 501C + 1560 (Mn \cdot S) - 82.8Mn - 47.9 \qquad [3.2]$$

was developed for autogenously TIG welded C:Mn steels containing 0.15–0.3% C, but having a relatively limited range of 0.025–0.038% P, so that phosphorus does not appear in this formula. Despite the interaction between Mn and S, Mn is regarded as beneficial within the range examined (0.31–1.69%).

Morgan-Warren and Jordan[5] covered a wider range of alloyed steel compositions, also using autogenous TIG welding, and developed the formula:

$$CSF = 42C + 847S + 265P - 10Mo - 3042(O) + 19 \qquad [3.3]$$

The steels contained 0.08–0.4% C, 0.4–1.5% Mn, 0.1–1.5% Si, 0.2–2% Mo, with up to 0.03% S, 0.03% P, 2% Ni, 5% Cr, 0.5% V, 6% Co, 2.1% Cu and 6.7% W. Other than the special case of manganese, molybdenum appears as the only alloying element which affects solidification cracking, although most people acknowledge that nickel is harmful in amounts greater than the 2% included in this study.

For electron beam welding, a cracking index (CI) has been developed for simple C:Mn steels, based on the composition of the steel being welded, and related to the ease of producing cracking in a standard test:[9]

$$CI = 52.6C + 1972S + P(4268C - 285) - Mn(1135S - 9) - 21.1 \quad [3.4]$$

The steels contained 0.09-0.31%C, 0.26±0.02% Si, 0.77-1.84% Mn, 0.005-0.055% S, 0.005-0.058% P with other elements <0.1%. All the important elements for this type of steel were varied over a significantly large range except Si. Although the equation has a family resemblance to other solidification cracking formula, guidance is not yet available on values to avoid for different welding situations.

Solidification cracking and the welding process

Translating the ideal weld shape factors (i.e. a weld pool which is not too deep or shallow, and which solidifies uniformly from root to cap) into welding parameters suggests that, where weld metal compositions are likely to crack, high welding speeds should be avoided and welding currents kept low enough to avoid high depth/width ratios. In avoiding a deeply penetrating weld pool, the dilution of parent steel is minimised, and this is usually advantageous because (except in some modern, very low carbon steels) the parent steel is usually of a composition more susceptible to cracking than the weld metal.

Depth/width ratio

Although the effect of varying depth/width ratio on cracking has not been fully explored, some guidance can be given for submerged arc welds, using the UCS values obtained from Eq. [3.1]. For butt welds with a ratio of about 0.8, cracking is not apparent until a UCS value of approximately 23 is reached. However, the weld root run illustrated in Fig. 3.17 cracked with a UCS of only 13, but its depth/width ratio probably exceeded 1.5. This example highlights the importance of controlling root run welding procedures, especially on the second side of a butt weld. There is a natural tendency to deposit the second side root run with a high current in order to ensure good root fusion and lack of welding faults. The risk of overdoing this is obvious from the figure: the current had been deliberately increased because it was known that a large root face had to be penetrated.

For T-fillet welds, it has been found that decreasing the depth/width ratio from 1.0 to 0.8 beneficially increased the critical UCS value by approximately 9 UCS (Fig. 3.18). Even further increases in the acceptable UCS value are apparent with MMA and TIG welding, where much higher UCS values can be tolerated without solidification cracking in

Solidification cracking

3.17 Solidification crack in root pass, second side of submerged arc weld in 60 mm thick plate, in which the welding current was increased considerably in order to achieve penetration.

weld beads with depth/width ratios of about 0.5. When changing welding parameters to reduce the risk of cracking by altering the depth/width ratio in fillet welds, it is important to determine this ratio by measuring from a line through the weld toes, rather than from the centre of the face of the weld, as shown in Fig. 3.4, in case the convexity of the weld is altered by the change in welding procedures.

The convexity of a weld bead can be increased by a limited extent to reduce the likelihood of cracking (Fig. 3.11). However, too convex a bead shape may be undesirable on other grounds, for example the difficulty of tying-in adjacent beads or the adverse effect on fatigue and other types of cracking due to the higher stress concentration factor.

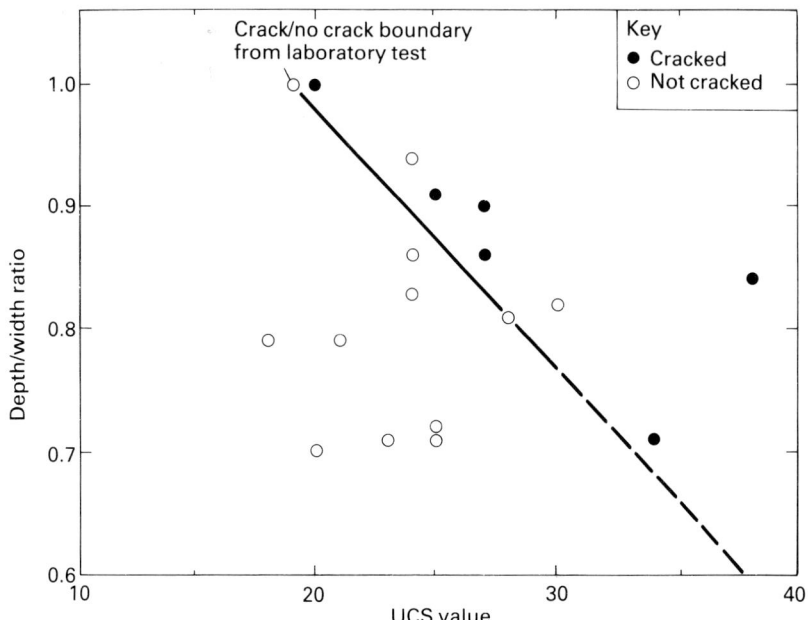

3.18 Results of tests on commercial submerged arc fillet welds in which D/W was varied and the results related to the UCS (Eq. [3.1]) of the weld metal).

The avoidance of a very shallow weld pool is as much a matter of welding technique as of selecting the correct parameters. One cause of the cracking of such shallow welds is that inadequate fusion in the root, with resultant slag particles, impairs the efficient removal of heat in the same way that a root gap does. The risk of such defects – and cracks – will be increased if the weld does not have an adequate platform to be deposited on, as was the case in the weld illustrated in Fig. 3.19.

Joint preparation and root gaps

For the faster welding process, such as submerged arc, narrow steep-sided preparations are best avoided, particularly in the root, unless slower processes (e.g. MMA) are used for the root itself. Back gouging of the first side root in preparation for the second side (which will not be welded with the slower process) should also be as wide and rounded in the base as possible. Root gaps should be minimised where possible; this means that the gap actually existing at the time of weld should be narrow (cf. Fig. 3.7). To ensure a continuing narrow root gap, particular care should be taken with tacking and clamping arrangements, as root gaps can open up

Solidification cracking

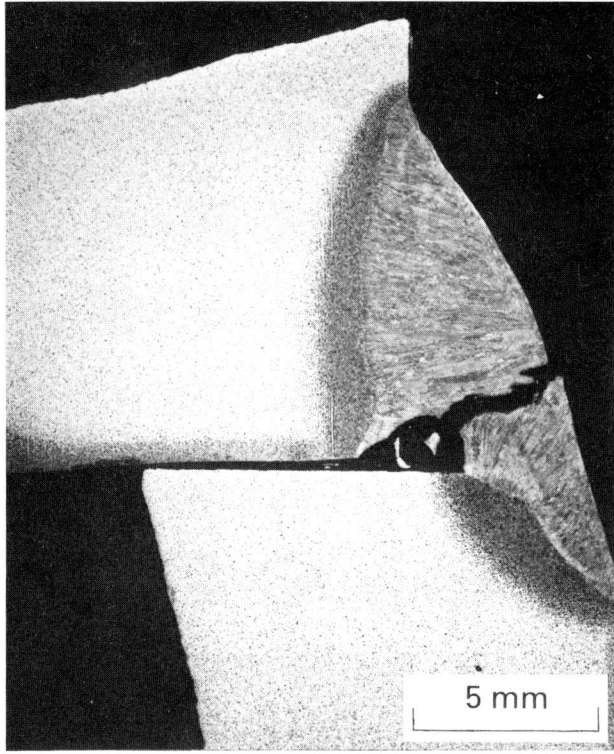

3.19 Solidification crack in cored wire MAG weld; owing to inadequate support for the weld, a shallow wide weld was inevitable and root defects resulted.

during the welding of long seams and this can lead to cracking towards the end of the weld. Sudden fracture or melting of tack welds throws a sudden strain on the solidifying weld which can accentuate this problem.[10] Care in clamping is particularly important in welding higher strength steels, as the forces to be resisted increase with the strength of the steel.

When welding sections of different thickness it is unwise to try to complete the joint with too large a root pass (or to try to make a single-pass weld where this is only just practicable), because there is a serious risk of the thinner component melting back, opening up the root gap and inducing cracking,[11] as in Fig. 3.7.

Other parameters
Weld size is a minor parameter compared with depth/width ratio, but increasing weld size in terms of the weld cross-section tends to increase

Weldability of ferritic steels

the risk of cracking. However, when weld size is changed, the alteration in welding parameters needed to give the change will usually have other effects, e.g. a change in dilution and hence weld metal composition, which will outweigh the simple effect of changing the weld size.

Weld pool length is a major factor. It has long been known that increasing the length of the weld pool from an oval shape (typical of TIG welding) to a long tear-drop shape (more typical of MAG and submerged arc) helps to promote cracking. This is because, as the weld pool is lengthened, the weld metal grains tend to grow more inwards and forwards than directly upwards, thus increasing the feeding distance for the molten weld metal to counteract solidification contraction. If, however, the pool is lengthened by using multiple arcs, then the adverse influence of a long weld pool is nullified by providing a new source of molten weld metal for feeding the solidification shrinkage and also by reducing the dilution of parent steel into the weld.

Welding speed, because of its effect on depth/width ratio and weld pool length, has a marked effect, increasing speed giving an increased risk of cracking.

Weld strain is difficult to visualise and to quantify, because the strains of concern are the local strains on the weld pool at a late stage in its solidification. These strains result from both large scale physical movement (i.e. distortion) and locally induced solidification and contraction strains. Such strains are continually changing as the weld is made. The local strain on the weld results from contraction of the weld metal during its

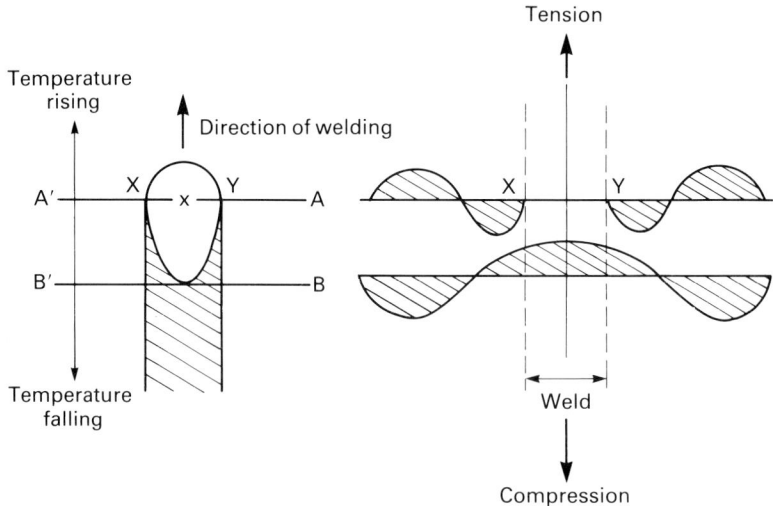

3.20 Stress distributions about moving heat source during welding.

solidification, induced by the restraint imposed by cooler parts of the joint as shown schematically in Fig. 3.20.

Restraint is needed to prevent the joint ahead of the welding arc opening up to such an extent that the weld shape would be affected, even allowing the weld to drop through. This opening up promotes solidification cracking, particularly in very long welds. In addition, however, the parts being joined would lift if not restrained; this lifting will reduce the tendency to crack, whereas restraining the tendency to lift will tend to promote cracking. After the first run has been deposited, the underlying weld metal restrains the upward contraction to some extent so that it is unlikely that increasing the restraint at this stage to prevent further distortion will have much effect on cracking.

Plate thickness is a minor variable; an increase in plate thickness, particularly for fillet welds, is likely to increase the risk of cracking because of the increased restraint. If the carbon content of the steel itself has been increased to maintain its strength as the thickness is increased, this will further increase the risk of cracking.

Parent metal strength also has a minor effect on cracking, an increase in strength increasing the risk of cracking because of the greater restraint offered by the stronger steel. This effect is noticeable, even if the weld metal strength is unchanged, which supports the view that the effect is due to the stronger parent steel being less able to yield to accommodate thermal strains.

Welding sequence and practice is most important; in multipass (and also two pass) butt welding, cracking is most likely in the root pass of the second side to be welded. This is partly because of the increased restraint but also because of a strong tendency to increase welding currents for this pass to ensure good inter-run fusion (Fig. 3.17). Little is known about the correct sequence to minimise cracking in multipass butt welds. It is believed that a balanced technique should be used, although it is probable that the deposition of several runs on the first side reduces the risk of cracking of the second side root run, provided distortion can be adequately handled.

A particularly difficult situation occurs when depositing simultaneous double fillet welds (e.g. T-joints and other stiffeners). If the arc on the first side leads by 100–200 mm (particularly in submerged arc welding), the first fillet may be solid and contracting and pulling on the second side bead when the latter is at a late stage of solidification, and thus most vulnerable to cracking.

Preheat has somewhat complex effects on solidification cracking. If the joint is clamped sufficiently to avoid distortion, preheat is beneficial, particularly if it is general rather than local. In less well-restrained structures, preheat can *add* to the level of distortion and thus increase the risk of cracking, particularly if it tends to open up the joint ahead of the welding arc. Very high levels of preheat will tend to increase dilution of parent steel into the joint, and this will usually increase the risk of cracking.

Although **grain refinement** of the as-deposited (columnar) weld metal grain structure is potentially capable of reducing the risk of solidification cracking, techniques for such refinement are as yet not available and are likely to be very difficult to achieve in any steel. Refinement of the grain size in the solid state has no effect on cracking.

Control and avoidance of cracking

Control of **weld metal** composition is not always the easiest option, particularly with high dilution processes such as submerged arc welding, because with them the weld metal composition depends more on the parent steel composition and the flux system (often selected to achieve toughness and/or weldability) than on the wire. In submerged arc welding, it is recommended to aim for UCS values (Eq. [3.1]) in the root runs below 25 UCS for butt welds and below 19 UCS for fillets. Higher values can be tolerated for other welding processes, probably in the order MAG (spray), MAG (dip), self-shielded FCAW, MMA and TIG. Experience is needed to decide on limits for these processes, although Eq. [3.2] and [3.3] may be used when TIG welding the higher carbon engineering steels.

In the parent plate, the elements carbon and sulphur are the dangerous ones, especially in the root region where dilution of the parent steel into the weld is high. Manganese and (to a limited extent) silicon are beneficial and the routine scrutiny of mill sheets for these elements is recommended where there is a known risk of cracking.

The effect of the composition of the wire on that of the root region is quite small when using high dilution welding processes. Nevertheless, good practice is to use wires of low C, S and P contents, together with the highest Mn content compatible with the mechanical properties required. It should also be borne in mind that excessively high strength levels (which could result from high weld metal manganese contents) can give rise to other problems, such as high residual stresses and an increased risk of weld metal hydrogen cracking; such problems are potentially more serious than solidification cracking, which is usually easy to detect and to remedy.

The flux also influences weld metal composition. The so-called acid fluxes, because they remove carbon and give moderately high silicon levels, are beneficial in avoiding cracking, provided that they are not so 'acid' that excessively high silicon levels (i.e. >0.6%) result. The acid fluxes can, however, increase weld metal sulphur and phosphorus contents, and

Solidification cracking

basic flux systems should always be used where high sulphur levels, such as are found in free-cutting steels, are likely.

Dilution control can be a problem with the high dilution processes because the welding parameters used for the root regions (i.e. high current levels), which are normally used to ensure good root fusion and lack of defects, also lead to high dilution levels. Good practice would be to ensure that root gaps and root faces are both small. However, the use of small root faces is incompatible with economical requirements to minimise the amount of weld metal used. Another disadvantage of a small root face is the risk of burn through when welding with a high current. One way to avoid this is to use a welding process of inherently low penetration, such as MMA or TIG, for the initial root runs.

After the root runs, dilution of parent steel into the weld is usually less, so that solidification cracking away from the root is indicative of something at fault, such as the use of the wrong wire, extensive penetration into the side of the preparation (e.g. Fig. 3.21), or a very marked change in weld bead shape as a result of changed welding parameters or inter-run grinding of defects.

High deposition and multipower submerged arc techniques can also be used to control dilution and, hence, weld metal composition. It is common

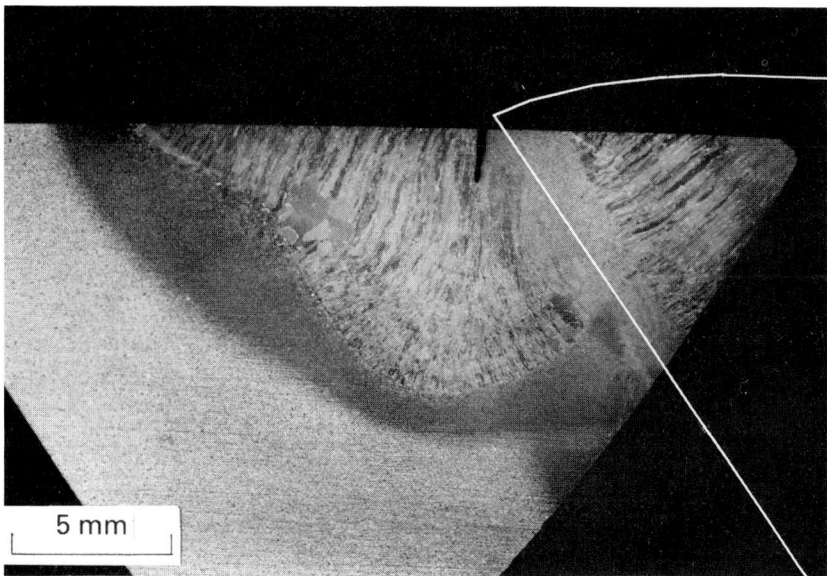

3.21 Centre-line solidification crack in a submerged arc weld run which melted into the parent plate, resulting in an unexpectedly high UCS value (Eq. [3.1]). Section shown is of a boat sample from the cracked weld; white line indicates weld preparation.

for the leading arc to use d.c. positive polarity in order to achieve good penetration, but to follow this with one or more arcs (up to three are common in pipe welding, four are sometimes used and experiments have been carried out with six) using a.c., d.c. negative or a combination of these. The arc functions in triple arc submerged arc welding are shown schematically in Fig. 3.22; it is important to ensure that the second arc is close enough to the first to avoid any cracks remaining in the lower part part of the weld bead due to the first arc.[10]

Other ways of increasing the amount of added weld metal without increasing current and dilution include the use of metal powders, chopped wire, hot or cold wire additions and increasing the nozzle/workpiece distance (stickout). Whichever high deposition welding technique is used, the available options are sketched in Fig. 3.23; these range from the same deposition rate with decreased current to increased deposition with the same current.

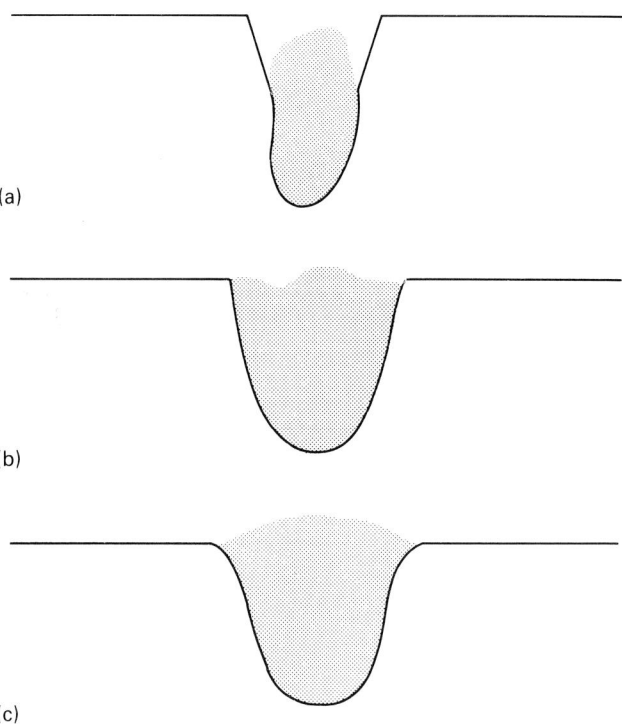

3.22 Arc functions in multipower submerged arc welding (schematic): (a) leading arc weld bead shape; (b) bead shape from leading and middle arcs; (c) final weld bead shape.

Solidification cracking

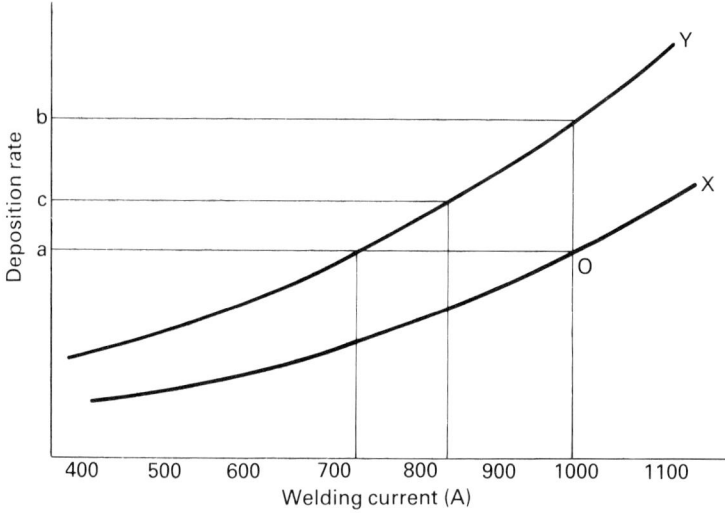

3.23 Schematic approach to control of dilution and heat input using high deposition welding technique, curve Y. Conventional technique is curve X and original welding condition, point O. Alternatives are: (a) same deposition rate with reduced welding current; (b) increased deposition rate with same welding current; (c) compromise with increased deposition rate but reduced current.

Control of **solidification pattern** is often a more successful way of controlling solidification cracking than attempting to improve the weld metal composition. Differences in solidification pattern usually accompany several of the methods of reducing dilution discussed in the previous section. Reducing the welding current is the easiest way of reducing the depth/width ratio, although it inevitably leads to a loss of productivity and, if carried to excess, can lead to such defects as lack of fusion (which may even bring back solidification cracking as a result of root defects plus too shallow a weld pool). Reducing welding speed also has similar beneficial effects on cracking to reducing welding current – and is also coupled with a loss of productivity.

Changing polarity, if this is possible, from d.c. positive to a.c. to d.c. negative, and the use of increased stick-out will both reduce weld depth/width ratios without reducing deposition rates. Also, where possible, the use of cold metal additions (hot or cold wire at the back of the weld pool or metal powder additions with submerged arc) tend to reduce depth/width ratios without penalising efficiency (Fig. 3.23).

In addition to altering the welding parameters, it is also possible to reduce the steepness of the weld preparation, although this usually has economic penalties, either by increasing the amount of weld metal needed or by the increased machining costs if a compound preparation is used.

Avoidance of root gaps is also important to achieve a weld shape resistant to solidification cracking. Where root gaps cannot be avoided, backing strips are used. Although no supporting work is known, it is likely that the use of high conductivity backings, such as steel or copper, will give less risk of cracking than ceramic or other non-metallic backings.

Control of **strain** by using lower strength parent steel or weld metal is rarely within the control of the fabricator. Nevertheless he or she should be aware of the *increased* risks when using the higher strength materials. Distortion during welding, which may occur as a result of preheating, inadequate restraint or the fabrication sequence, is certainly under the fabricator's control.

Preheating causes most distortion when applied locally, and a narrow band is heated. Consideration should always be given to using very low hydrogen consumables to avoid the need for preheat, otherwise wide areas, preheated with electrical heating mats, are preferred. If preheat is essential and is causing solidification cracking, it may be possible to use subsidiary heating units to counteract distortion produced by preheating mats.

Restraint must be considered in two senses. Firstly, sufficient restraint must be applied when welding the first pass to avoid the two pieces moving apart and thus cracking. Once the first pass has been completed, further restraint hinders free contraction and thus makes cracking more likely.

When welding the first run of very long welds (i.e. >20 m or so), very high forces are needed to prevent the weld gap opening, and the use of multipower techniques is advantageous, so that the leading arc achieves adequate penetration and the trailing arc(s) adds weld metal of low crack susceptibility and completes the pass with a good weld shape, i.e. slightly convex (Fig. 3.22c). Run-off blocks also play their part in avoiding any of the crater cracking at the end of the weld run remaining in the joint itself, providing that they are sufficiently long to accommodate the weld pool completely; this may be longer than 300 mm when multiarc-welding. It is also important that tack welds and clamping do not fail suddenly when welding the first seam, as the sudden shock is more likely to give cracking than the slow application to the solidifying weld of a similar strain.[10]

The restraint offered by the rest of the fabrication usually builds up during welding and can lead to cracking if the welding sequence is not selected to minimise this build-up. Nevertheless, it may be necessary to reduce welding currents for particularly critical seams (especially if sensitive steels are involved) or even change to a less sensitive welding process, such as **MMA** or **TIG** if the bulk of the welding is by submerged arc or MAG, for example.

Solidification cracking

Remedial measures

If cracking is discovered before a joint is too deeply buried by later weld runs, it is usually a simple matter to repair it. When automatic welding processes have been used, repair is often by means of **MMA** or **TIG**, the slower speed of which, coupled with an appropriate choice of consumables, will make solidification cracking during the repair much less likely. A long solidification crack can be repaired by gouging it out and re-welding with the original process, provided the reasons for the cracking are sufficiently well understood for the welding parameters to be correctly modified.

In practice, it is likely that many solidification cracks occur in root runs and are unknowingly 'repaired' by melting out when the second run is deposited, without anyone being aware of their brief existence. However, if one is aware of such cracks, their repair by gouging and re-welding (followed by modification of the welding procedure for subsequent similar welds) is a safer option than hoping that they will always be removed by the next weld run; such removal depends very much on consistent adherence to the welding procedures.

Once solidification cracking has been detected, it is important to establish the likely reason for the cracking, so that the necessary remedial measures will be both appropriate and economical. A common cause of such cracking is that the welding procedures have deviated from what was originally used successfully. If restoration of the original welding parameters fails to eliminate cracking, some fairly simple remedial measures are possible:

1. Change welding parameters to reduce penetration, and hence depth/width ratio and dilution (unless the cracking is associated with a very low depth/width ratio).
2. Reduce welding speed to give a shorter weld pool.
3. Improve weld metal composition by reducing C and S levels within the wire specification.
4. Improve clamping for cracking in the first pass, reduce root gaps, using gap filling techniques if necessary.
5. In severe cases it may be necessary to re-qualify the procedure by using, for instance, MMA for the root run(s) or buttering one half of a joint involving a high carbon steel.

Check list

Design

Check that weld preparations are not too steep; that root faces are not so large that they require very high currents to penetrate; check whether steel and weld metal strengths are no greater than have been used previously, as this itself will increase the risk of cracking; check that there is enough solid metal at the joint to hold a weld pool of adequate size (Fig. 3.19) without having to reduce current levels excessively.

Composition and weld bead shape
Check that weld metal compositions, particularly in the root run, are not excessively high in C, S and P when welding automatically; the UCS formula, Eq. [3.1], should be used with submerged arc welding, Eq. [3.2] and/or [3.3] for TIG welding when welding engineering steels; allow for such factors as restraint, high strength steels, poor (concave) weld shape and root gaps based on previous experience. For other processes, Eq. [3.1] may be used, with adjustments to critical UCS levels in the light of experience.

Use welding procedures to minimise dilution and depth/width ratio for root runs, i.e. current and welding speed not too high, or use appropriate multipower submerged arc techniques; ensure that multi-arc techniques lead to a single weld pool; ensure that arc angles do not lead to excessive heating of the thinner steel being welded; check that root gaps are small.

Restraint
Check that clamping and tacking for root runs are adequate for the strength and thickness of steel being welded; ensure that a welding sequence does not lead to excessive restraint building up during welding.

Examination
During multipass welding, check weld craters for signs of cracking. If cracks are unexpectedly found, carry out careful examination for surface and sub-surface cracks.

Liquation cracking

Liquation cracking, sometimes known as hot tearing, is a relatively minor type of cracking in ferritic steels. It can give problems in oxy-acetylene welding of thin sections, where the time of heating is relatively long. In the arc welding of medium carbon low alloy steels, its presence can increasing the risk of hydrogen cracking by providing pre-existing defects at which cracking can initiate. It has also led to welders failing their qualification tests when unsuitable steels have been used, because welds containing liquation cracks can give premature failure in the HAZ in bend tests. Steels for such testing should, therefore, be selected to be unlikely to produce liquation cracks, i.e. they should be of low sulphur and carbon contents.

Liquation cracking occurs in the high temperature region of the HAZ when the steel contains sufficient impurities for significant liquation of the eutectic between these impurities and iron to occur.[12] On cooling, the

Solidification cracking

liquid formed at the high temperature grain boundaries is unable to accommodate tensile strains caused by the contraction, and small cracks result. The cracks are identical in their formation with weld metal solidification cracks and, if significant segregation of the responsible impurities is present in the parent steel, liquation cracks can continue from segregate bands in the parent steel into the weld metal as solidification cracks (Fig. 3.24), provided that the weld metal is susceptible to solidification cracking. In submerged arc welds and those of similar profile, cracking often occurs in the so-called bay area, Fig. 3.25, where the finger penetration joins the upper part of the weld bead, because this is the region of the HAZ which is at a high temperature for the longest time.

Although it is possible that some of the solidification cracking formulae discussed earlier could be used to forecast the likelihood of liquation cracking, the diagrams due to D. J. Widgery reproduced in ref. 13 and here in Fig. 3.26, are useful for assessing the likelihood of liquation cracking in relation to the S, C and Mn contents of the parent steel. Although it is likely that P and the Mn/Si ratio can also influence liquation cracking, their roles are minor and can usually be ignored.

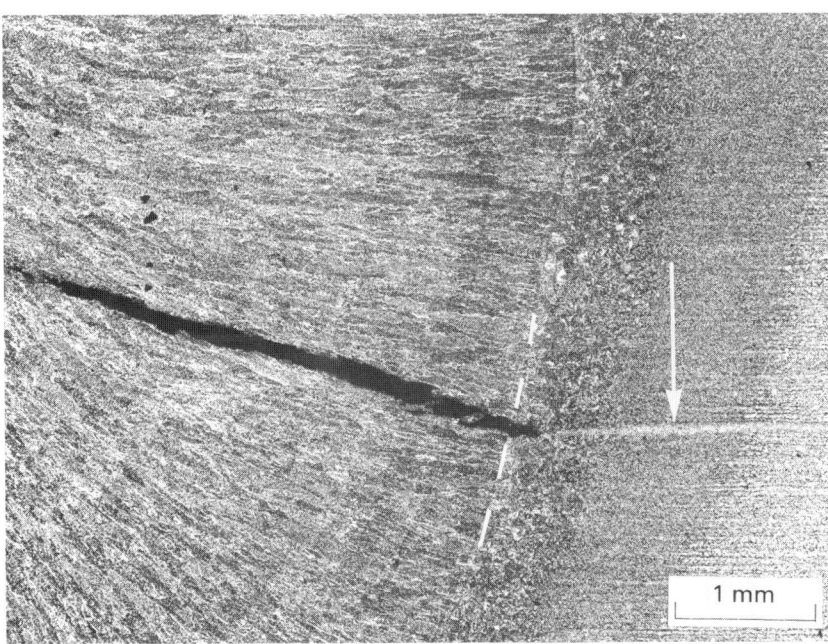

3.24 Section of submerged arc weld showing segregate band (arrowed) from which a liquation crack and a solidification crack have initiated; fusion boundary is shown by a broken line.

88 Weldability of ferritic steels

3.25 Liquation crack (arrowed) in bay area of submerged arc weld bead.

It can be seen from the diagrams that liquation cracking is not possible with sulphur levels below about 0.010% S at any level of carbon, but with 0.2% C, there is a risk of cracking with >0.010% S and a strong likelihood with >0.05% S, unless the Mn/S ratio exceeds 40 (this last combination of circumstances is unlikely for most steels as it would imply a Mn content of >2% when 0.05% S was present). Hence free-cutting steels, containing anything up to 0.35% S (BS970: Part 3: 1991: 230M07) will usually suffer liquation cracking when welded and should not be used for welder qualification tests where bend tests are required.

Because liquation cracks are restricted to the high temperature HAZ and are aligned roughly perpendicular to the fusion boundary, they are never more than 1–2 mm long. By themselves they are insignificant for most practical purposes, particularly as their tips are blunt like solidification cracks and unlike most low temperature cracks. However, they can increase the likelihood of subsequent hydrogen cracking (and probably also reheat cracking in susceptible steels), and for this reason it is not good practice to select steels with a deliberate sulphur addition when there is a risk of hydrogen cracking during welding. The risk of low ductility and toughness in the high temperature HAZ can also be important, particularly in high heat input welds, in which liquation cracks are likely

Solidification cracking

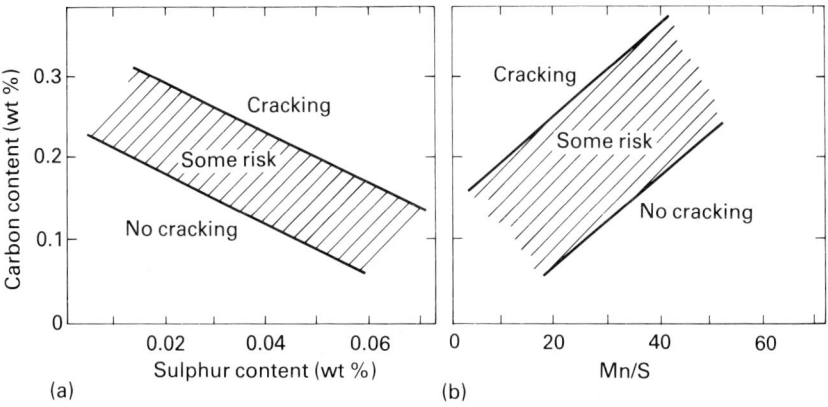

3.26 Relationship between steel composition and occurrence of liquation cracking in HAZs of welds made with heat inputs of 2 kJ/mm.

to be larger than if the heat input is low. In machined components, liquation cracks may also initiate fatigue cracking if present in areas of high stress.[12]

Liquation cracks are dependent on the steel composition itself so that little can be done to minimise their presence. Welding on to the susceptible steel using a low heat input is mildly beneficial, as it slightly reduces the size of any cracks formed because of the reduced size of the high temperature HAZ. However, there is no evidence that a low heat input reduces the risk of liquation cracking occurring. One technique of minimising the risk of tearing is to use weld metals of low solidification temperatures. If a low heat input is used, a weld metal freezing temperature of ~1100 °C would be needed,[12] but at high heat inputs stainless steel or nickel alloy consumables may have enough time for the liquid weld metal to be drawn into the liquation cracks as they form, thus healing them.

As with solidification cracks, high sulphur contents (particularly when associated with medium to high carbon and low manganese levels) promote cracks, and these are associated with MnS films along prior austenite grain boundaries. Because the MnS inclusions in the steel start dissolving at the very high temperatures involved, chains of very small inclusions appear in a micro-section close to the major MnS inclusions (Fig. 3.27). However, if the region containing liquation cracks is opened and examined by fractographic techniques, the small chains of inclusions are seen to be sections through a eutectic structure along the prior austenite boundaries (Fig. 3.28); these are sometimes described as chrysanthemum-shaped.

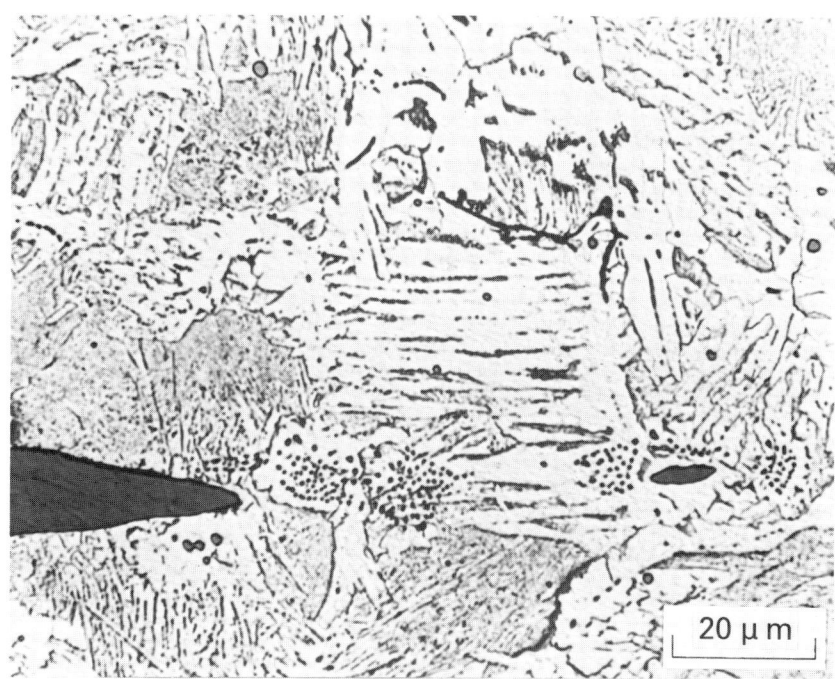

3.27 Partial dissolution of a manganese sulphide particle (known as burning) in the HAZ of an electroslag weld.

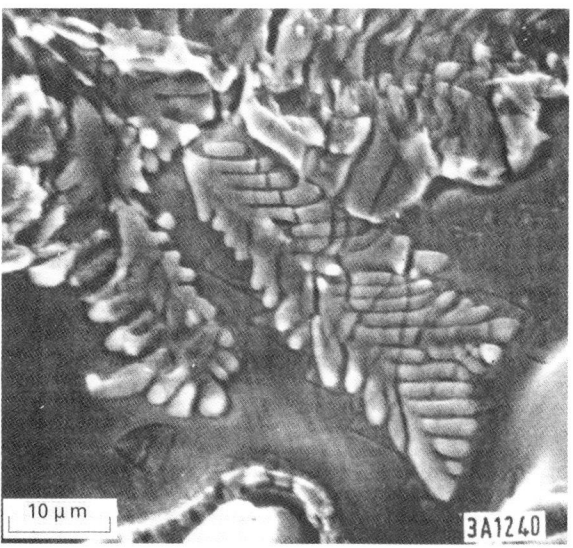

3.28 SEM of liquation crack showing 'chrysanthemum'-shaped eutectic, essentially of Fe and (Fe,Mn)S, in a stainless steel.

Solidification cracking

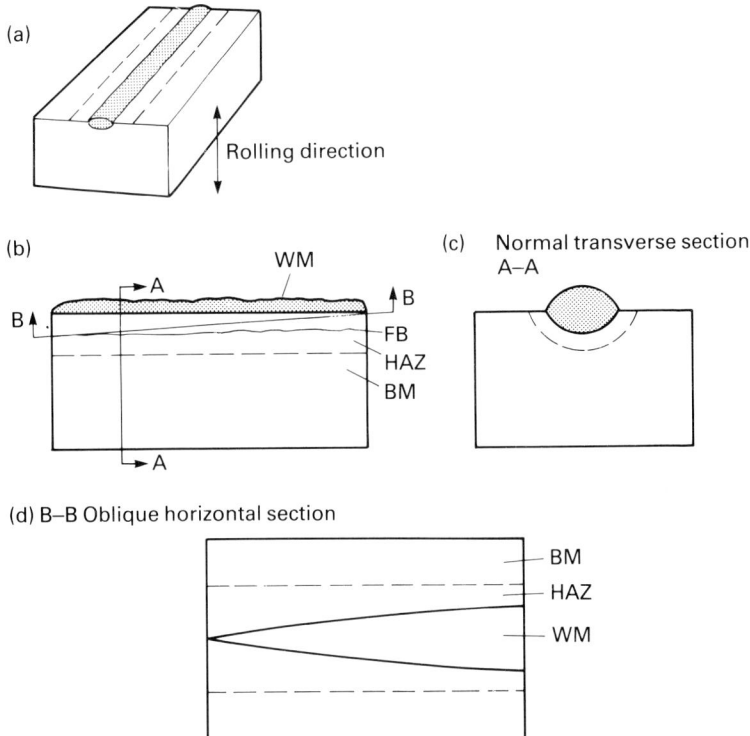

3.29 Taper sectioning used when examining for liquation cracks. BM, base metal; WM, weld metal; FB, fusion boundary; HAZ, heat-affected zone.

The most useful test for assessing liquation cracking is to make a bead-on-plate MMA deposit on the steel of interest and to examine a plan section made at a taper to the weld surface (Fig. 3.29), thus examining all regions of the HAZ from root to cap. This technique is preferred to simple transverse sections, as the latter can miss most liquation cracks present.[12] Examination of the taper sections should be carried out at a magnification of at least ×300 after normal polishing and etching.

Copper pick-up

Akin to liquation cracking is the penetration of copper into the HAZ (and occasionally the weld metal) when the copper, usually from a contact tip, inadvertently comes into contact with the workpiece during automatic welding. If small amounts of copper touch the molten weld metal there is no problem, as copper dissolves directly into the weld pool and, in small quantities, is not harmful. However, copper has a freezing point of

1083 °C, several hundred degrees below that of steel, and large amounts of copper in the weld pool, or smaller pieces coming into contact with the hot HAZ (or weld metal) while it is still above 1083 °C, will melt and penetrate the HAZ and/or weld metal by a liquid metal penetration mechanism. Although seeming possibly less harmful than a crack, such penetration is not desirable, particularly as it is often accompanied by cracks which are like liquation cracks, but which may be appreciably longer (Fig. 3.30).

Small amounts of copper penetration can also result when shavings of copper from the contact tip (or possibly also as debris from copper coated welding wires) land on the HAZ while it is still hot. There have also been occasions where such penetration has occurred in weld metal during submerged arc welding. These latter have not been fully explained, although it is likely that the copper particles have taken sufficient time to work their way through the flux burden and slag to the weld metal surface for the weld to have completely solidified and cooled to a temperature still in excess of 1083 °C. Similar instances of cracking have occasionally been noticed in MMA deposits when copper has been alloyed into the weld metal from the electrode covering, e.g. when welding weathering steels.

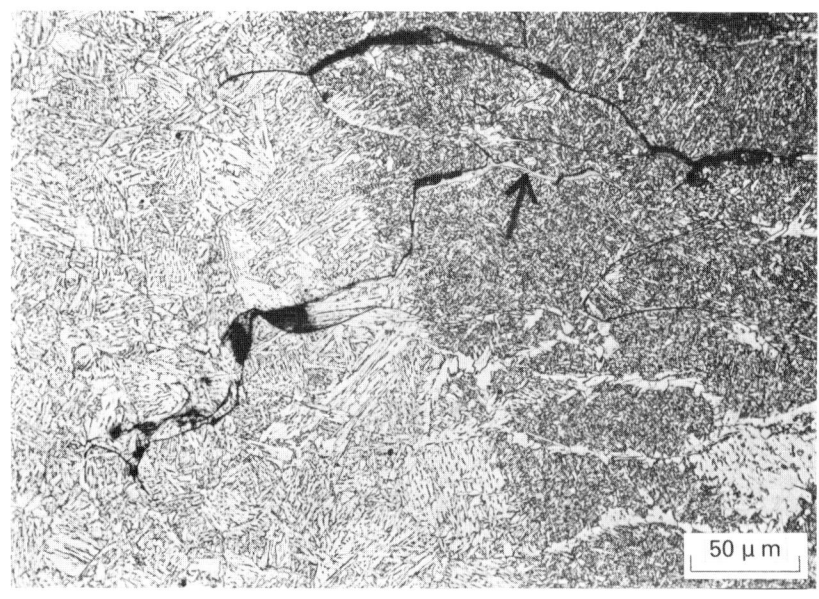

3.30 Cu penetration (arrowed) and cracking in HAZ and weld metal as a result of partial melting of contact tip.

Solidification cracking

Ductility dip cracking

This is a type of cracking which is theoretically possible at elevated temperatures when the ductility of face-centred cubic metals is reduced as a result of harmful amounts of certain impurities. The temperature range for the ductility dip in ferritic steels is within that in which face-centred cubic austenite is the stable phase. Experiments[14] have shown that, even with susceptible ferritic steel compositions, the amounts of strain needed to give cracking are beyond those present in normal welding, unless very large strains are superimposed on the weld as it is cooling. Furthermore, no cases of ductility dip cracking have been found, either in service or in tests to produce various types of weld cracking in the laboratory, with the possible exception of test welds made under fluctuating stresses (see Chapter 11). It is concluded, therefore, that ductility dip cracking is not a problem of any importance in the welding of ferritic steels.

Apart from welding under fluctuating stresses, the only other problems within the ductility dip range have occasionally occurred when hot working welded parts in which the minimum ductility temperatures of parent steel and weld metal were so different that it was difficult to select a temperature suitable for both materials. However, such instances have not been documented in the published literature, hence no guidance can be given as to the impurity levels or temperatures involved.

Detection and identification

Detection

Many solidification cracks are detected by visual examination, when the cracks are well developed and open to the surface. Most sub-surface centre-line cracks are detected by radiography, provided they are also well developed and roughly aligned with the beam. For intermittent, sub-surface cracks, as in Fig. 3.2(c), ultrasonic examination is necessary.

Liquation cracks are much harder to detect – indeed the only sure way is by destructive sectioning. However, ultrasonic examination may reveal multiple fine reflections close to the fusion boundary, and this could provide a clue to their possible presence.

Identification

Well-developed solidification cracks are easy to identify in either polished weld sections or by fractography; when located on the weld centre-line they can usually be recognised visually or even on radiographs (Fig. 3.31). Because of their small size, liquation cracks require some form of microscopic examination for their identification. A solidification crack which has been broken open will be oxidised if it was open to the surface

94 Weldability of ferritic steels

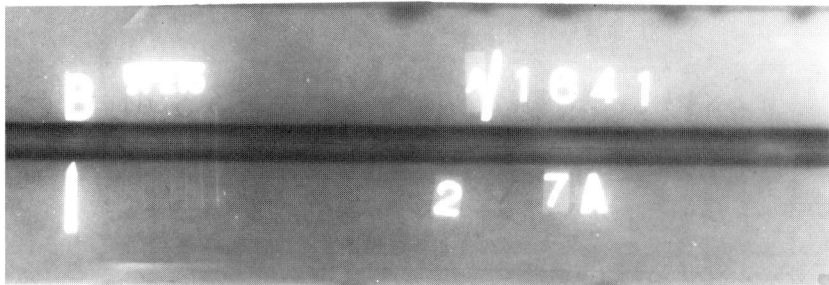

3.31 Radiograph of a centre-line solidification crack in a submerged arc weld.

when it formed. If not, a slight difference in texture of the fracture surface may be apparent, with some rounding of the fracture surface detail being apparent at low magnification in comparison with the rest of the surface.

Light microscopy of sections of solidification cracks shows them to lie along the solidification boundaries. Usually, because of contraction following solidification, the cracks are wide open. If the plane of sectioning coincides roughly with that of the dendrite arms, the sides of the crack can be seen to be undulating (Fig. 3.9 and 3.32(a)), matching the dendritic pattern seen in fractographic examination, described later. Etching to reveal the solidification structure (e.g. in SASPA NANSA* or at low power in a modified ferric chloride etch †) shows the relation to the solidification structure, as in Fig. 3.32(a), where a line of segregation can be seen beyond the tip of the crack. In Fig 3.32(b) the relation to the solidification structure can be seen, although conventional etching in nital had originally shown the first run to have a fully equiaxed structure as a result of heating by the second pass.

The location of cracks in the weld cross-section and their orientation both provide useful clues to the identification of solidification cracks. Such cracks are never transverse to the weld, either perpendicularly or at 45° to the surface; such transverse cracks are usually weld metal hydrogen cracks (Chapter 5). The most usual location is along the weld centre-line, although this will be the heat centre, as in Fig. 3.7, rather than the geometrical centre. So-called interdendritic cracks may originate from the flare angle or bay region, where there is a distinct finger penetration (Fig. 3.33) or as cracks additional to the centre-line crack (Fig. 3.3(a), 3.9, 3.13). The junction between the two parts of the weld bead locally delays heat

---

*100 ml saturated aqueous picric acid (SASPA), 1 g sodium dodecyl benzene sulphonate (NANSA SS60 wetting agent).
†13 g hydrated $FeCl_3$, 1 g $CuCl_2$, 13 ml HCl, 300 ml water, 250 ml methyl alcohol. Immerse weld for a few seconds.

Solidification cracking

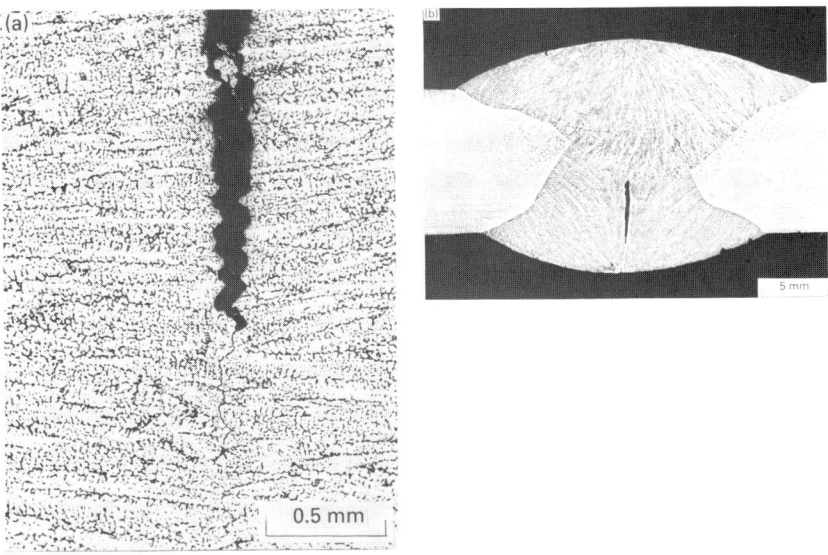

3.32 Sections through solidification cracks in submerged arc welds etched to reveal segregation pattern: (a) centre-line crack with heavy segregate beyond crack tip, SASPA–NANSA etch; (b) crack in first submerged arc run, modified ferric chloride to show solidification structure in a run in which nital etch showed a refined transformation structure.

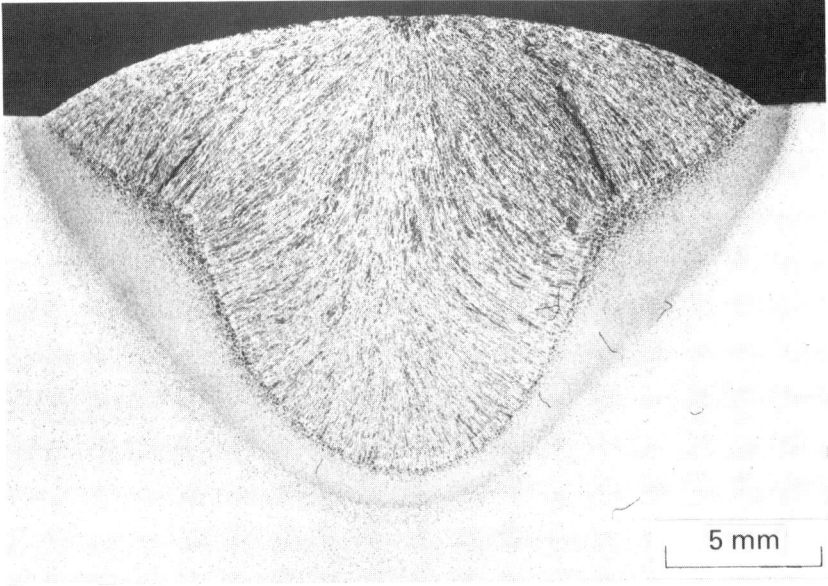

3.33 Solidification crack in submerged arc bead from flare angle.

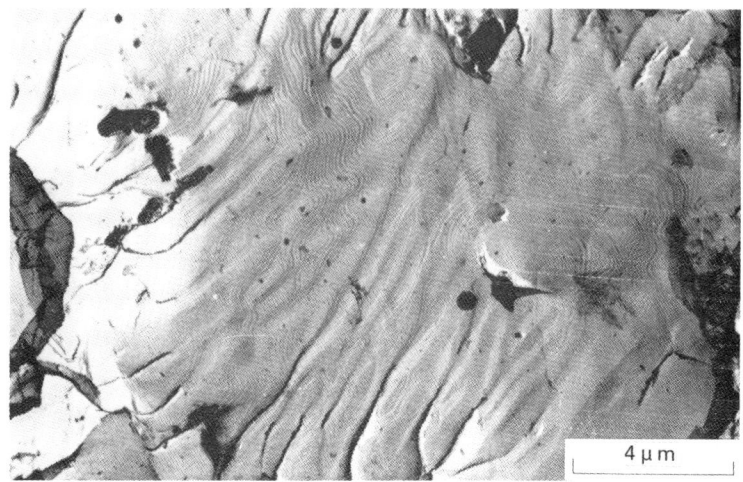

3.34 Thermal faceting on replica of solidification crack surface examined on TEM.

extraction, as can be seen from the wider HAZ, and thus makes feeding of any incipient cracks difficult. Cracks can also initiate in susceptible weld metals from isolated liquation cracks at segregate bands, as in Fig. 3.24.

Occasionally, if formed late in solidification from material with a low susceptibility to cracking, the cracks are very narrow and undulating and less recognisable as solidification cracks (Fig. 3.2(a)).

Electron microscopy can be used to examine buried, i.e. unoxidised, solidification and liquation crack surfaces, but the finer detail is irretrievably lost if the surfaces are oxidised. Scanning electron microscopic (SEM) examination is adequate for most details, although it will not show thermal striations or faceting, which can be found on the surfaces of unoxidised solidification and liquation cracks, and for which transmission electron microscopic (TEM) examination is necessary (Fig. 3.34). Segregation associated with cracking is normally beyond the surface analytical capabilities available on many SEMs; true surface analytical techniques, such as Auger and ESCA (electron spectroscopy for chemical analysis), are required to detect such segregation. However, second phase particles in the eutectics associated with cracking can be seen by both SEM and TEM and occasionally in the light microscope.

In the SEM, well-developed solidification crack surfaces show a typical well-rounded dendrite structure (Fig. 3.3(b) and 3.35). Even at low magnification, the linear pattern of the dendrite tips is unmistakable (Fig. 3.3(b)) and the ends of dendrites growing inwards and be distinguished

Solidification cracking

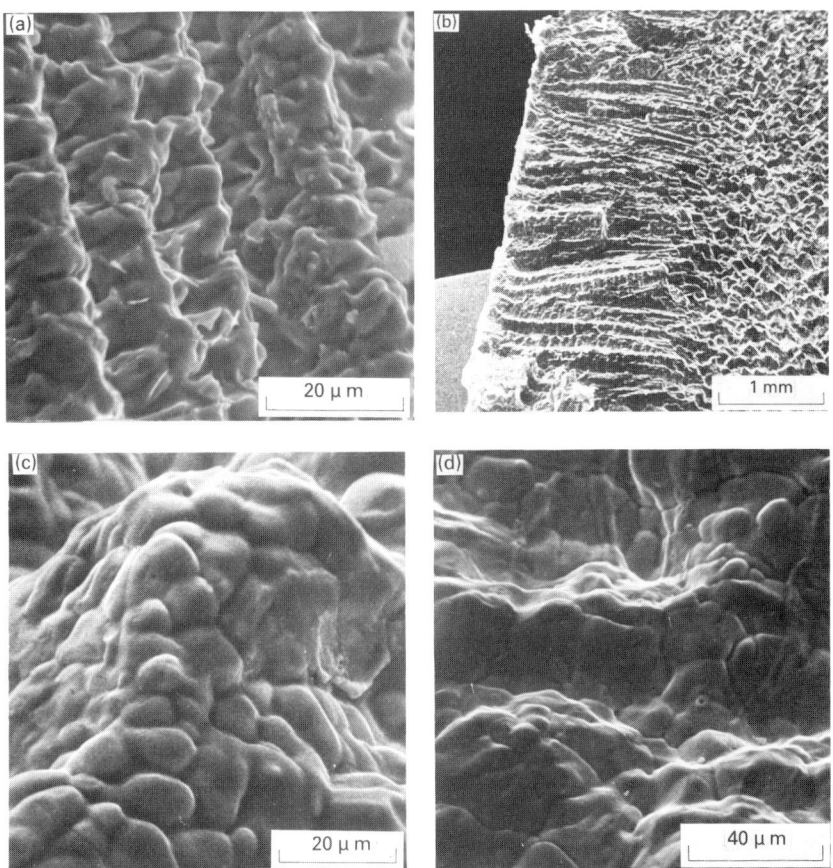

3.35 Typical solidification crack surfaces in SEM: (a) well-developed crack; (b) upwards dendrite growth from backing bar on left, inwards growth from fusion boundary on right; (c): detail of inwards growth; (d) detail of upwards growth.

from cracking parallel to the dendrite growth direction (Fig. 3.35(c) and (d)). Less well-developed cracks form at lower temperatures, when the spaces between the dendrites are more nearly filled with solid metal and the crack surfaces are considerably flatter (Fig. 3.36). Even so, the surfaces are slightly undulating. Very late-developing solidification cracks (Fig. 3.2(b)) may contain a mixture of solidification cracking and the dimpled ductile fracture (microvoid coalescence) produced when the specimen was broken open. Fracture just beyond the tip of a solidification crack can show linear markings from the segregated region (Fig. 3.37). Oxidised solidification crack surfaces show these features less well (Fig. 3.38), particularly if the oxidation is heavy (Fig. 3.38(b) and (c)).

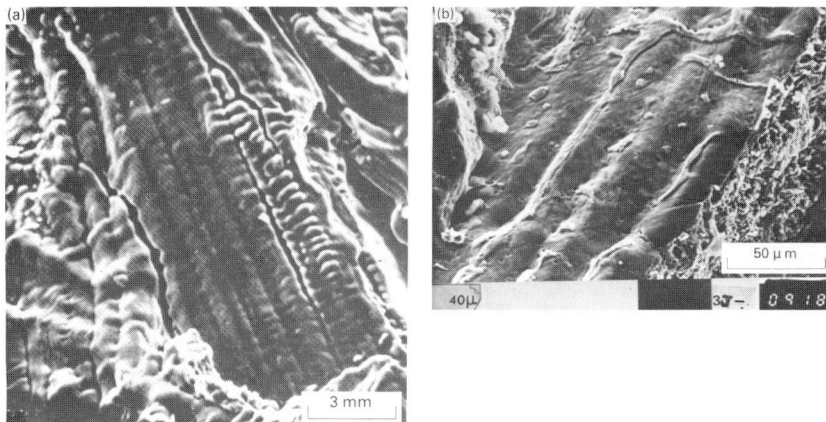

3.36 Less well-developed solidification crack surfaces in SEM: (a) flatter surface with slight dendritic markings; (b) smooth undulating surface.

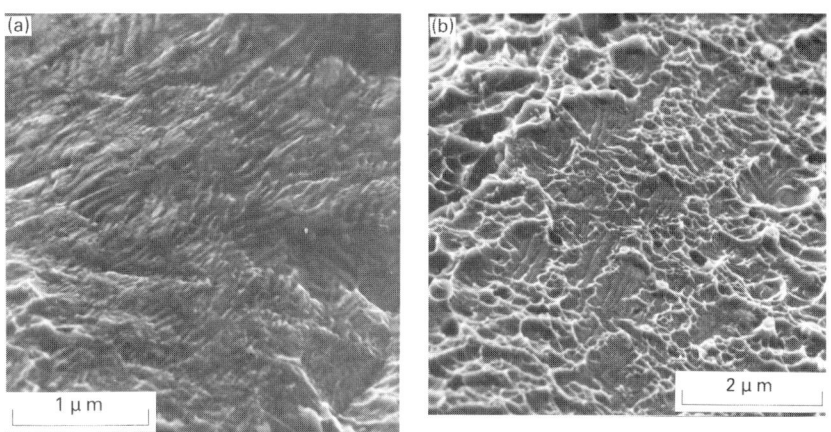

3.37 Striations associated with less well-developed solidification cracks: (a) on flattish crack surface; (b) in microvoids in segregated region ahead of crack tip.

Occasionally, spikes are seen on fracture surfaces, as in Fig. 3.38(a). These were believed to be bridges of solid metal formed across a solidification crack and then pulled down as the crack faces were pulled apart. More recent work has shown them to lie opposite to cavities on the other side of the crack and that they are segregates rich in phosphorus.[15] With SEM examination, signs of MnS or other inclusions are quite rare, although an example is shown in Fig. 3.39. Small globules of lead seen on

Solidification cracking

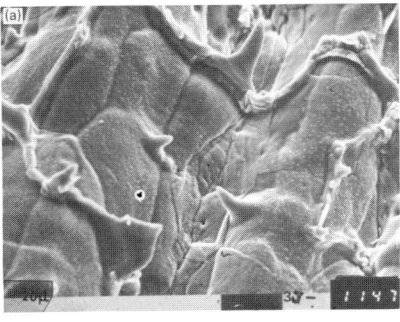
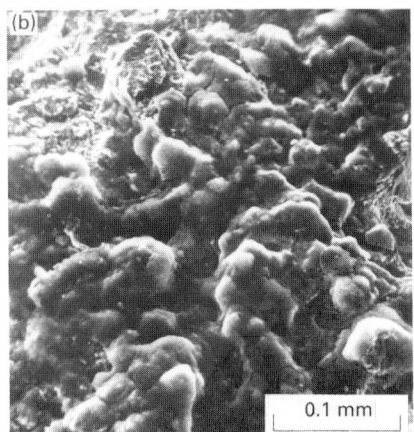
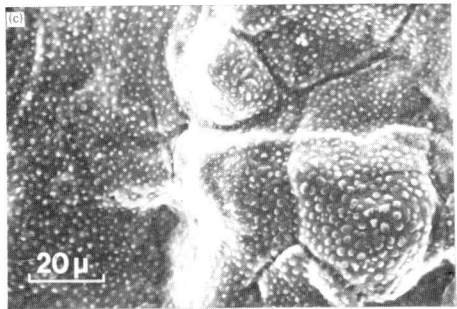

3.38 Oxidised solidification crack surfaces in SEM: (a) less well-developed surface with spikes; (b) well-developed crack with small spikes; (c) well-developed open crack.

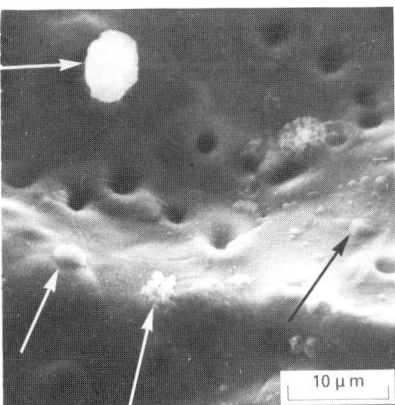

3.39 Solidification crack surface in SEM showing MnS particles (arrowed).

100　　Weldability of ferritic steels

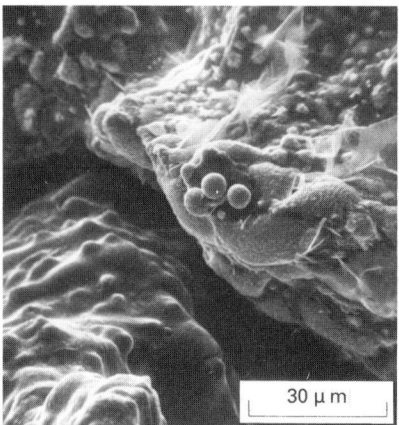

3.40　Solidification crack surface in SEM, three small particles in centre were identified as rich in lead.

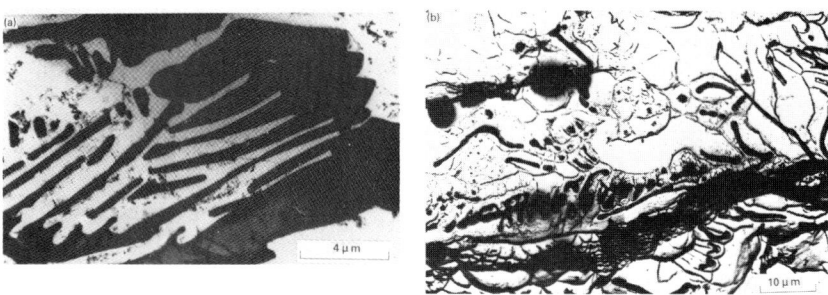

3.41　Crack surfaces in TEM showing: (a) Nb_4C_3 on solidification crack surface; (b) α-MnS in weld metal ahead of crack tip.

a submerged arc crack surface are shown in Fig. 3.40, although it is not known whether they took any part in the formation of the crack, or whether they formed subsequently.

Using the TEM, it is possible to detect and identify (by local area diffraction) constituents that took part in solidification or liquation cracking. Compounds, usually in eutectic form (Fig. 3.41), such as MnS, Nb_4C_3 and Ti(C.S) have been identified by these means, often in the segregated region just beyond the crack tip, and have made it possible to identify the cause of cracking. The detection of thermal faceting on solidification crack surfaces in the TEM (Fig. 3.34) is evidence that the surface has been present at a high temperature; however such striations have also been found on cold (e.g. hydrogen) crack surfaces that have been heated (e.g. by PWHT) subsequent to their formation.

References

1. Dixon, B.F., 'Weld metal solidification cracking in ferritic steels – a review', *Australian Weld. J.*, 1981, **24**(4), 23-30.
2. Ohshita, O., Yurioka, N., Mori, N. and Kimura, T., 'Prevention of solidification cracking in very low carbon steel welds', *Weld. J.*, 1983, **62**, 129s-136s.
3. Nakagawa, H., Matsuda, F. and Senda, T., 'Effect of sulphur on solidification cracking in weld metal of low carbon and low nickel alloy steels', *Trans. Jap. Weld. Soc.* 1974, **5**(1), 39-45; **5**(2), 18-33, 84-89; 1975, **6**(1), 3--16.
4. Bailey, N. and Jones, S.B., 'The solidification cracking of ferritic steel during submerged arc welding', *Weld. J.*, 1978, **57**, 217s-231s.
5. Morgan-Warren, E.J. and Jordan, M.F., 'Quantitative study of the effect of composition on weld solidification cracking in low alloy steels', *Metals Technol.*, 1974, **1**(6), 271-278.
6. British Standard BS5135: 1984, 'Arc welding of C and C-Mn steels', BSI, London, 1984.
7. Davis, M.L.E., Pargeter, R.J. and Bailey, N., 'The effects of titanium and boron additions to submerged-arc fluxes', *Metal Construction*, 1983, **15**, 338-344.
8. Huxley, H.V., 'The influence of composition on weld solidification cracking in C: Mn steel', *Metallurgia*, 1970, **82**, 167-174.
9. Russell, J.D., 'The development of an electron beam weld solidification cracking test', *Welding Res. Intl.*, 1979, **9**(6), 1-21.
10. Fujita, Y., Terai, K., Yamada, S., Suzuwa, R. and Matsumara, H., 'Prevention of end cracking in automatic one-side welding', *J. Soc. Naval Architects Jpn*, 1973, **133**, 267-274. (TWI Translation No 330.)
11. Bailey, N. and Roberts, C., 'Maraging steel for structural welding', *Weld. J.*, 1078, **57**, 15-28.
12. Boniszewski, T. and Watkinson, F., 'Examination of hot tearing in the weld HAZ of ferritic steels', *Brit. Weld. J.*, 1964, **11**, 610-619.
13. Baker, R.G. and Watkinson, F., 'The assessment of cracking problems', TWI Conference on 'Weldability of Structural and Pressure Vessel Steels', London, Nov, 1970.
14. Keville, B.R. and Cochrane, R.C., 'Strain-induced hot cracking in ferritic weld metal', in *Proceedings of the TWI Conference 'Trends in Steels and Consumables for Welding'*, London, Nov, 1978, TWI, Cambridge, 1979.
15. Dixon, B.F., 'Cracking in the Transvarestraint test', *Metal Construction*, 1984, **16**, 86-90, 156-160, 232-237.

4 Lamellar tearing

Description

Lamellar tearing is confined to plates (and perhaps extrusions) when they have inadequate ductility in the short transverse direction (also known as the through-thickness or Z-direction). Steel plate can contain a population of non-metallic inclusions sufficient to reduce its short transverse ductility to a considerable extent. When such steel is welded, tearing or cracking can result if the ductility in the short transverse direction is not sufficient to accommodate the welding contraction strains. Such strains result from welding stresses, which are at a maximum in the short transverse direction when a fusion boundary is parallel to the plate surface (Fig. 4.1), i.e. to the planes into which inclusions have been rolled to their maximum dimensions.

Such stressing occurs when a weld in a rigid structure is made on the surface of a plate as shown in Fig. 4.1. The stresses are set up when the free contraction of the weld on solidifying and cooling is restrained by the rest of the structure. The tearing process takes place in two distinct stages (Fig. 4.2). Firstly, non-metallic inclusions 'decohere', i.e. they separate from the matrix of steel in which they lie and thus form minute cracks between themselves and the steel. The second stage is for the minute cracks at the decohered inclusions to join up, jumping from one plane in the steel to another to link up with similar decohered inclusions on adjacent planes. It should be noted that the decohesion of inclusions can sometimes occur as a result of cold working operations (particularly when they involve bending) and PWHT.

Unlike hydrogen cracking, discussed in the next chapter, the stresses needed to produce lamellar tearing are long range. Also in contrast to hydrogen cracking, cracking usually extends well beyond the visible HAZ into the parent plate and it cannot occur in weld metal. Tearing may also start at a higher temperature than is possible for hydrogen cracking (perhaps as high as 300 °C).

Lamellar tearing 103

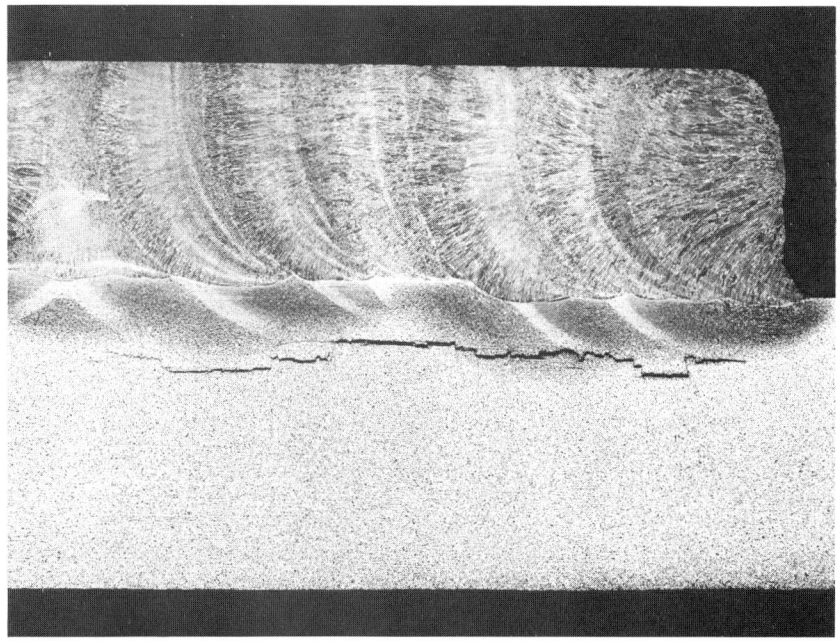

4.1 Lamellar tear under a restrained weld made on the surface of a plate with inadequate ductility in the through-thickness direction.

Despite the need for low ductility in the steel for cracking to occur, lamellar tearing occurs by a ductile and not a brittle fracture mechanism, the cracks being initiated by inclusions in the base plate which have been rolled and extended by the steel rolling process. Nowadays, grades of plate containing low levels of inclusions are available which are very resistant to lamellar tearing. Because these have good ductility in the short transverse or Z-direction, they are frequently known by a name which includes the letter Z, e.g. 'Z quality' or 'Z grade'.

Lamellar tears, unlike hydrogen cracks, form directly on cooling, although they may extend when welding elsewhere on the structure if this welding builds up and increases the long range stresses. Cracks are usually seen at weld toes, and remedial measures usually involve a repair with buttering of the susceptible plate by soft weld metal. Although the inclusions in the weld metal may be much larger in number than in the plate, they are spherical, very small and not oriented parallel to any surface, so that their effect on ductility is small and isotropic.

Much of the present knowledge of lamellar tearing was obtained from research carried out at TWI in the early 1970s. Details of these investigations, including a bibliography up to 1974, are given in ref. 1.

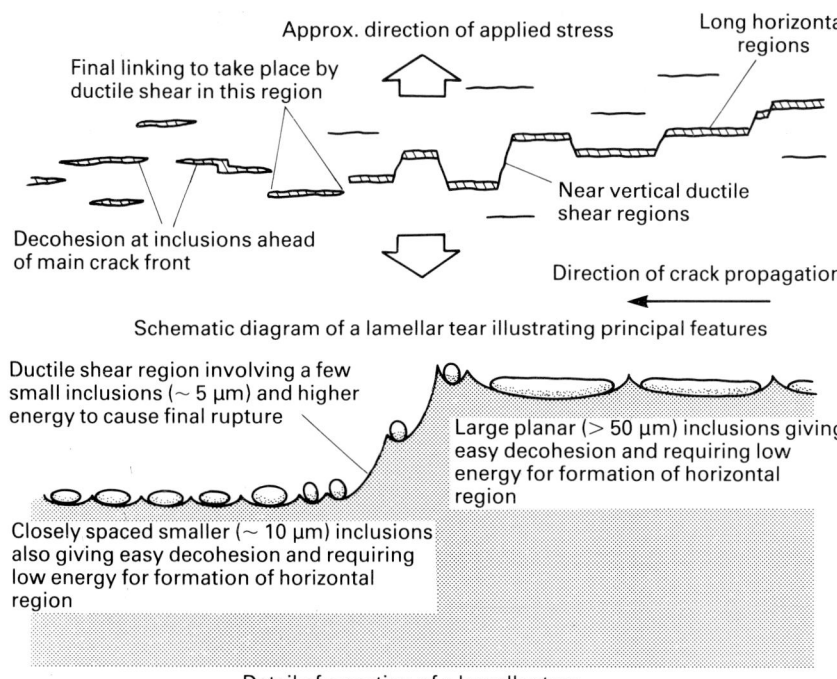

4.2 The lamellar tearing process.

Role of inclusions

The most important aspects of inclusions in relation to lamellar tearing are their form, number and distribution. Their type is of less importance, unless it influences the form of the inclusions. For example, manganese sulphide (MnS) inclusions are much more readily rolled down to thin inclusions of large surface area than those of the alumina (Al_2O_3) type. Similarly, the type and form of sulphide inclusions can be altered (inclusion shape control) by treating the steel with rare earth metals, calcium or other metals in order to produce complex sulphide inclusions which are much less deformable than MnS at steel rolling temperatures.

In castings (as in weld metals), non-metallic inclusions are of spherical form so that ductility is isotropic, and very low values are not possible in any orientation, provided the steel is above its transition temperature. Forgings are much more uniformly hot worked than plate so that inclusions are less flattened and extended, ductility is relatively uniform and forgings are not prone to lamellar tearing. Steel plate, and possibly extrusions, however, are heavily hot worked in one direction (Fig. 4.3) and are therefore at risk.

Lamellar tearing

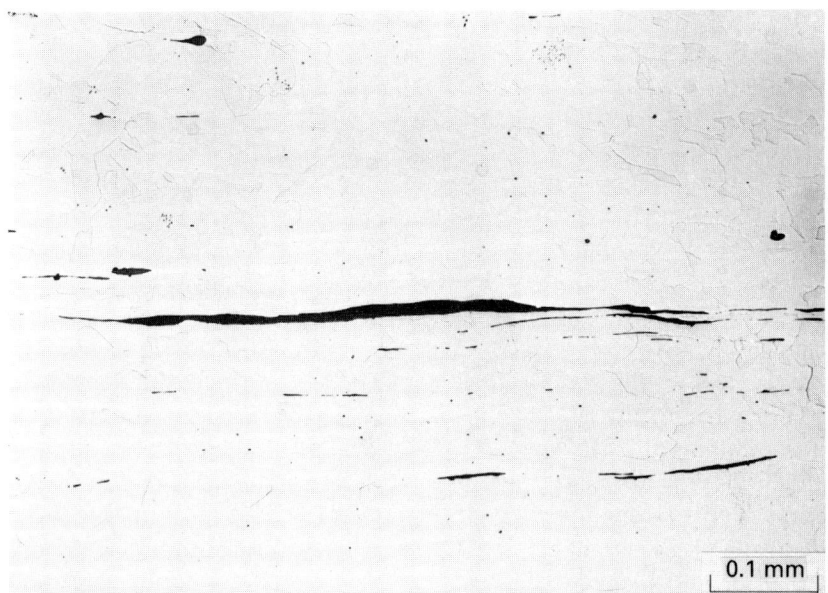

4.3 Typical large silicate inclusions.

An increase in the number of inclusions usually increases the risk of lamellar tearing, particularly if size distribution and position are not significantly altered. The spatial distribution of inclusions is also important. If most of the inclusions are just below the plate surface, lamellar tearing will be very easy, particularly if welded with a technique which does not penetrate far enough to melt through the inclusion-rich region, as in Fig. 4.1. Inclusions concentrated along the plate centre-line, as is likely in continuously cast plate, require a high degree of restraint and a relatively long-range stress system to be damaging. However, if such concast plate is relatively thin and/or the welding process is penetrating, then a central region of inclusions can be very harmful, particularly if it is associated with heavy segregation likely to give lower toughness than the bulk of the steel.

Even when centrally segregated, inclusions do not lie in a single plane but in parallel planes separated by fractions of a millimetre. The second stage of lamellar tearing depends on the ease with which the first stage cracks can jump from one of these planes to another.

The ease with which the first stage of lamellar tearing (decohesion of inclusions from the steel matrix) occurs, depends on how well the inclusions are bonded to the steel matrix. This depends, in turn, on the relative coefficients of thermal expansion and the ease of deformability of inclusions and steel.

Manganese sulphide inclusions tend to contract away from the steel matrix when the steel cools from high temperature during its manufacture, and thus decohere readily. Other types of inclusions contract less than steel, which therefore tends to contract on to the inclusions. Aluminium-treated steel plate (provided it has not been produced with inclusion shape control) contains compact angular Al_2O_3 inclusions and separate leafy MnS. The MnS inclusions easily decohere and, if the population is suitable, readily produce lamellar tearing. Silicon-killed and other types of steel contain inclusions which are mixtures of manganese sulphides and silicates. Because these mixed inclusions have a mean thermal expansion coefficient closer to that of steel than MnS by itself, they decohere less readily than MnS alone, so that a given inclusion population should be less likely to start a lamellar tear in a Si-killed steel than a similar population of inclusions in a steel which is Al-treated.

Even though plate may be cross-rolled, the compressive deformation always occurs in the direction which undergoes a reduction in thickness. As a result, deformable inclusions, such as manganese sulphides and silicates, finish up shaped like broad leaves, being of large surface area and thin cross-section. Their widths depend on the degree of cross-rolling to which the plate has been given. Such thin inclusions naturally occupy a much larger area in planes parallel to the plate surface than the more compact inclusions in castings and forgings, or the less deformable alumina or calcium oxide/silicate inclusions in Al- or Ca-treated steels after rolling.

In steels generally, the inclusion content depends mainly on two compositional parameters, its oxygen and its sulphur contents, although the precise content (in terms of volume or weight percentage) and size distribution will depend on the actual oxides and sulphides present and the thermal and mechanical history of the steel, i.e. on the way the steel has been made.

Furthermore, very low oxygen contents in steel (i.e. in Al-treated steels) tend to give Type II sulphide inclusions, which are more flattened, and therefore more damaging for a given sulphur content, than the Type I sulphide inclusions found in the older Si-killed and semi-killed steels. Most current steels are made by continuous casting, which requires an aluminium addition, and hence ensures low oxygen contents, together with Type II MnS inclusions, unless inclusion shape control additions have been made, in which case more complex sulphides of higher melting temperature will be present. Even for one type of plate steel, in terms of composition and deoxidation practice and other treatments, the susceptibility to lamellar tearing depends on other factors, such as the degree of rolling from the cast ingot to the finished plate, the position of the plate in the original ingot and the distribution of the inclusions within the plate.

Lamellar tearing

The **degree of rolling** increases the flattening of the inclusions and their final potency as initiators of lamellar tearing. Unless the ingot size has been selected in relation to the plate size to minimise rolling, the most susceptible plate thickness is probably between about 25 and 65 mm. It is difficult to give very heavy rolling reductions when making plate thicker than this range, whereas thinner plates tend to be more flexible and can accommodate some of the through-thickness stressing by distortion of the whole plate. However, cases of lamellar tearing in plate as thin as 8 mm are known to have occurred in very rigid structures.

The **position of the plate in the original ingot** is perhaps less important now than in the past, because of the advent of continuous casting. In ingots the distribution of inclusions is far from uniform, as a result of the different types of segregation which can occur. Such variability in inclusion distribution inevitably leads to a considerable scatter in the results of whatever tests are used to assess lamellar tearing behaviour, and it is unwise to scrap or accept a plate or a batch of plates on the results of a single test or test series taken from one location.

The **distribution of inclusions within the plate** has a profound effect on the likelihood of lamellar tearing. If relatively small fillet welds are made on the surface of thick plate (e.g. 100 mm), the presence of very severe centre-line segregation will not be of much consequence. The long range stresses, at a high level close to the fusion boundary, will have fallen to a relatively low level at the centre of the plate and any distortion will have been accommodated in the relatively clean steel close to the surface.

In thinner plate (e.g. 20 mm thick), with a larger weld size, the weld will penetrate an appreciable fraction of the distance to the plate centre-line, so that stresses will still be high in that region and tearing is possible. A high density of inclusions near the plate surface is likely to give lamellar tearing problems, regardless of plate thickness, but such a population is less likely with continuously cast than with ingot cast steel.

Despite all these factors leading to variability of tearing behaviour, rough correlations for Al-treated steels have been provided to estimate the effect of sulphur on the risk of lamellar tearing, as assessed by the short transverse tensile test. This test, described in a later section, gives a value of the short transverse reduction of area (STRA); the effect of sulphur on this value[2] is shown for different thicknesses of plate in Fig. 4.4. Under conditions of highest likely restraint, a minimum value of 20% STRA is usually recommended, and Fig. 4.4 shows that this can be achieved comfortably if the sulphur content of the steel is kept below 0.005%.

The individual points on the figure represent TWI results, mainly on ingot cast steels, and these mainly fall within scatter bands which had been proposed for this type of steel in an IIW document.[3] It is significant

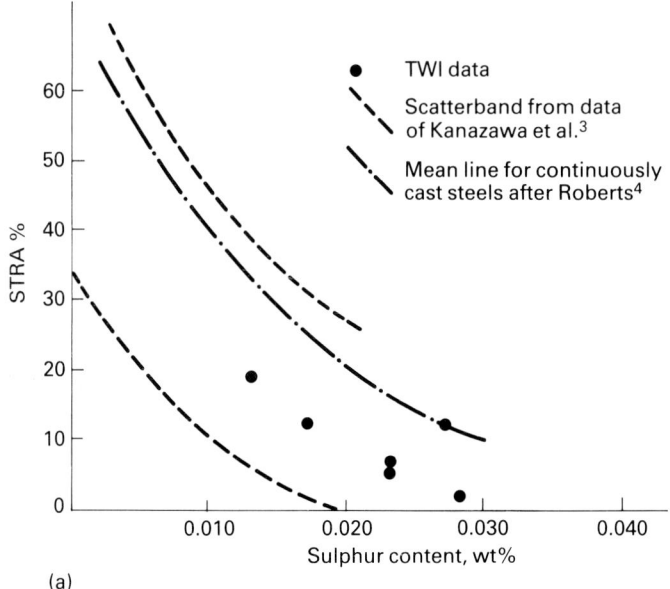

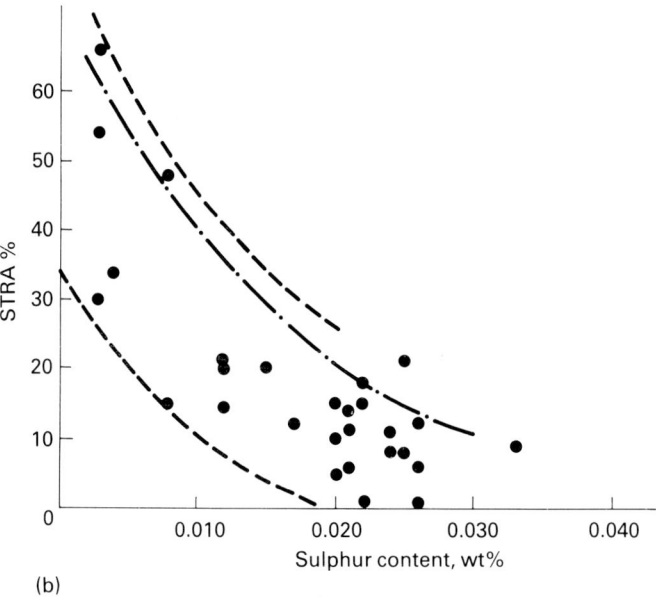

4.4 Influence of plate sulphur content on STRA values in short transverse tensile test: (a) plate <12.5 mm thick; (b) plate 12.5–50 mm thick; (c) plate >50 mm thick.

Lamellar tearing

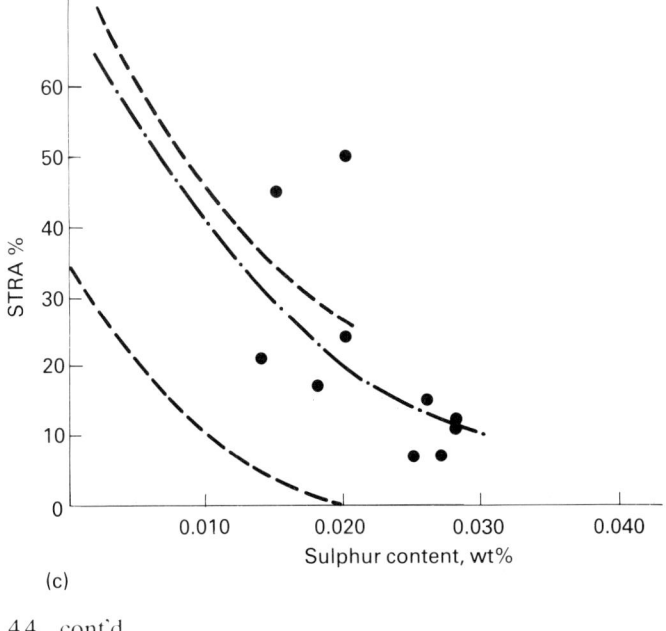

(c)

4.4 cont'd

that the only values outside the proposed scatter band lie above it (i.e. on the safe side) and are for thick-sectioned steels, which tend to be less prone to lamellar tearing because of the normally lower degree of rolling given. The mean values for continuously cast steels given by Roberts[4] lie just below the upper scatter band. This supports the reportedly good behaviour of such steels (except, presumably, where the stress fields associated with any deeply penetrating welds approach any centre-line segregates). It should be emphasised that the results in Fig. 4.4 relate to normal steels and *not* to those with shape control additions.

Susceptible joint types

Lamellar tearing may occur in both fillet and butt welds. The susceptible joint types are those in which one of the fusion boundaries is parallel to the rolling plane of the plate. The risk of tearing depends on the quality of the plate, and normally a high degree of restraint is necessary in a fabrication before any tearing occurs. Nevertheless, if the plate quality is very poor, lamellar tearing can occur (and has occurred) in such lightly restrained fabrications as lifting lugs and fillet-welded I-beams.

With high to moderate restraint present, three types of joint are most likely to give rise to lamellar tearing problems: they are T-fillets, T-butts and

110   Weldability of ferritic steels

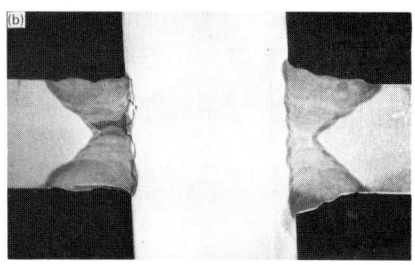
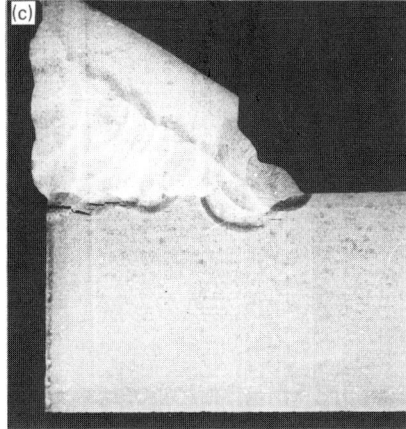

4.5 Weld configurations susceptible to lamellar tearing: (a) T-fillet; (b) T-butt (cruciform joint); (c) corner joint.

corner joints; examples are illustrated in Fig. 4.5. The situations in which these joint types are likely to occur are discussed below.

Set-through nozzles and other penetrators (Fig. 4.6) may be of the full or partially penetrating butt type, with or without reinforcing fillets, or they may be fillet-welded. Tearing is likely if the penetrator, such as a set-through nozzle, is made from plate of poor quality with regard to lamellar tearing. The risk can be reduced by using set-on nozzles (Fig. 4.7), by using a grade of plate resistant to lamellar tearing (so-called 'Z quality'), or by using forged or cast inserts. Set-on nozzles, although apparently the cheapest option, are likely to require the use of compensating plates – themselves likely to pose a risk of lamellar tearing on the vessel itself, albeit less severe than with set-through nozzles.

Stiffeners (Fig. 4.8) are a potent source of lamellar tearing, because the tearing can occur in the stiffeners themselves if the structure is of the egg-box or of the stiffened diaphragm (Fig. 4.8(c)) type, especially as stiffeners themselves are not often considered to be important members of a struc-

Lamellar tearing 111

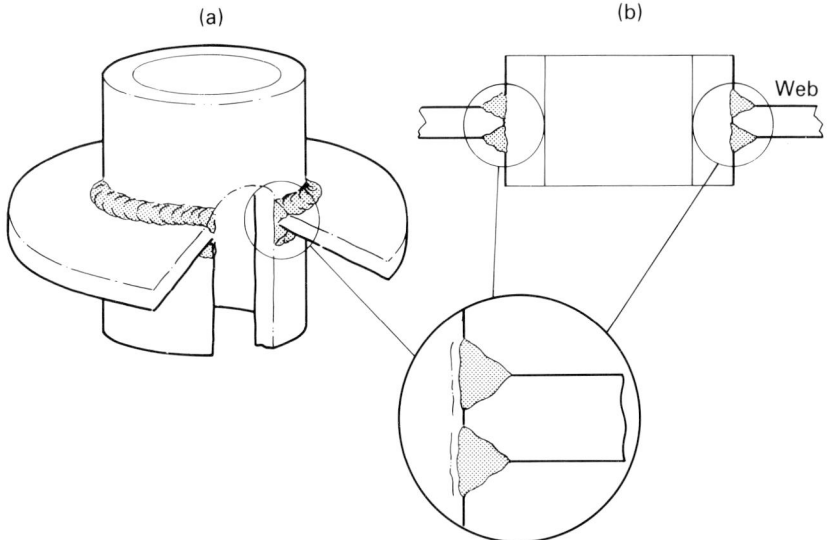

4.6 Joint types susceptible to lamellar tearing: (a) set-through nozzle; (b) Vierendeel girder.

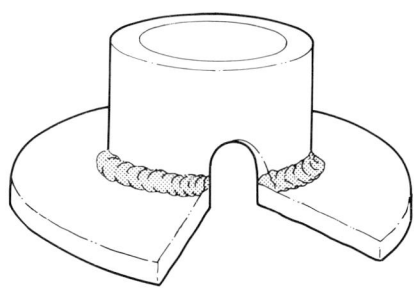

4.7 Set-on nozzle, less prone to lamellar tearing than set-through type.

ture. Stiffeners inside cylinders (Fig. 4.8(a)) are particularly likely to cause tearing in the cylinder itself, and nowadays steel plate of a lamellar tearing-resistant grade is nearly always specified for such duty, although forgings or castings are equally suitable.

Joint design
Although nowadays it is common to specify grades of steel resistant to lamellar tearing where the risk of tearing is known to be present, the susceptibility of different joint types can be varied by altering their configuration.

T-joints with simple fillets give fewer problems than joints with full

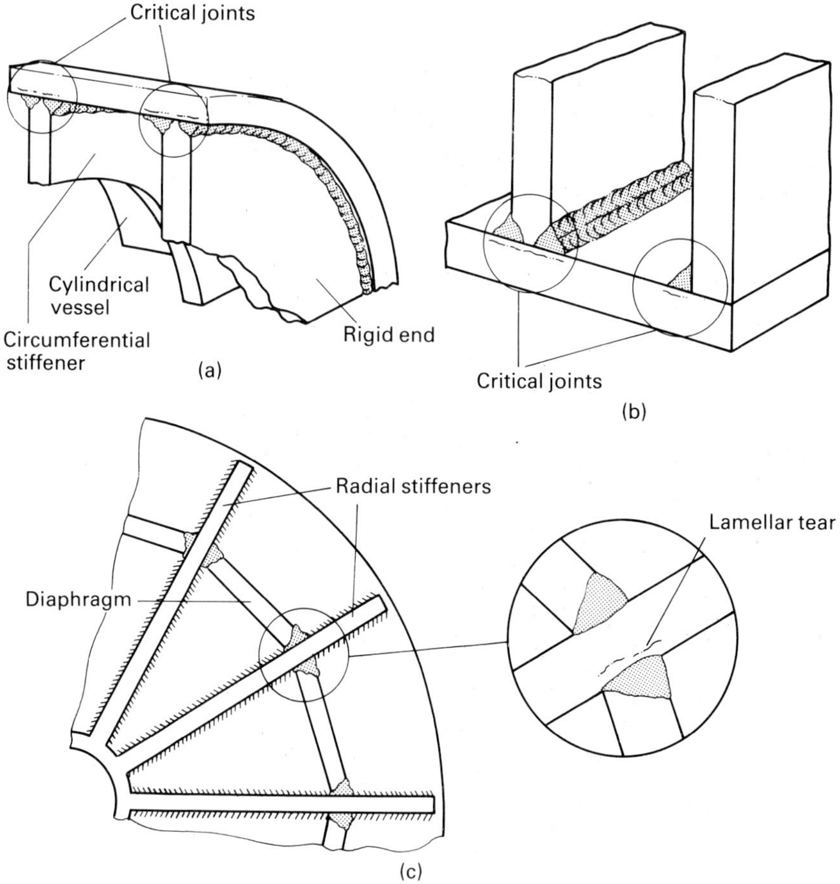

4.8 Types of stiffener liable to lamellar tearing: (a) circumferential stiffener; (b) joints in box structure; (c) radial stiffeners.

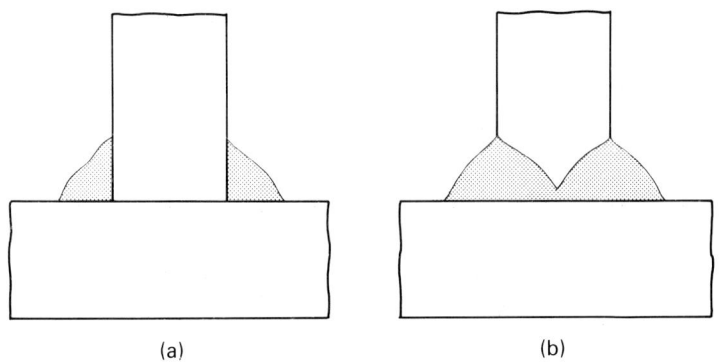

4.9 T-fillets (a) are less susceptible to lamellar tearing than T-butt welds (b).

Lamellar tearing 113

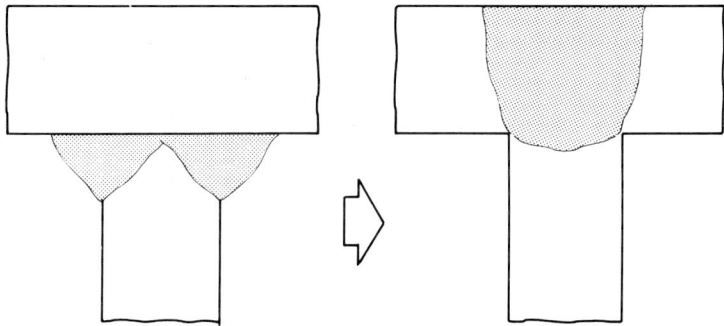

4.10 Conventional vertically welded T-joints may be replaced with electroslag or similar welds with no risk of tearing.

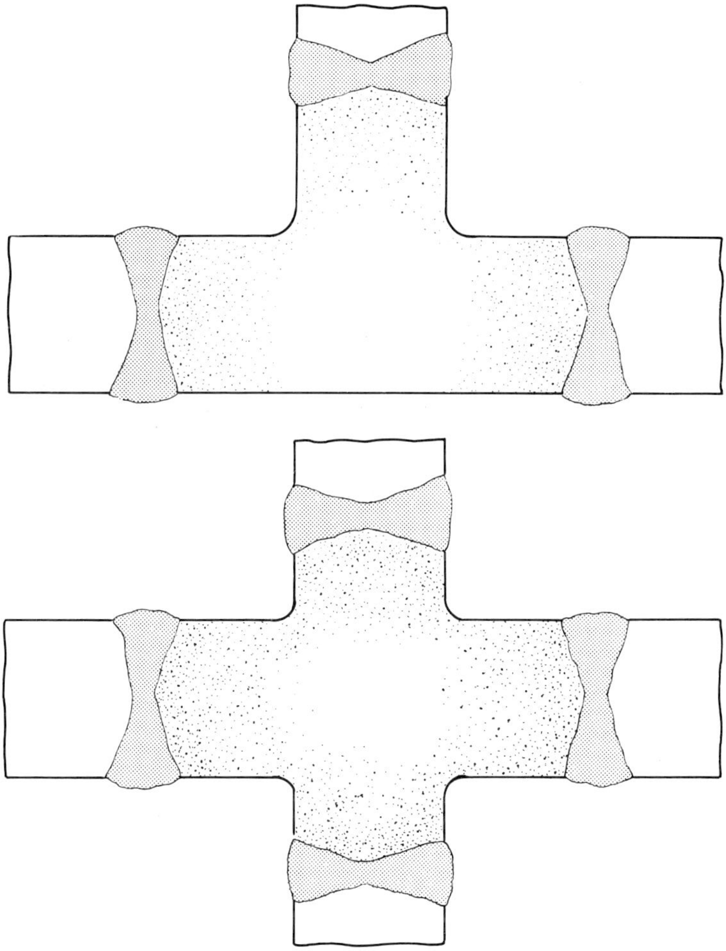

4.11 Castings or forgings can replace conventional T-joints with no risk of tearing.

penetration (Fig. 4.9). The most severe form of the T-joint is the cruciform joint, particularly where full penetration welds are employed (Fig. 4.5(b)). In both fillet and full penetration joints, a balanced, double-sided weld provides less risk of tearing than a single-sided weld. The improvement is largely due to the smaller amount of weld metal deposited, but some risk of tearing still remains, even with a T-fillet joint welded with small fillets in a balanced manner. It is possible to re-design this type of joint, possibly using electroslag or electrogas welding, to eliminate all risk of lamellar tearing (Fig. 4.10). The use of forgings or castings is also possible, although this can increase the amount of welding, as shown in Fig. 4.11.

Corner joints (Fig. 4.12) can also give rise to lamellar tearing if suitable steel grades are not used. The risk can be reduced by putting the joint angle on the surface of the plate at risk, instead of on the end of the other plate. However, this is more costly, particularly if a large angle is used, as it increases the amount of weld metal needed to fill the joint.

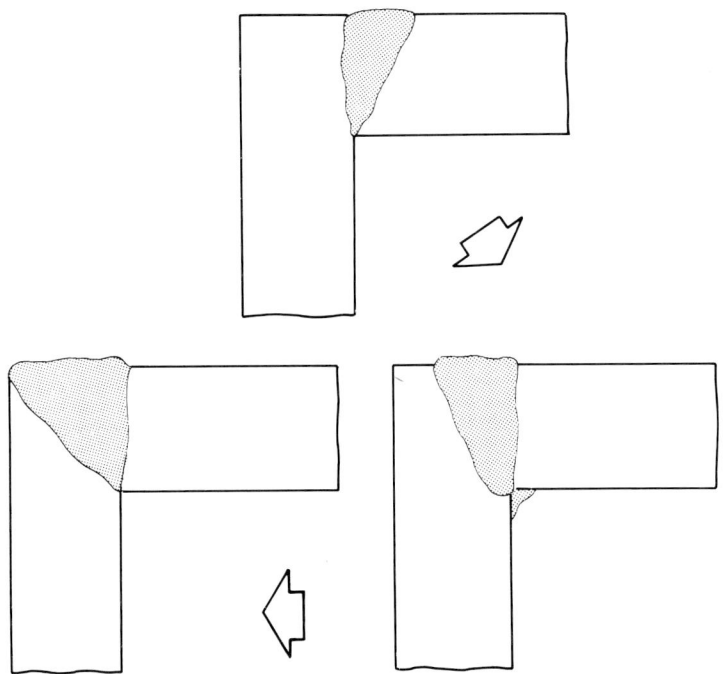

4.12   Bevel on corner joints modified to reduce risk of tearing.

Lamellar tearing

Other factors

Steel plate type
Many of the factors associated with steel type were discused earlier. These can be briefly summarised:

1 Only plate (and possibly extrusions) are prone to lamellar tearing.
2 Al-treated steel is more susceptible than a similar steel that is Si- or semi-killed.
3 Inclusion shape control, low sulphur contents and (usually) continuous casting reduce the likelihood of problems.

One factor not discussed earlier is the inherent strength of the steel. Tests have suggested that with very high strength steels (i.e. with minimum yield strengths of about 900 N/mm^2) low STRA values can occur with sulphur contents of 0.005% S and less,[5] so that the yield strength of plate itself is a factor to consider when assessing the lamellar tearing susceptibility of high strength plate steels. Certainly, lamellar tearing problems occurred during the development in the UK of weldable steels of 550 N/mm^2 yield strength, before lamellar tearing behaviour was adequately understood.

Direction of residual stress
To reduce the risk of tearing, it can be helpful to reduce the angle on the plate at risk from 90°. This is most easy to carry out in corner joints (Fig. 4.12). This technique is only practicable on the type of corner joint illustrated, as its use on the other types of susceptible joints would involve gouging out or machining into the major component of, for example, a T-joint, which would not be acceptable in most circumstances.

Hydrogen
Although lamellar tearing and hydrogen cracking (Chapter 5) are normally distinguishable, it is known that hydrogen introduced during welding (or even present in the steel to be welded), can increase the risk of tearing, because it reduces the ductility of the steel. Nevertheless, not all the techniques that can reduce the risk of hydrogen cracking are of benefit in combating lamellar tearing. Preheating (or increasing the preheat temperature) is not always a good option, particularly when a set-through nozzle or penetrator is being welded into a hole in a vessel. In such circumstances, the increased temperature gives an increased risk of a significant temperature difference between the two components being welded and, hence, of increased stresses on cooling after welding. The use of consumables of lower hydrogen potential or post-heating may well be of

benefit, as these options are not known to increase the risk of lamellar tearing.

With continuously cast plate of relatively thin section, any centre-line segregates are likely to lie close to the visible weld HAZ. This proximity to the HAZ (making diffusion of weld hydrogen to this area easy), coupled with their increased composition (making them more susceptible to hydrogen cracking) as a result of segregation, makes such regions particularly prone to lamellar tearing, hydrogen cracking or a combination of the two.

Welding parameters

Apart from the influence of hydrogen discussed in the previous section, welding variables have a minor influence on lamellar tearing. However, the use of lower strength weld deposits is helpful, although the scope for variation is often limited. Nevertheless, basic manual electrodes of low strength have been developed to help to cope with lamellar tearing problems. These have yield strengths around the 350 N/mm^2 level. The use of lower strength consumables is, however, *not* recommended where it would mean increasing the fillet weld size to maintain the strength of a joint.

The use of high heat input processes reportedly gives less risk of cracking, but changing heat input within a single process is said to be without effect on the likelihood of tearing. Indeed, reducing heat input within a process can actually have a beneficial influence, when it is reduced in order to decrease the size of a fillet weld under which tearing is likely.

Welding technique

Although varying the actual welding parameters is likely to have little effect on lamellar tearing, the technique with which a joint is made can have a profound effect, and special methods are used both to avoid tearing in difficult situations and to repair tearing which has occurred. The two techniques that have been used successfully are buttering and balanced welding. Among those tried unsuccessfully are the peening of weld runs at an intermediate stage and intermediate PWHT for stress relief. The latter may, however, have been unsuccessful because it produced decohesion of inclusions (i.e. the first stage of lamellar tearing) to a sufficient extent as to increase the incidence of unacceptable ultrasonic indications and not because of the increased amount of actual lamellar tearing.

The simplest method of buttering is to deposit two layers of weld metal (preferably of low strength) on the surface at risk (Fig. 4.13(a)). This produces a surface of high resistance to lamellar tearing on which the joint can be completed in a manner which removes the suspect region of plate to a region of somewhat lower stress. If the plate is of very poor quality

Lamellar tearing

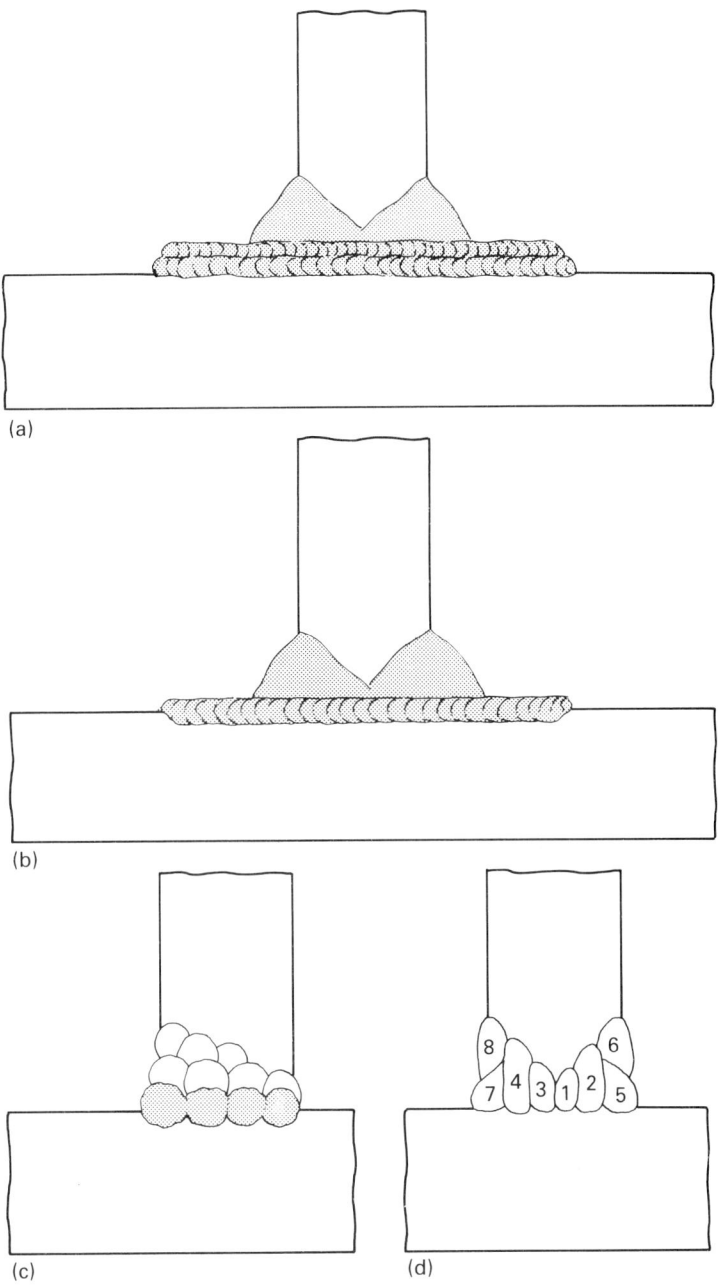

4.13 Risk of lamellar tearing can be reduced by: (a) buttering before welding; (b) removing susceptible plate material and buttering; (c) *in situ* buttering; (d) balanced welding when tearing originates from original weld root of type shown in (c).

close to the surface, this buttering technique may not be successful and it may be necessary to groove out the plate and fill the groove with weld metal, as in Fig. 4.13(b), before proceeding. Figure 4.13(c) shows how a more limited buttering may be incorporated into the joint. This technique may also be used to effect a repair of a torn weld; it aims to avoid increasing the weld size, and hence the amount of metal influenced by high through-thickness stresses.

A balanced welding technique is particularly recommended for T-butt joints, as shown in Fig. 4.13(d). Here, several runs are alternately made at each side of the joint so that the joint is built up in a symmetrical manner.

Any repair of lamellar tearing will involve removal of the cracked area, cleaning by gouging or grinding to clean metal (preferably clean with regard to harmful non-metallic inclusions), buttering with weld metal to the original plate profile before welding of the joint itself is started, and re-making the joint.

Control and avoidance of cracking

Selection of joint design

The selection of joint design is inevitably an interactive process with the selection of steel quality. If high quality, 'Z grade' plates, forgings or castings are to be used, the selection of joint type need not consider lamellar tearing. If prime quality grades of plate are not selected, then the fabrication must be assessed to consider the degree of restraint, particularly during the later stages of welding the component. If restraint is likely to be high, joints should be selected that avoid residual stresses being applied directly perpendicular to the plate surface, bearing in mind that the severity of the situation is increased going from two-sided T-fillets to single-sided T-fillets to T-butts to butt-welded cruciform joints. Corner joints cannot be directly fitted into this scale; they can also be made less severe by angling the preparation as shown in Fig. 4.12.

Selection of steel

Together with an assessment of the joint types present, the grades of steel plate to be used should be selected, bearing in mind that only plates (and possibly extrusions) are likely to give problems, and that the stronger the steel, the cleaner it has to be to avoid any risk of lamellar tearing. If machinable or free-cutting grades of steel plate, particularly with very high sulphur contents (i.e. >0.05% S), are to be used, welds should never be deposited on the plate surface without buttering with at least two layers of weld metal, and then only if the level of restraint is low.

Lamellar tearing

Because most steels are currently cast continuously, it is assumed that the steel will have been Al-treated, and that its resistance to lamellar tearing may be poor if it contains more than about 0.005% S. Although steels with such low sulphur contents may sometimes increase the risk of HAZ hydrogen cracking (Chapter 5), hydrogen cracking is easier to overcome than lamellar tearing, and the low sulphur levels have advantages in other directions, such as better HAZ and weld metal toughness. A fuller appreciation of tolerable sulphur levels in relation to restraint levels is given below in the section dealing with destructive control tests.

The welding process

The welding process has little direct influence on lamellar tearing, the higher heat input processes are said to provide less risk. On the other hand, small overall weld sizes reduce the level of long-range stresses and give less risk of tearing. Low hydrogen processes are of some advantage, but high preheat may be harmful, particularly when welding set-in nozzles and other penetrators.

Control tests

Non-destructive control tests

Considerable use is made of ultrasonic examination of plate to assess its suitability for applications where there is considered to be a risk of tearing. These tests are often called for in specifications and can also be used on plate of variable quality in order to select areas which should or should not be used for weld details prone to lamellar tearing. If these tests are used on plate that has been cold worked, particularly by cold-forming techniques, it must be realised that cold-forming operations can cause decohesion of inclusions, and thus enhance their detectability to ultrasonics. This phenomenon of decohesion has, in the past, led to doubts as to the usefulness of ultrasonics for assessing plate quality. It is possible for plate to pass the initial ultrasonic examination when manufactured and later give rise to apparently unacceptable ultrasonic reflections after cold-forming or even a PWHT. In such cases it is advisable to carry out small-scale tests, such as the short transverse tensile test (see below), before rejecting such plate.

Destructive control tests

Two types of test have been used to investigate lamellar tearing, the small-scale tests intended to assess plate quality, and the larger tests intended to reproduce lamellar tearing in the laboratory in a controlled manner. The mechanical property tests to examine plate susceptibility to lamellar tearing are mainly considered here.

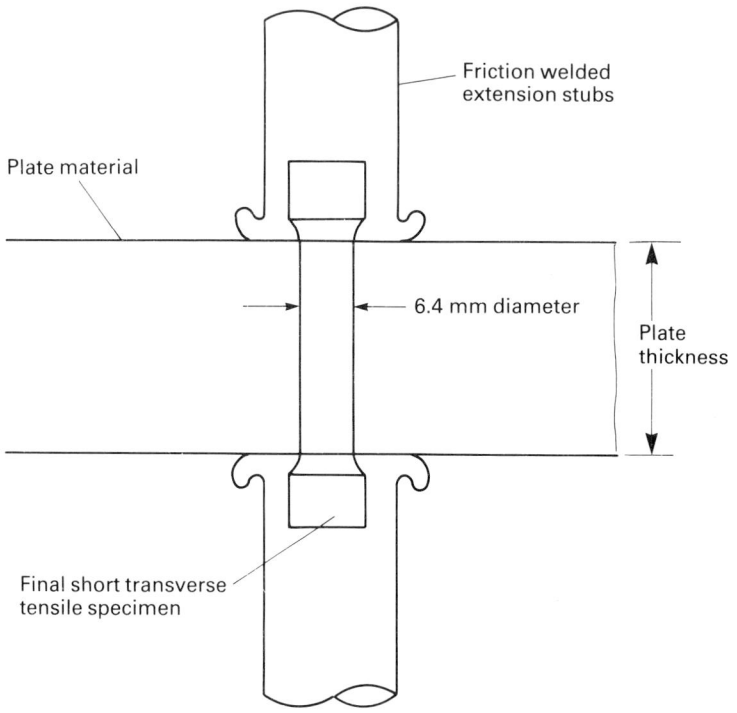

4.14 Extraction of short transverse tensile test piece after extensions have been friction-welded in place.

The **short transverse tensile** test is the simplest of these tests which gives a quantitative result, and which is included in the British Standard for welding structural steels.[6] The principle of the test is illustrated in Fig. 4.14. The number of specimens required is usually laid down in the appropriate specification, but it should be borne in mind that through-thickness ductility is a very variable parameter, both in one area of a piece of plate and from place to place in that plate. Several specimens should therefore be taken, preferably from several locations.

Extension pieces are first welded (preferably by friction welding, but certainly by a very low heat input process) on to the opposite plate surfaces at the intended locations of each test specimen, and the specimen blanks cut out. For plate thinner than 10–12 mm, it is advisable to normalise the test blanks, as the HAZ of the friction weld becomes an increasingly high proportion of the gauge length as the plate thickness is reduced; this will consequently adversely affect the as-welded through thickness ductility measured. For freshly manufactured plate, a hydrogen removal heat treatment of a few hours at 150–250 °C may be advisable to avoid any hydrogen (remaining from the steel melting process) reducing ductility, particularly if the steel is of thick section and has not been given a separate normalising or tempering heat treatment. Otherwise, no heat treatment is necessary, as the HAZ of the friction weld (1–2 mm in width on plate thicker than 10–12 mm) is too narrow to affect the test results significantly.

Test specimens are machined out, normally with a gauge diameter of 6.4 mm, and a gauge length sufficient to include the full plate thickness if this is practicable. For very thick plate, it may be more convenient to extract two or more specimens, slightly overlapping, to cover the full plate thickness. Tensile tests are then carried out at ambient temperature, the reduction of area (RA) measured and the location of the fracture noted. The tensile strength may also be measured, but this is only significant where there may be some doubt as to the origin of the test specimens, or when the results are very poor.

The assessment of the results depends on STRA values which have, in the past, given problems of lamellar tearing in practice with normal structural steels. Figure 4.15 shows that STRA values in excess of 20% have been found to give no lamellar tearing, even in the most highly restrained situations. It should be noted that plate steels tested in the longitudinal and long transverse directions are regularly capable of giving 65% RA values, but to achieve these levels in the short transverse direction requires a very low inclusion content and sulphur contents well below 0.005% S.

A minimum STRA value of 15% has been found to give freedom from lamellar tearing in all except the most highly restrained situations. That is to say, 15% minimum STRA would not guarantee freedom from tearing in highly restrained nozzles, but would be acceptable for box fabrications and other less severely restrained cases. A minimum of 10% STRA would only ensure freedom from tearing in lightly restrained T-joints, such as web to flange joints. Values of 5% STRA and lower would give high risks of tearing when welding lightly restrained, but large section joints, such as lifting lugs, on to plate. Although the latter may not sound particularly serious, failure of such a joint in service could lead to loss of life.

Inevitably, the scatter of STRA results requires some care in their interpretation. Uniformly high or low values, within the designations in Fig. 4.15, are unambiguous. However, an odd low result, or a scattered level of values, could require either more testing to clarify the situation, or the ultrasonic examination of the plate, so that poor areas can be detected and used only where risks of lamellar tearing are very low. Correlations between STRA results and plate sulphur contents has already been discussed in Fig. 4.4.

The **slice bend** test is a semi-quantitative test, which allows the short

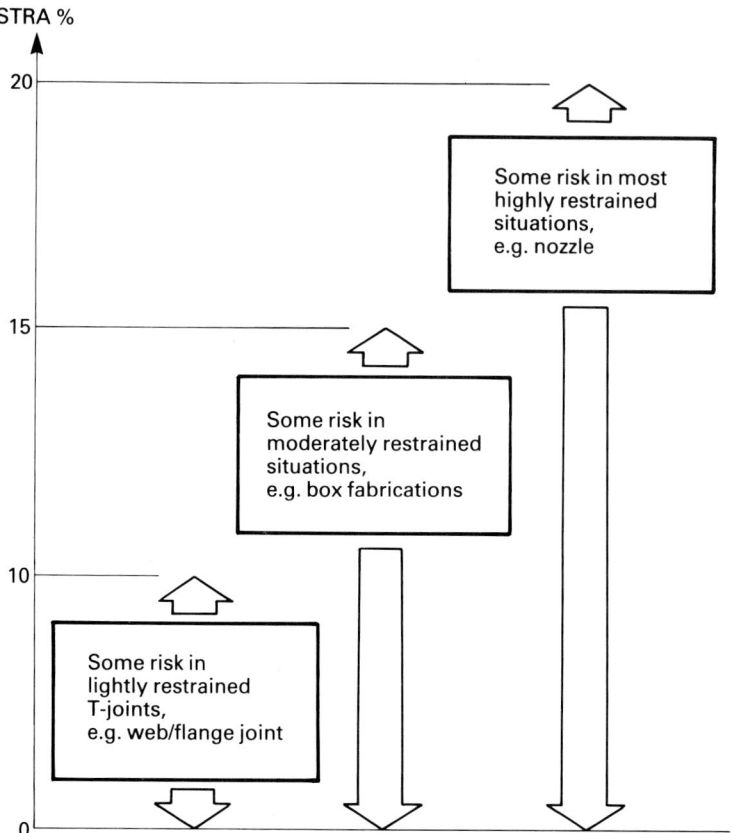

4.15 Risk of lamellar tearing in relation to reduction of area in short transverse tensile test.

transverse ductility to be assessed from the bend angle to the onset of cracking. The completed test provides a macrosection through the plate thickness showing the distribution of harmful inclusions, and how easily inclusions on different planes link up to form a continuous tear. The test was intended to overcome early objections to the short transverse tensile test that, without the use of the friction-welded endpieces included in Fig. 4.14, it was not possible to assess plate near to its surfaces, i.e. where lamellar tearing is most common.[1] A further advantage claimed is that a bigger volume of steel is sampled in a single test. On the other hand, the test is more complicated and probably more expensive than the short transverse tensile test.

The sequence of manufacture of test specimens developed at TWI is illustrated in Fig. 4.16 (an alternative version of the test is given in ref. 7).

Lamellar tearing

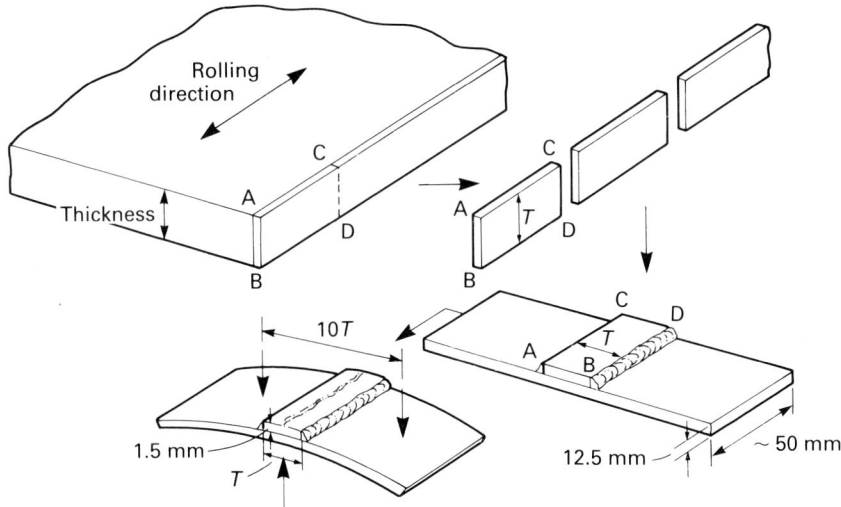

4.16 Manufacture, testing sequence and dimensions of TWI slice bend test.

A thin, 1.5 mm, slice through the thickness of the test plate is fillet-welded to a standard bend test specimen 12.5 mm thick and about 50 mm wide. Welding is by MMA using 2.5 mm electrodes, although TIG (using a suitable filler) would be equally suitable. Good quality is needed for this weld; adequate fusion, no lack-of-fusion defects, no crater cracks and no overlapping on to the upper surface of the test specimen. After welding, the test surface is ground to a controlled surface finish. For research purposes, 32 CLA may be needed, but for routine testing 125 CLA is adequate.

Test specimens are tested in three-point bending, with a span 10× the original test plate thickness (assuming the full thickness is being tested) in order to keep the bending moment reasonably constant over the full test thickness. Bending is continued until the first signs of cracking are seen by eye. Good quality plate may not crack, and failure will then occur in the attachment weld. In such cases the quality of this weld needs to be checked; if defects are found, the test should be repeated.

The deflection at the onset of cracking can be related to the surface strain and this, rather than the bend angle, is the parameter best related to lamellar tearing behaviour.[1] Certainly cracking at 2% surface strain or less, as in Fig. 4.17, is cause for concern. Cracking at 2% surface strain is very roughly equivalent to 5% STRA in the short transverse tensile test.

The **tab** test is a simple qualitative workshop test that can be carried out in the field, as it requires no laboratory equipment. Various forms have been proposed, and the version developed by TWI[8] is illustrated in Fig. 4.18.

4.17 Cracking in slice bend test after ~1% strain.

Lamellar tearing

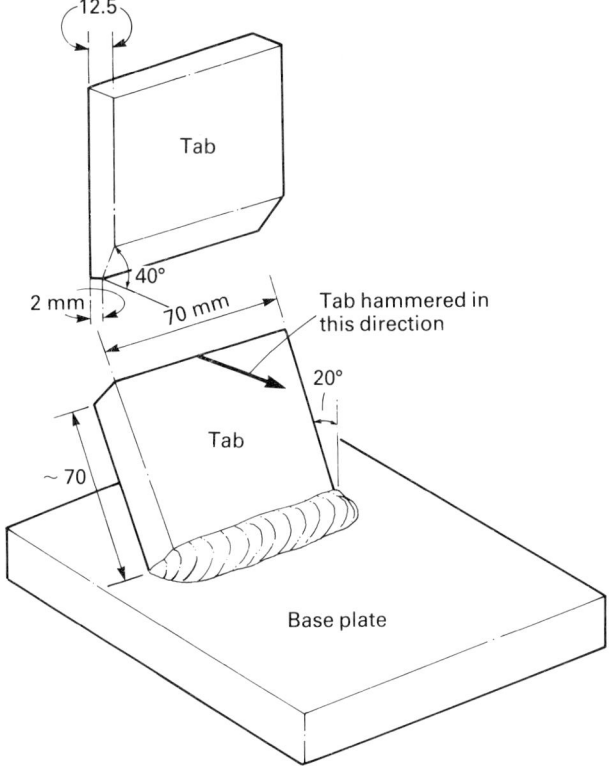

4.18 Sketch of tab test.

A mild steel plate approximately 70×70×15 mm, with a 40° angle preparation and a 2 mm root face, is angled back about 20° and welded to the plate to be tested, welding on to the surface of interest. To minimise hydrogen from the welding operation adversely affecting the results, it is recommended that a minimum preheat of 150 °C is used together with very low hydrogen basic electrodes (i.e. dried, if necessary, at 400–450 °C), appropriate to the plate being tested. Two runs are deposited, both with 4 mm diameter electrodes, leaving the root gap (resulting from the use of the 2 mm root face) to start fracture when the specimen is tested. The purpose of the second run is to temper and improve the toughness of the HAZ of the first, to avoid obtaining misleading results.

To test, the tab is hammered off with blows on the underneath of the angled plate. Plate susceptible to lamellar tearing will show a predominantly woody fracture (Fig. 4.19) typical of lamellar tearing. Plate which is of good quality will usually fracture through the weld metal and give a fine, usually ductile, fracture.

4.19 Typical woody fracture surface in tab test revealing susceptibility to lamellar tearing.

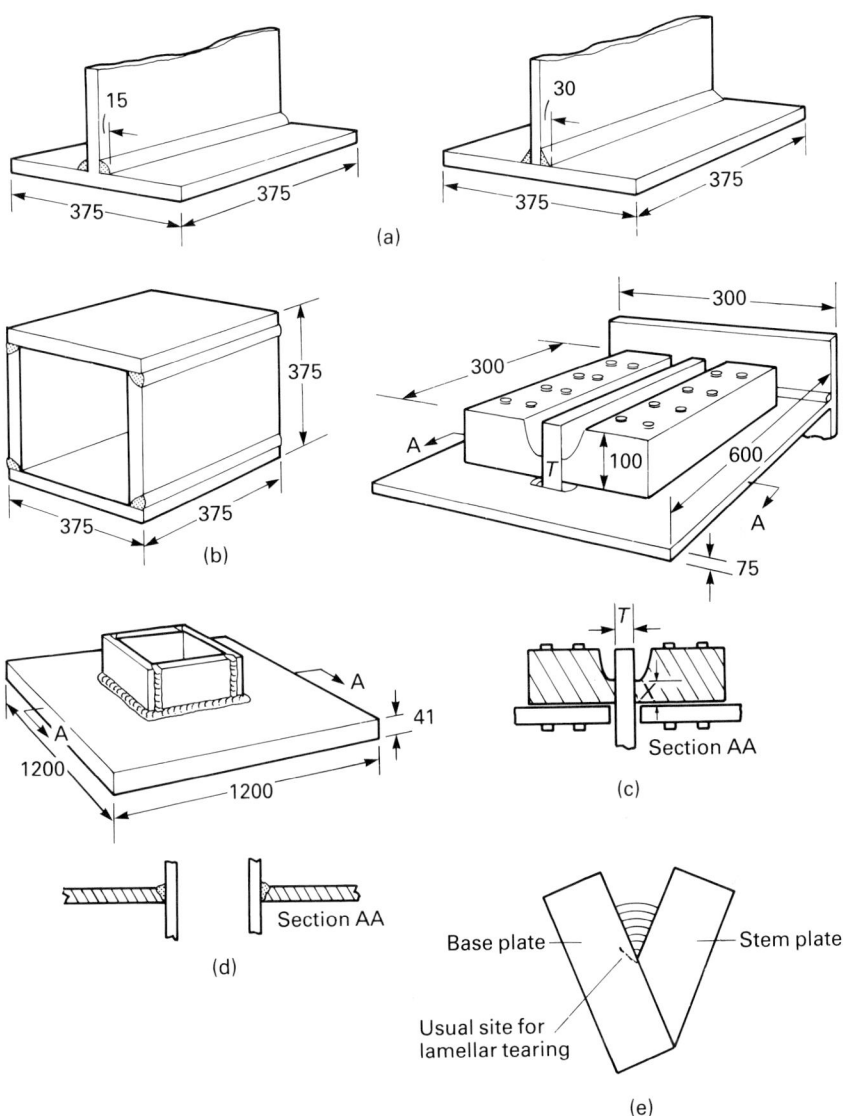

4.20 Examples of lamellar tearing testpiece design: (a) T-joints; (b) box section; (c) double corner joint; (d) box section through plate; (e) Cranfield test.

When interpreting the results, two facts should be remembered. Firstly, the test is not capable of differentiating between moderately poor and very poor plate; the transition from woody to fully ductile fracture takes place at the equivalent of 12–20% STRA in the short transverse tensile test. Secondly, the test only samples a surface layer no more than 2–3 mm thick. If heavy welds are being made on to fairly thick plate with high restraint, segregation deeper than this can give lamellar tearing, but would not be detected in the tab test.

All **large-scale** tests which are intended to reproduce lamellar tearing in such a way that, for example, the controlling factors can be studied, are relatively expensive, because of the need to build a high degree of restraint into the testpiece. These include the Cranfield and window tests.[7] However, as shown in ref. 1, tests can rapidly be devised to meet the needs of the particular problem under investigation, the quality of the plate and the restraint needed. Some examples of these tests are illustrated in Fig. 4.21, from which their areas of application can readily be appreciated.

Detection and identification

Detection

Any surface technique, i.e. visual, magnetic particle inspection (MPI) and dye penetrant examination, is capable of revealing lamellar tears open to the surface. Sub-surface cracking can only be detected by a suitable ultrasonic examination technique; radiography is of little use.

Some care is needed in the interpretation of the results of ultrasonic examination. If the conditions are too sensitive, it is possible to detect clusters of inclusions, or even bands of different microstructures. To be sure of identifying lamellar tearing, the position of any suspect region should be checked to confirm that the joint is of a susceptible type and that the boundaries of the suspect region are related in a sensible way to the likely position of a lamellar tear. (The solidification cracks illustrated in Fig. 3.12 were originally incorrectly identified as lamellar tearing!) If the plate is originally suspect, the position of any doubtful areas should be noted, and their ultrasonic response checked after welding to see whether there has been a change or not.

Some of any increase in response could be due to the decohesion of inclusions from the steel matrix – such behaviour is also possible after PWHT and cold working – and a careful examination should be made for tears which have broken surface, as these would provide positive confirmation of cracking. Any ultrasonic examination of thick, dirty plates is liable to be difficult because heavy attenuation of the ultrasonic signal is likely.

Lamellar tearing

4.21 Typical woody fracture produced by steel susceptible to lamellar tearing.

4.22 Magnified view of short transverse fracture surface of steel susceptible to lamellar tearing.

130　Weldability of ferritic steels

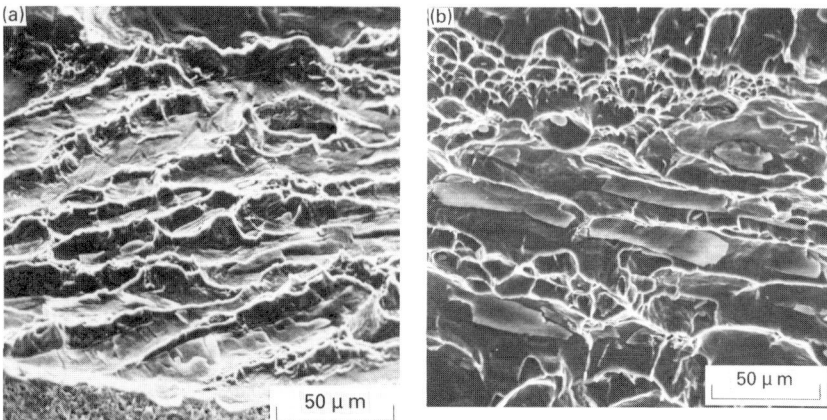

4.23　Type-II MnS inclusions on fracture surfaces of steels susceptible to lamellar tearing: (a) from Fig. 4.22, steel with 7% STRA; (b) steel with 16% STRA.

Identification

Of the types of cracking liable to be encountered when welding steels, lamellar tearing is usually the easiest to identify. If an open crack surface is available for visual examination, the characteristic woody fracture surface (Fig. 4.21), is unmistakable. This structure is reflected in low power fractographic examination in the SEM (Fig. 4.22), where the large inclusions, whose decohesion from the matrix starts lamellar tearing, can be clearly seen, usually separated by walls of shear fracture. Examination at higher magnification in the SEM (Fig. 4.23) will reveal more detail of the inclusions (which can be identified, if needed, by selective area analysis (Fig. 4.24)).

Examinations of prepared sections by standard macro- and microscopical techniques will show the path of the tear alternately running along long flattened inclusions (usually of the MnS or mixed silicate type) and then cutting across to inclusions in different planes. Examples of macrosections of typical tears are illustrated in Fig. 4.1 and 4.5 and in the microsection in Fig. 4.25. It will be seen that in all the examples shown, lamellar tearing is not confined to the visible HAZ but always extends into the parent plate. Its general direction is parallel to the nearest fusion boundary. Lamellar tearing can only be positively identified if cracking is found to extend *outside* the visible HAZ.

Lamellar tearing

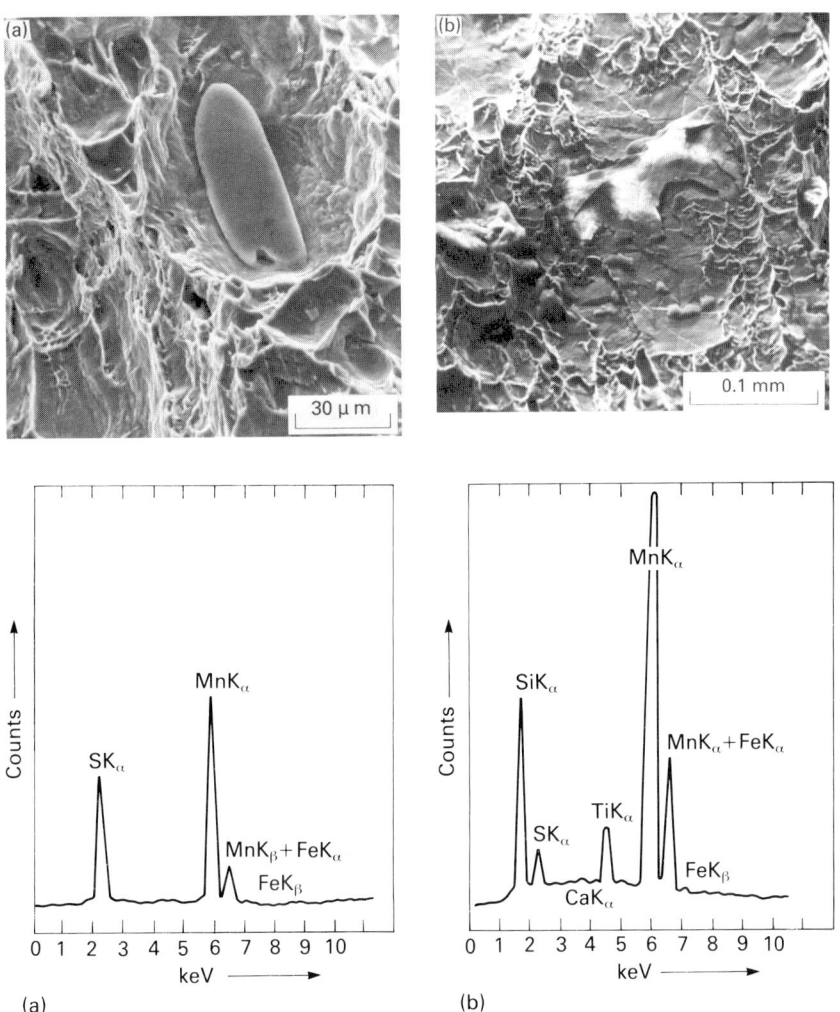

4.24 Typical inclusions associated with lamellar tearing and SEM analysis traces: (a) Type-I MnS; (b) complex silicate.

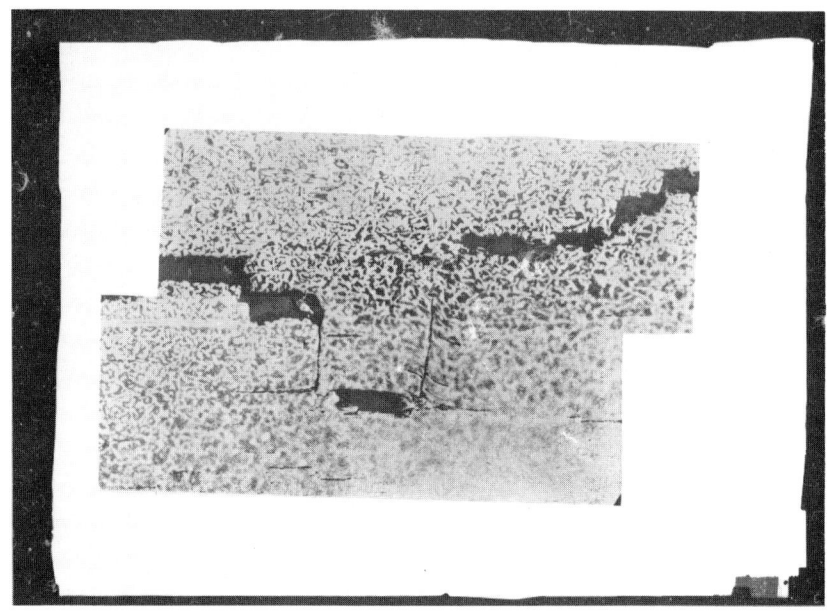

4.25 Mosaic of microsections showing wide cracks along inclusion bands and narrow cracks with plastic deformation between them.

References

1 Farrar, J.C.M. and Dolby, R.E., 'Investigations into lamellar tearing', TWI Report Series, TWI, Cambridge, March 1975.
2 Davey, T.G., 'Sulphur content and lamellar tearing susceptibility', *TWI Res. Bull.*, 1979, **20**, 169–171.
3 Kanazawa, S., Kawamura, K., Yamato, K., Haze, T., Inoue, T., Fukuda, I. and Niube, T., 'Lamellar tearing resisting steels and the directions for the use of them', IIW Doc. IX-873-74, IIW, 1974.
4 Roberts, J.E., 'Development of normalised structural steels', Rosenhain Centenary Conference, 1975, Sept., The Royal Society, London, 1976, 277–288.
5 Bailey, N., 'Aspects of the weldability of HY 130', *Metal Construction*, 1970, **2**, 339–344.
6 British Standard BS5135: 1984, 'Arc welding of C and C–Mn steels', BSI, London, 1984.
7 Jubb, J.E.M., 'Lamellar tearing', *WRC Bull.*, No. 168, Dec. 1971.
8 Davey, T.G. and Dolby, R.E., 'The tab test for lamellar tearing susceptibility', *TWI Res. Bull.*, 1978, **19**, 170–173.

5 Hydrogen cracking

Description

Hydrogen cracking has been, and still is, the most likely cause of metallurgical problems when welding ferritic steels. As such, it is the subject of a companion book, *Welding Steels Without Hydrogen Cracking*,[1] which contains detailed recommendations for ferritic steels. The present chapter is, therefore, less detailed than those on the other types of cracking and describes how hydrogen cracking occurs, its detection and identification and the principles of the methods taken to avoid it. In addition, other interactions between hydrogen and steels are mentioned, several of which are discussed more thoroughly in Chapter 10 and elsewhere.

Hydrogen cracking occurs because hydrogen embrittles ferritic steels at and near ambient temperature by a mechanism, the details of which are still obscure. One hypothesis is that the atoms of hydrogen, which are interstitially dissolved in the ferrite lattice, interfere with the movement of dislocations (i.e. the imperfections in the atomic lattice which allow plastic deformation). Because hydrogen atoms in steel move more slowly as the temperature is reduced, they can only move at the correct speed to interfere with dislocations within a limited temperature range of one or two hundred degrees above and below normal ambient temperature. Furthermore, the interference only occurs when the steel is deformed at slow strain rates; hydrogen embrittlement is not normally perceptible in high speed tests, such as the Charpy impact test.

This slow movement of hydrogen in steel means that hydrogen cracking is a slow process. Time is needed for hydrogen to diffuse to high stress concentrations, and to move to the crack tip as the crack slowly grows. Cracking appears to be particularly slow and protracted if the conditions are borderline for cracking, so that not all of the hydrogen needed to start cracking is immediately available at the correct site. In effect, there is a race between hydrogen diffusing locally to possible sites of crack initiation and hydrogen diffusing on a larger scale out of the steel and out of harm's way. A practical consequence of this slow crack growth is that time is needed before any cracks can grow to a detectable size and a structure can be inspected and declared free from cracks.

Cracking may occur in HAZ, weld metal or both. For cracking to occur, four simultaneous conditions are necessary:

1 Sufficient hydrogen must be present.
2 The microstructure must be susceptible to hydrogen embrittlement.
3 Sufficient stress must be present.
4 The temperature must be within the sensitive range while all the other conditions are operative.

Each of these topics will be discussed separately; the way they interrelate will become apparent in a later section dealing with the techniques of avoiding hydrogen cracking.

Hydrogen

Sources of hydrogen

Hydrogen must be assumed to be always present during welding, although the amounts can be controlled. Because hydrogen can diffuse within (and escape from) steel at ambient temperatures, special methods have been devised to measure the amount of hydrogen put into a weld by welding. The usual test consists of a single weld bead deposited under exactly controlled conditions and, because it is cooled rather faster than in most real welds, the results tend to be higher than those measured (with some difficulty) from actual welds. Whereas hydrogen contents of most metals (including steels) are usually expressed as parts per million (ppm) by mass, the hydrogen contents of welds are normally given in terms of millilitres of hydrogen at STP per 100 grams of weld metal (mL/100 g). For comparison:

$$1 \text{ mL}/100 \text{ g} \equiv 0.89 \text{ ppm}$$

The weld metal referred to may be either that added to make the test (the **deposited metal**) or the amount of **fused metal**, i.e. deposited weld metal plus melted parent steel. The relationship between the two depends on the welding process and the welding conditions (i.e. the dilution of parent steel into the weld): for TIG welding without filler, the deposited metal would be zero; for submerged arc welding, about 30% of the total; and for MMA about 70%.

A further complication is that, whereas some tests measure the **total hydrogen** content, others measure the **diffusible hydrogen**, i.e. the amount of hydrogen that diffuses out of the sample at ambient temperature (defined as 25±5 °C), the difference being the **residual hydrogen**. The distinction is made because it is believed that only diffusible hydrogen can cause hydrogen embrittlement and cracking. In order to classify welding consumables, the following hydrogen levels, related to the diffusible hydrogen content of deposited metal, are in common use:

High: >15 mL/100 g;
Medium: >10 but ⩽15 mL/100 g;
Low: >5 but ⩽10 mL/100 g;
Very low: ⩽5 mL/100 g.

In addition, an ultra-low level (⩽3 mL/100 g) has been proposed but, as the value has not yet been agreed, it has not generally been used in this chapter.

Three sources of hydrogen must be considered:

1 The welding consumable.
2 The atmosphere.
3 The parent steel.

In most circumstances, the first is the most important, but sometimes cracking can occur because significant amounts of hydrogen have originated from one of the other sources.

Hydrogen from the welding consumable
The highest levels of weld hydrogen result from moisture, combined water and other hydrogen-containing compounds in electrode coverings or fluxes. These decompose in the high temperatures of the welding arc to give free hydrogen which can dissolve in the molten steel. In MMA electrodes, particularly of the cellulosic and rutile electrode types, this hydrogen is necessary for the correct welding behaviour of the electrodes and should not be reduced by drying or baking.

Basic electrodes, and most submerged arc fluxes, do not give high levels of weld hydrogen, but need careful storing and, if necessary, re-drying or baking (the term baking is often used for higher temperature drying, e.g. from 250 to 450 °C – possibly higher for fluxes) in order to maintain suitably low moisture contents to give low and very low hydrogen levels. Recent developments have led to improvements in the resistance of electrode coverings and fluxes to the pick-up of moisture, but care is still needed, particularly when welding in humid ambient conditions. Fluxes in cored wires vary considerably in their hydrogen levels from very low to medium. Fully sealed cored wires give very low hydrogen levels and are also resistant to pick up of moisture. Some, but not all, wires of the unsealed type contain fluxes resistant to the pick-up of moisture.

Current solid wires normally give very low weld hydrogen levels, although this was not always the case. Solid wires are unlikely to deteriorate on storage, unless conditions are so bad that they actually rust.

Hydrogen from the atmosphere
When expected hydrogen levels are very low, the contribution of moisture picked up from the atmosphere during welding and broken down to hydrogen in the welding arc should not be ignored. It is likely that the absolute level of humidity is important, so that welding in a hot humid climate is more likely to increase weld hydrogen levels significantly (i.e. by 1 or 2 ml/100 g) than welding in a moist temperate or cool climate. Another circumstance where the humidity of the atmosphere is important is when welding underwater in a hyperbaric chamber, where the high pressure in the chamber increases the partial pressure of moisture.

Hydrogen from the parent steel
This is frequently neglected, because the hot working and heat treatments used to produce the steel often reduce hydrogen levels to insignificance. This is not always the case, however, particularly if the steel is of heavy section, and welding is being carried out away from a free surface that had been present during heat treatment. For example, if a large forging is heat treated and then bored out and welded within the bore, sufficient hydrogen can be picked up from the parent steel to give cracking problems, even when welding with a very low hydrogen process. Problems from this source have been recorded with both TIG and friction welding, both of which are normally regarded as giving ultra-low hydrogen levels.

A far more usual source of hydrogen pick up from the steel is from surface moisture resulting from condensation, grease and oil, rust or paint. Some paints, e.g. weld-through primers, are acceptable as they give known, very small increases in weld hydrogen level.

Corrosion products (i.e. rust) are also typical sources of hydrogen pickup when steels are welded after they have been in service or have been stored badly. More insidious sources are if the steel has been in sour (H_2S) service or in high temperature, high pressure hydrogen service. In these cases steel can pick up sufficiently high levels of hydrogen to give cracking during welding, unless it is suitably heated to diffuse out dangerous amounts of hydrogen, or if tests and/or experience shows the practice to be safe.

Behaviour of hydrogen in steel
Molten steel (including weld metal) has a high solubility for hydrogen. The equilibrium solubility depends on the concentration or partial pressure of hydrogen above the molten steel (Fig. 5.1); it falls as liquid steel cools (Fig. 5.2) and then falls sharply as the steel solidifies.

Hydrogen cracking

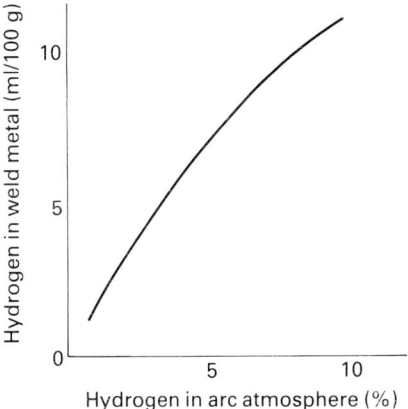

5.1 Variation in solubility of hydrogen in molten weld pool at about 1900 °C with hydrogen content of atmosphere surrounding welding arc.

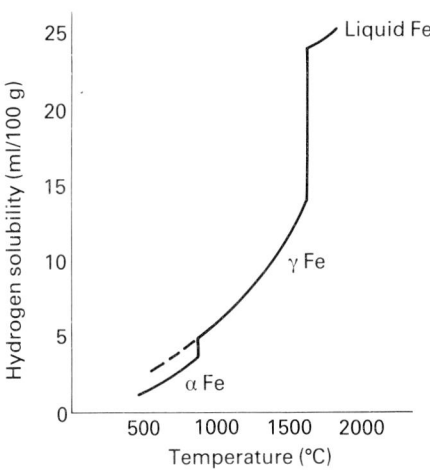

5.2 Solubility of hydrogen with temperature in a typical weld metal.

As the high temperature phase (austenite) cools, the equilibrium solubility of hydrogen (with an external hydrogen pressure of 1 bar) falls still further, reaching a value of about 5 ml/100 g at 800 °C. When the steel transforms to ferrite, hydrogen solubility again falls sharply and continues to fall, reaching extremely low levels (a small fraction of a ml/100 g) in ferrite at ambient temperature. However, in a cooling weld, equilibrium is never reached, so that the amount of hydrogen retained in a solidified weld can reach quite high values if cooling is rapid – values as high as 80 ml/100 g have been reported for welds deposited from cellulosic elec-

138 Weldability of ferritic steels

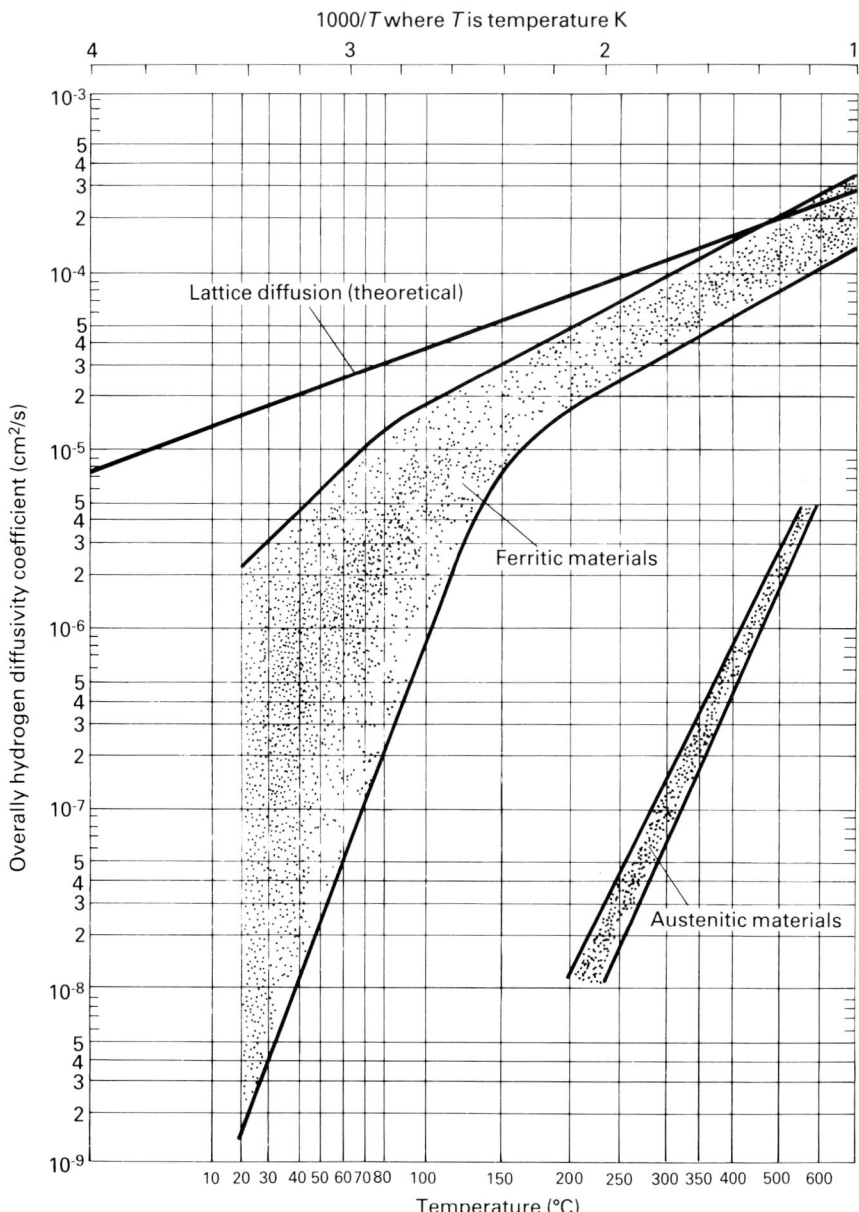

5.3 Variation of overall diffusivity coefficient of hydrogen in steel with temperature.

Hydrogen cracking

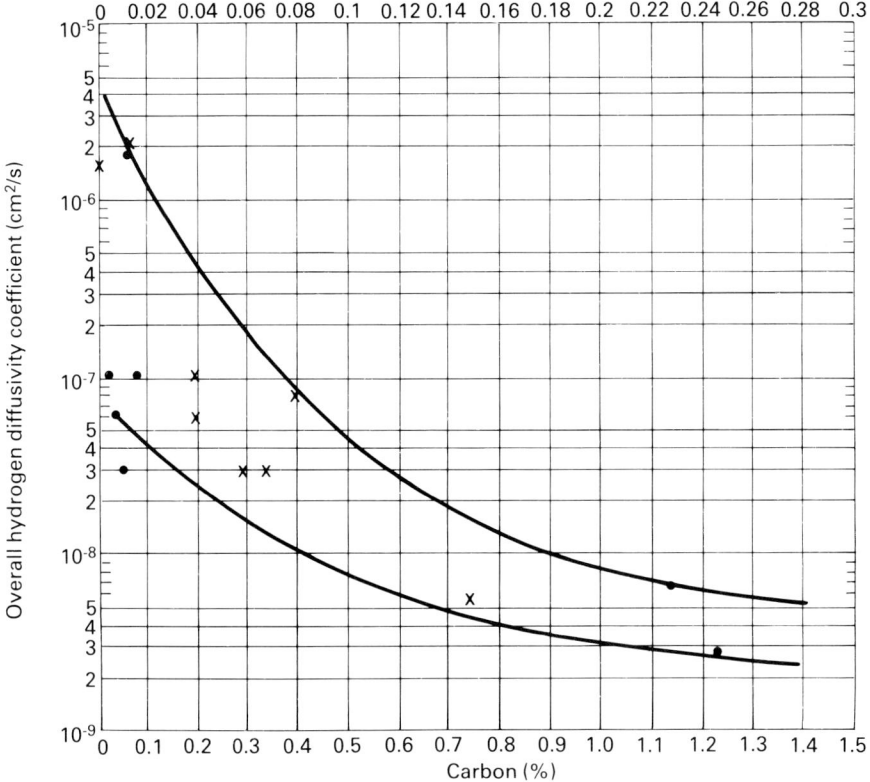

5.4 Influence of carbon, ×, and sulphur, ●, on overall diffusivity of hydrogen in steel at 20 °C.

trodes (which give much higher hydrogen levels in the arc atmosphere than are shown in Fig. 5.1).

To understand how hydrogen escapes from steel on cooling, it is necessary to understand how the diffusion rate of hydrogen in steel varies with temperature. A partial plot in Fig. 5.3 shows that, as expected, the diffusion rate slows considerably as the temperature falls.

It also shows that hydrogen diffuses much more slowly in austenite than in ferrite. The diffusion rate in austenite below 500 °C is similar to that in ferrite at ambient temperature, whilst at ambient temperature, the rate is so slow that hydrogen will stay in austenite indefinitely, whereas ferrite needs to be cooled below about −70 °C to retain its hydrogen. This slower diffusion rate means that when a steel (or weld metal) is at a high temperature in the austenitic condition, it not only has a high solubility for hydrogen (provided hydrogen is in the atmosphere in equilibrium with the steel) but any hydrogen will only move very slowly. Thus, even if the furnace atmosphere is free from hydrogen (so that the equilibrium solubility is less than that shown in Fig. 5.2), hydrogen will not diffuse out of thick sections very quickly.

Figure 5.3 also shows that below about 200 °C, the diffusion rate in ferrite begins to deviate below the theoretical line for lattice diffusion. This results in a spread of diffusion rates at ambient temperature of about three orders of magnitude! This disturbing result is largely a result of second phase particles (carbides and non-metallic inclusions, particularly sulphides) in the steel (Fig. 5.4). Near ambient temperature, these particles act as traps for hydrogen – some temporary, some permanent – which slow its diffusion by a very large margin and making the removal of hydrogen a very slow process at and just above ambient temperature.

Susceptibility to embrittlement

The susceptibility of a steel to hydrogen embrittlement – parent steel, weld metal or HAZ – depends on the inherent toughness of the steel. The tougher the material, the greater is the content of hydrogen needed to embrittle it sufficiently to cause it to crack. As in many other materials, toughness usually decreases as strength, or hardness, is increased. This is true for HAZs and generally true for parent steels, but some weld metals are exceptional.

The likelihood of (or susceptibility to) hydrogen cracking in weld HAZs increases as their hardness increases. Cracking usually occurs in the as-deposited HAZ, so that its hardness before it is tempered and softened by succeeding weld runs, or by PWHT, is the value of interest. Hardness measurements made on completed multipass welds are, therefore, of no use in assessing the likelihood of HAZ cracking in the weld root as the weld is being made. Also, because the hardness usually varies within any HAZ, it is the maximum hardness that is of importance, not the mean value.

Heat-affected zone hardness depends on several factors. The HAZ is best considered as steel that has been heated to austenite and quenched. Unlike a conventional austenitising heat treatment, the HAZ close to the fusion boundary is coarsened as a result of heating to a temperature near to its melting temperature – up to perhaps 1500 °C – which coarsens the prior austenite grain size and thus increases its hardenability. Unlike steel quenched by immersion in water or oil, the HAZ is back-quenched by the steel which has not been heated by welding, so that the thicker the section, the faster the cooling (unlike when steel is cooled externally during heat treatment).

When a HAZ is cooled at a sufficiently fast rate, its hardness will be

Hydrogen cracking 141

that of martensite of the same composition, and will depend solely on its carbon content (Eq. [1.6] and [1.8]). The cooling rate needed to achieve a fully martensitic microstructure depends on the amount of alloying elements present, as well as on the carbon content. The higher the alloy content, the slower will be the cooling rate at which martensite is still formed and the greater will be the **hardenability** of the steel. For any particular welding situation, in terms of hydrogen level, restraint, preheat and/or interpass temperature and post-weld cooling conditions, there is a particular maximum HAZ hardness at which there is a risk of cracking; this is termed the critical hardness.

The ease with which a HAZ can develop a particular hardness under a particular cooling regime can be related to a single compositional parameter, usually termed the **carbon equivalent** (CE). The most common of these (although it strictly relates to the risk of cracking, rather than hardenability pure and simple) is termed the IIW carbon equivalent:

$$CE_{IIW} = C + \frac{Mn}{6} + \frac{Cr + Mo + V}{5} + \frac{Ni + Cu}{15} \qquad [1.2]$$

Values of CE_{IIW} below 0.42 denote a steel which is easy to weld without hydrogen cracking, whereas steels with CE_{IIW} values above 0.5 are difficult. Although other CE formulae have been proposed, the only one of these which is widely used is the Japanese P_{cm} formula (Eq. [1.3]), which is particularly applicable to low carbon steels; it places more weight on carbon itself than the CE_{IIW} formula, and includes a term for boron.

Application of the carbon equivalent is modified by the inclusion population in a steel, as inclusions can affect hardenability and susceptibility to cracking. This was first appreciated when hardenable steels of low sulphur contents were found to have cracked, while similar steels of higher sulphur level did not. The effect was found to be due, in part at least, to non-metallic inclusions (particularly sulphides) acting as nuclei for the formation of ferrite at temperatures higher than those at which it would have formed in the absence of such nuclei. Having formed at higher temperatures, the ferrite was softer than any of the lower temperature transformation products, so that removing the inclusions made the steel harder. However, in other situations, sulphides can be damaging if they are sufficient to form liquation cracks (Chapter 3), which are believed to be capable of nucleating hydrogen cracks; they are also a potent cause of lamellar tearing (Chapter 4).

At one time, segregation in steel in terms of considerable variations between ladle and ingot composition, or differences in the compositions of ingots from the same cast, caused considerable problems in devising safe welding procedures without analysing every piece of steel before it

142 Weldability of ferritic steels

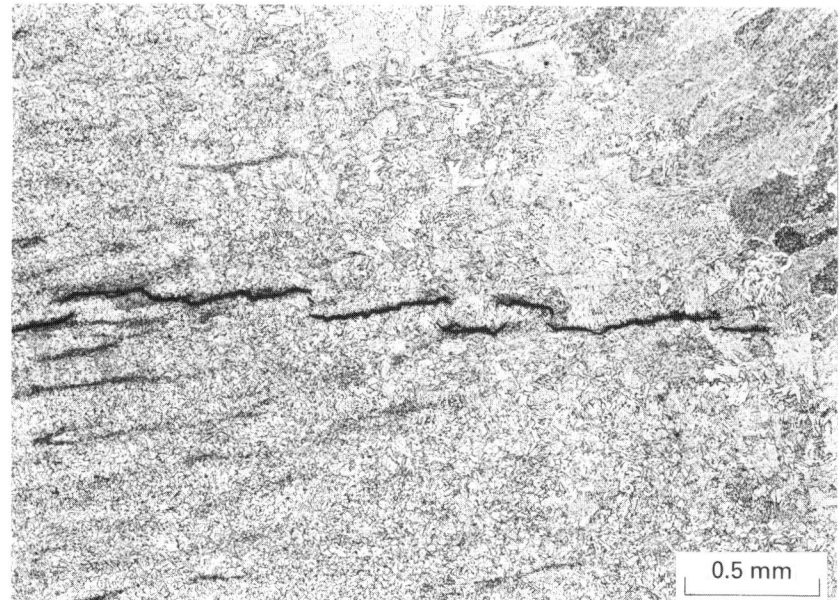

5.5 HAZ hydrogen cracking in the segregate-rich (dark-etching) regions at the centre of a Concast plate. The weld is at the top right.

was welded. Nowadays, these problems appear to have been replaced by other concerns. One is that some continuously cast (concast) plates can have regions close to the centre-line which are enriched in carbon and alloying elements to such an extent that, if conditions are somewhat marginal for hydrogen cracking for the bulk of the weld, cracking can occur in the segregated regions (Fig. 5.5). Such cracking is often relatively benign, in that it usually runs parallel to the plate surface and is well away from a free surface, so that it is not potentially harmful.

Another current concern with modern steel is that those casts that have been made using a high proportion of scrap can contain a high proportion of alloying elements. When not declared in the mill sheet, these residual elements can unexpectedly increase the CE_{IIW} by as much as 0.08, i.e. much higher than the level of 0.03 that was recommended to be added to the CE_{IIW} if only the C and Mn contents were known.[2]

To reduce the hardness level below a critical value in order to avoid cracking, it is necessary to control HAZ cooling by adjusting the temperature of the steel being welded (i.e. the preheat and interpass temperatures) and also the amount it is heated during welding by adjusting the heat input of the welding process. The cooling of a weld bead depends on these factors, as well as on the thickness of the steel being welded. The thickness for this purpose is commonly referred to as the **combined thickness** of the joint, that is to say the sum of the thicknesses of the heat paths through which the heat of welding flows away. This takes account of the greater severity of cooling from a conventional fillet weld compared with a simple butt weld, as shown by the diagrams in Fig. 1.1. It also takes account of parts of the joint whose thickness is tapered or which are short, by averaging each value over the 75 mm from the joint line.

The second factor that affects HAZ cooling, and hence microstructure and hardness, is **heat input**. This describes the arc energy distribution to the steel being welded in kilojoules per unit length of weld and can be calculated from the welding parameters, Eq. [1.1].

It is important that the arc voltage is measured near the arc, as voltage drops can occur in long welding cables. For multi-arc welding, a sum of the separate arc energies gives the required value for heat input, provided the arcs give a single weld pool. For different types of welding, the arc energy as calculated in Eq [1.1] should be factored to derive the heat input to the steel being welded, because different welding processes have different **arc efficiencies**. Factors of approximately 0.8 are applicable to MMA welding and the gas shielded metal arc processes; 0.6 is suitable for TIG welding and 1.0 for submerged arc.

Unfortunately two systems apply to the practical application of heat input values. The one used in Ref. 1 takes the arc energy calculated (without factoring) for MMA welding to construct tables and diagrams of 'safe' welding procedures. The use of TIG or submerged arc welding is then estimated by factoring the arc energy, as quoted in Chapter 1, page 10, to obtain values for use in welding diagrams. In the second system, diagrams and tables are constructed using heat inputs, i.e. the arc energy values *after* they have been factored using the factors in the previous paragraph. Hence, two otherwise identical diagrams may appear with scales of heat input differing by 20%!

The use of **preheat** also slows the cooling of a HAZ. It is important to measure the preheat correctly, i.e. at 75 mm from the joint line and preferably on the side which is *not* heated, immediately before starting welding. If this is not practicable, a time of 1 minute per 25 mm thickness should be allowed after removing the heating source and before measuring the temperature. If the whole component, rather than an area centred on the joint, is preheated (i.e. using *general* as opposed to *local* preheat), it may be possible to reduce the preheating temperature. In multipass welding, the **interpass temperature** is measured immediately before depositing the following pass.

By combining the concepts of carbon equivalent and cooling time or rate, it is possible to devise welding procedures aimed at keeping the HAZ softer than the critical level at which cracking can occur, and this is discussed later. However, for weld metals, particularly those that contain acicular ferrite in their microstructures, the situation is more complicated, because increasing the proportion of acicular ferrite in such microstructures can increase both strength *and* toughness. Furthermore, most weld metals are of compositions that do not form martensite as easily as in a HAZ. Thus, they show much smaller changes in hardness as cooling conditions are altered.

Stress to give cracking

In an as-deposited weld or weld bead, residual stresses up to the yield strength of the weld metal are nearly always present; these are tensile at and near the weld. Residual stresses are the driving force for hydrogen cracking. They can be intensified by stress concentrations, which may take the form of root gaps (Fig. 5.6), undercut (Fig. 5.7(a)) and associated intrusions at the weld toe (Fig. 5.7(b)), and stringers of inclusions in parent plate.

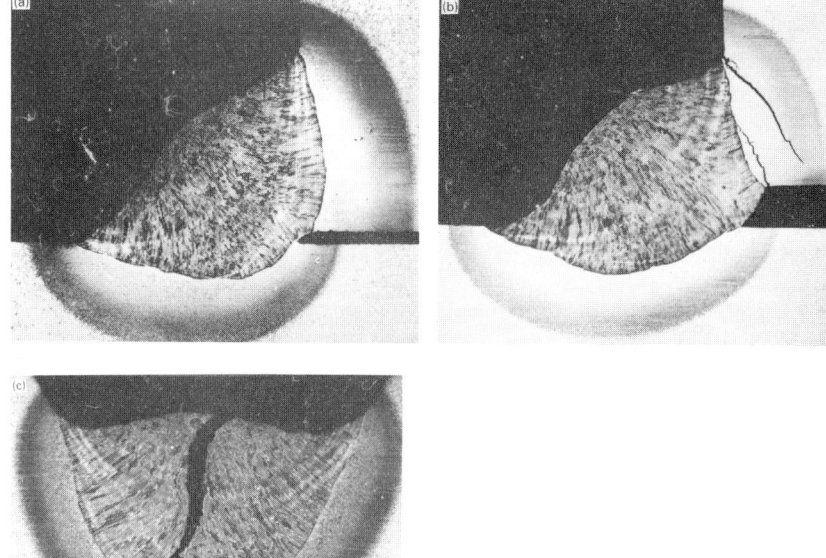

5.6 Effect of root gap on hydrogen cracking: (a) restrained fillet weld in low alloy steel with small root gap (<3 mm); (b) identical weld but with 1.6 mm root gap giving HAZ cracking; (c) root run of butt weld showing that cracking originates from sharp notch at right-hand side of 1.6 mm root gap.

Hydrogen cracking

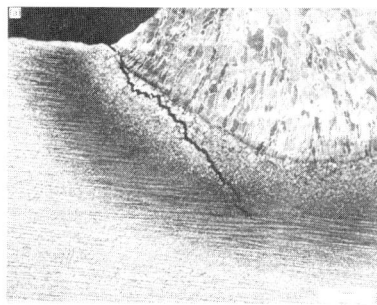 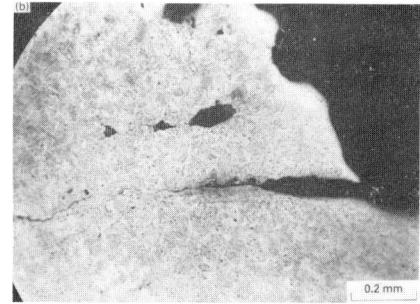

5.7 Stress concentrations at weld toe: (a) hydrogen crack in HAZ of C : Mn steel originating from undercut at weld toe; (b) defects at toe of fillet weld exaggerated by taper-sectioning specimen preparation technique.

Because hydrogen embrittles steel at slow strain rates, and residual stresses build up slowly, stress conditions are eminently suitable for hydrogen cracking. Also, when a small root run may be all that is holding two large pieces of steel together, root runs are at particular risk.

Little can be done to reduce stress levels to prevent hydrogen cracking. Peening is difficult to control and may cause damage, grinding of weld toes is expensive and can do nothing for the roots of fillet welds, and the choice of a low strength weld metal, giving lower residual stress levels, is not usually allowable.

Temperature of embrittlement

The effect of temperature on hydrogen embrittlement is illustrated in Fig. 5.8, which compares the results of notched tensile tests on simulated HAZ material containing hydrogen with those on similar specimens containing no hydrogen. Embrittlement is apparent from just above 200 °C to a temperature well below −100 °C. For steels whose HAZs are very susceptible to embrittlement, hydrogen cracking has been found to occur at temperatures as high as 190 °C, even when the hydrogen level was at the ultra low levels of TIG welding. For other less susceptible steels, the upper limit for cracking is likely to be somewhat lower, so that maintaining such a steel at 150 °C, or even lower, can be used to avoid cracking.

The possibility of avoiding cracking by maintaining steels at a temperature *below* the sensitive range for cracking is somewhat academic; steels are not very tough at such temperatures and hydrogen would escape so slowly that it could be unsafe to bring the joint back to normal ambient temperatures, even after several years.

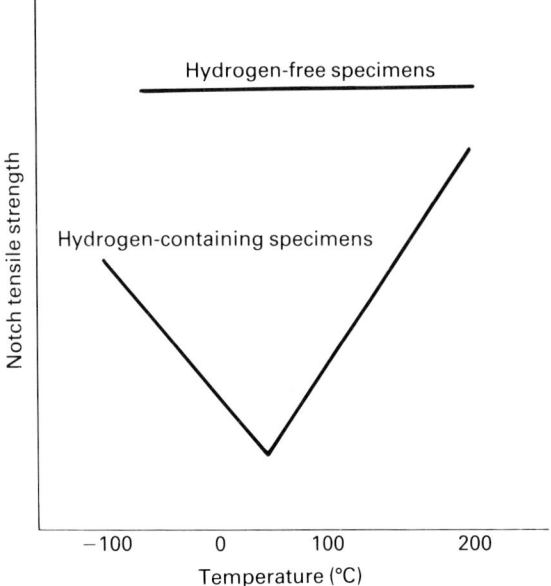

5.8 Effect of temperature on hydrogen embrittlement of simulated HAZ material illustrated by the results of notched tensile tests on similar specimens with and without hydrogen.

Techniques for avoiding cracking

Consideration of the four factors needed for cracking shows how it is possible to formulate strategies for its avoidance. As stress reduction is not a feasible option, the strategies revolve around control of hydrogen level, control of microstructure and avoiding the embrittling temperature range while the hydrogen content is high. The methods available are:

1 Direct control of hydrogen level.
2 Control of microstructure by control of cooling.
3 Temperature control method.
4 Control of microstructure by isothermal transformation.
5 Use of austenitic or nickel alloy consumables.

For a range of conditions, the safe welding procedures for one steel may, in fact, involve several of these methods of control, hence nomograms devised to derive procedures intended to avoid cracking tend to be complex.

The **direct control of hydrogen level** is the most important method of avoiding hydrogen cracking. As the hydrogen level is reduced from high to very low, higher hardness levels can be tolerated in the HAZ without recourse to expensive pre- and post-heating. Weld metal cracking in the so-called weldable C and C:Mn steels can usually be avoided by using consumables giving very low hydrogen levels. However, strict precautions are necessary, particularly with the welding process involving fluxes (MMA, submerged arc and cored wire welding), to maintain very low hydrogen levels, especially on site.

Manual electrodes and submerged arc fluxes need careful storing in warm stores and may need re-drying at relatively high temperature (up to 450 °C) if they are to be supplied for use in good condition. Very low hydrogen electrodes may not retain their hydrogen levels if welding is carried out in unusually humid conditions (i.e. in hot and humid climates or in hyperbaric chambers). Some batches or brands of cored wires may not retain their hydrogen levels if reels are kept on machines unprotected for several days in humid conditions.

Besides directly controlling the hydrogen level of the consumable when using low hydrogen procedures, it is also important to maintain control of steel cleanliness, i.e. freedom from rust, oil, grease, paint on the weld preparation, as well as being sure that the steel itself does not contain hydrogen.

The **control of microstructure by regulating cooling** is intended to ensure that the HAZ is not too hard. This is achieved by maintaining a sufficiently high heat input and, where this is not practicable (for example, in positional welding), by using preheat. However, preheat has two other benefits: it reduces the risk of cracking while the weld is actually at the preheat temperature and it allows more time for the weld hydrogen content to be reduced by diffusion at temperatures where diffusion rates are faster than at ambient.

Use of welding diagrams, such as the examples shown in Fig. 5.9, allow different combinations of preheat temperature and heat input to be selected to suit the circumstances. In addition, these diagrams show the benefits of reducing the hydrogen input to the weld. Diagrams of this type are available in ref. 1 for C:Mn steels with CE values from 0.32 up to 0.58; for other types of steel, different strategies are needed.

The **temperature control method** is mainly used for hardenable alloy steels whose HAZs are not amenable to softening by slow cooling and the use of preheat. Although primarily intended for alloy steels, the temperature control method can equally well be used for C and C:Mn steels of such thick section that they are fully hardened by welding.

The method relies on the relatively rapid diffusion of hydrogen out of the steel at temperatures at which the steel will not crack, i.e. above 150–200 °C. If the type of steel and its carbon content are known, the HAZ hardness can be determined from the lower part of Fig. 5.10. This can be used to estimate the minimum preheat and interpass temperature in the upper part. Some latitude is given in the figure so that account can be

148 Weldability of ferritic steels

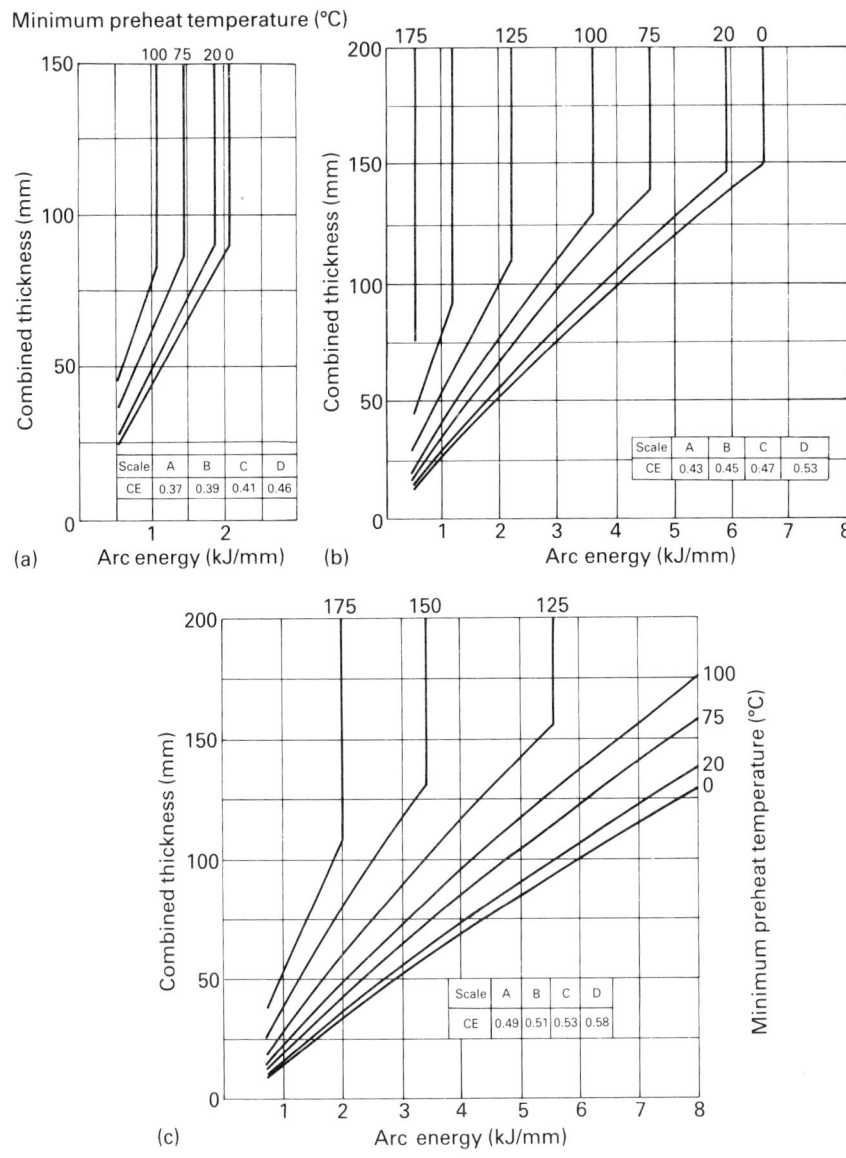

5.9 Examples of diagrams devised to avoid hydrogen cracking when welding steels of different CE and thickness using different hydrogen levels and heat inputs. (Source: Fig. 4.3b, 4.3e and 4.3h of ref. 1.)

Hydrogen cracking

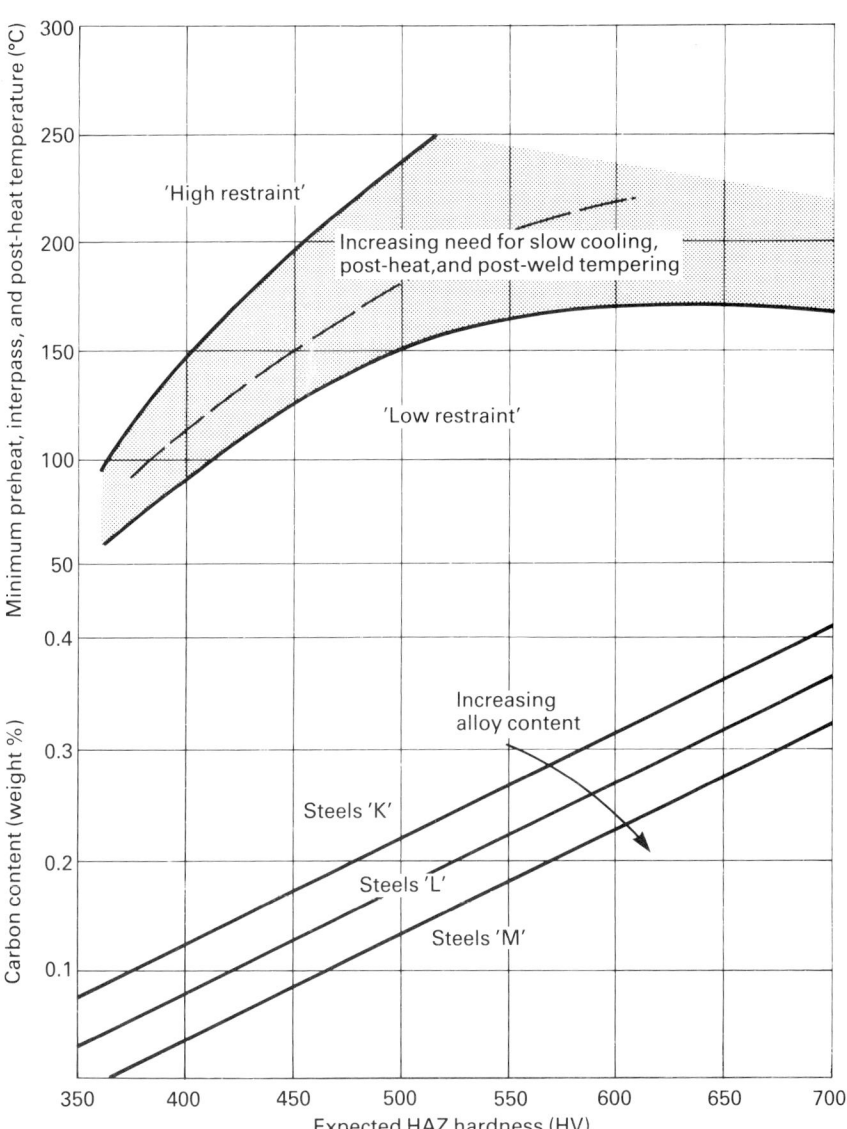

5.10 Diagram for welding alloy steels by the temperature control method (for explanation of steel types see ref. 1).

taken of factors such as differing restraint and hydrogen levels. For example, the lower curve is applicable to bead-on-plate deposits made with relatively low hydrogen levels; the upper curve is for higher restraint and higher hydrogen. Methods for estimating HAZ hardness are described in ref. 1 and 3.

If either the expected HAZ hardness or the restraint is increased, preheat alone will not be sufficient to avoid cracking; the joint will need to be post-heated at the preheat temperature or higher in order to allow sufficient hydrogen to diffuse out before it can be cooled out.

One difficulty which can arise when making multipass welds in 'difficult' steels is that the hydrogen level may build up during welding to a harmful level, particularly if the weld is short, and one run follows another with inadequate time for much hydrogen diffusion. In such cases, as well as minimising the hydrogen input to the weld, it is also helpful to allow a minimum interpass *time*. This can also be calculated from hydrogen diffusion data at the interpass temperature,[1] to allow sufficient time for diffusion of hydrogen out of each weld run before the next run is deposited.

When using the method, details of the steel need to be known so that the temperature selected is not so high that it tempers the steel and softens it, or is above the temperature above which the transformation to martensite is completed (the M_f temperature). Details of M_f temperatures can be found in appropriate reference books or calculated as described in ref. 1. If the weld is held above the M_f, the HAZ will retain some austenite at the preheat temperature, which will not lose its hydrogen until it cools to ambient temperature, transforms and leaves the hydrogen it contained free to cause cracking. Even if the steel is taken up to a tempering (PWHT) temperature on completion of welding without intermediate cooling, such retained austenite will not transform during the PWHT, but will retain most of its hydrogen until it is cooled and transforms.

Some steels may have M_f temperatures which are so low that they are well below the desired preheat level. In such cases, if neither of the methods described subsequently can be used, welding should be carried out at the preferred preheat temperature, using as low a hydrogen level as possible. On completion of welding, the temperature is reduced very slowly in order to give hydrogen from the transforming austenite as much opportunity as possible to escape before it can do any harm.

A disadvantage of the temperature control method is that it leaves a hard HAZ which, if the joint is not to be given PWHT, is likely to have poor toughness and poor resistance to some types of stress corrosion. One method of achieving some degree of tempering is to use the **temper bead** technique. This requires a weld bead to be deposited, as in Fig. 5.11, at a fixed and closely controlled distance from the weld toe. This bead tem-

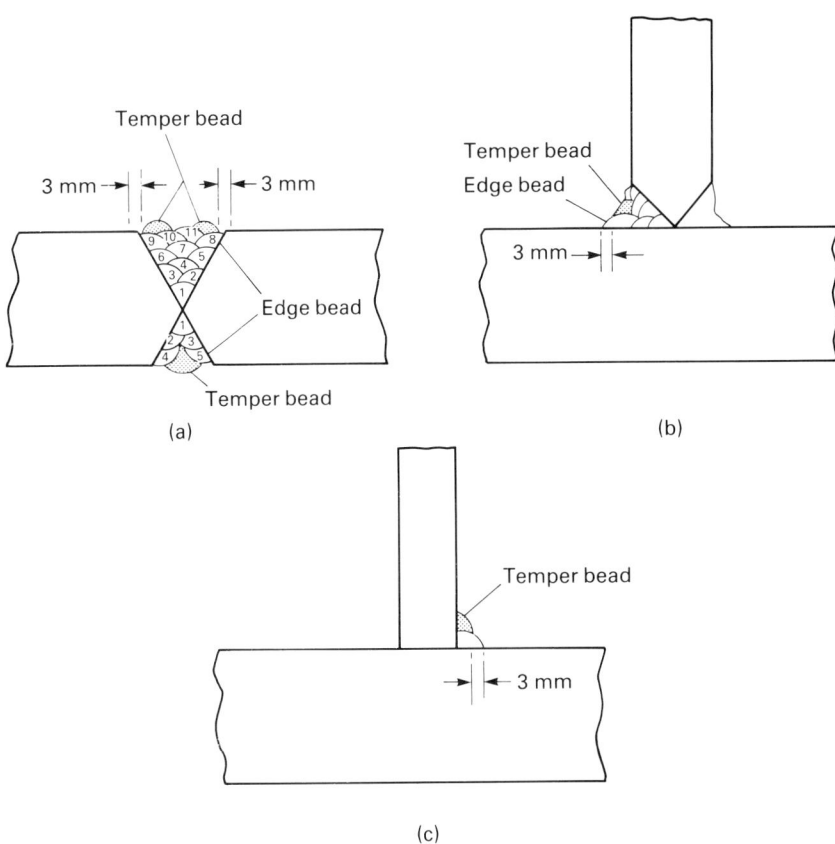

5.11 Temper bead technique for: (a) multipass butt weld; (b) multipass fillet weld where upstanding member is thinner than through member; (c) two-pass fillet weld.

pers the hard HAZ on the parent steel of the final weld run and leaves its own HAZ in less hardenable weld metal. If necessary, the temper bead can be ground off on completion. The method suffers from the disadvantage that if the temper bead approaches the weld toe too closely, it creates its own hard HAZ on the parent steel; also the underlying HAZ may still contain unacceptably hard regions. The next method provides an alternative technique of avoiding cracking and excessively hard HAZs.

The **control of microstructure by isothermal transformation** is intended to produce a softer HAZ than the previous technique, by carrying out the welding operation at a sufficiently high temperature for the HAZ to transform isothermally to bainite or a similar transformation product softer than martensite. For this method, it is necessary to use the isothermal

152 Weldability of ferritic steels

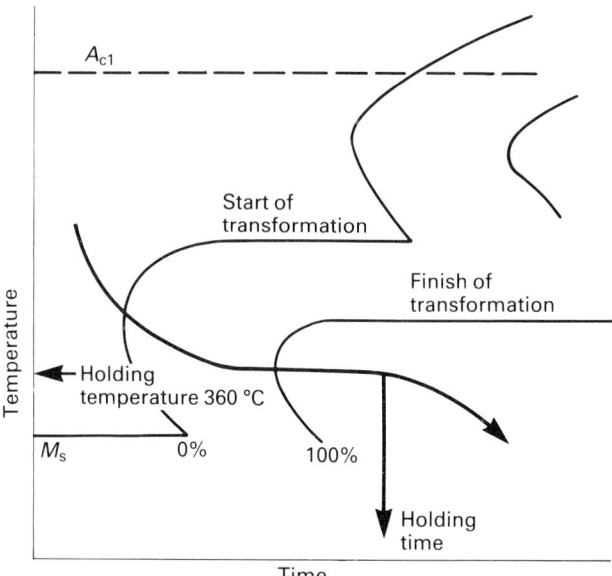

5.12 Typical isothermal transformation diagram of the type used with the isothermal transformation technique; unless the diagram was produced for steel that had been austenitised at above about 1250 °C, the holding times should be doubled.

transformation diagram for the steel being welded. Such diagrams can be found in standard reference works and will look something like Fig. 5.12.

The temperature selected should give a transformation to bainite in a reasonably short time, although it is advisable to use a time twice that indicated by the diagram, as the coarse regions of a weld HAZ will take longer to transform than the steel used for the construction of the diagram, which will probably have been austenitised at an appreciably lower temperature. This longer transformation time and a normal cool out should give sufficient time for damaging amounts of hydrogen to diffuse out of the joint.

Austenitic or **nickel alloy consumables** are used when the preheat levels necessary by other methods are unacceptably high, either because they would damage the steel (or nearby components) or are too high for the safety of the welders. With steels containing up to 0.2 to 0.3% C, preheat is not necessary, and for higher carbon levels 150 °C is adequate; welding alloy steels without adequate preheat can, as shown in Fig. 5.13, give rise to HAZ cracking. Figure 5.14 gives guidance on how the preheat level varies with carbon content and other factors when using austenitic stainless steel electrodes.

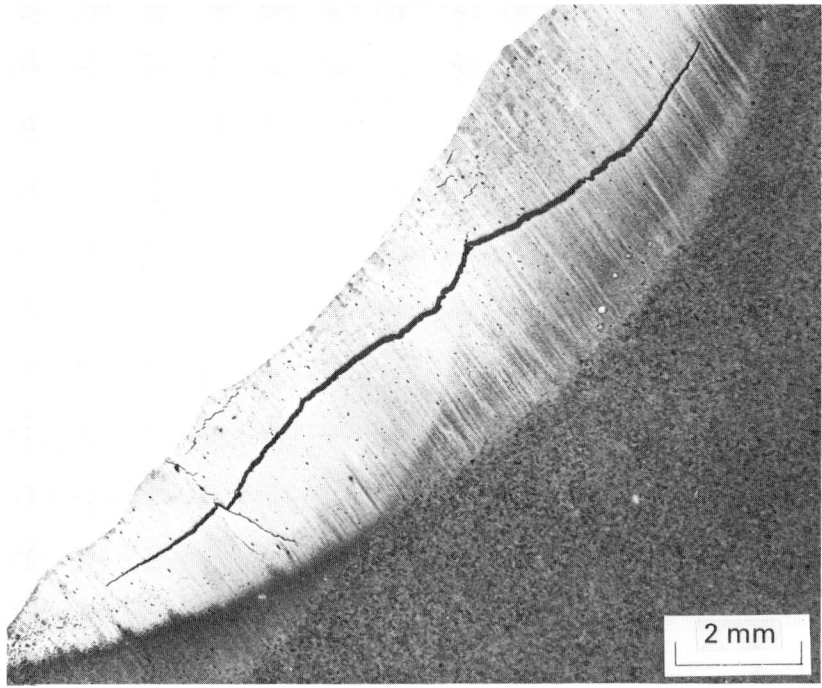

5.13 Hydrogen cracking in HAZ of low alloy steel welded with austenitic stainless steel electrodes with inadequate preheat.

The principle of the method is that both austenitic stainless steels and nickel alloys can dissolve appreciable amounts of hydrogen in the solid state, and are not normally susceptible to hydrogen embrittlement and cracking. During welding some hydrogen will inevitably diffuse into the parent steel HAZ while it is austenitic, but it rapidly diffuses back into the weld metal on transformation.

Some care is needed with the use of these consumables. The nickel alloys are prone to solidification cracking, particularly if they pick up sulphur from the parent steel, but otherwise they have advantages over the stainless steels. The latter are rather more prone to form hard martensite at and near the fusion boundary, as a result of incomplete mixing of parent steel and weld metal during welding, and this can give rise to cracking problems. Also, austenitic stainless steels differ more in their coefficients of thermal expansion from ferritic steels than do nickel alloys. Such differences can give rise to problems when large amounts of weld metal are being deposited; they also do not allow so much stress relief to occur during PWHT as do nickel alloy fillers.

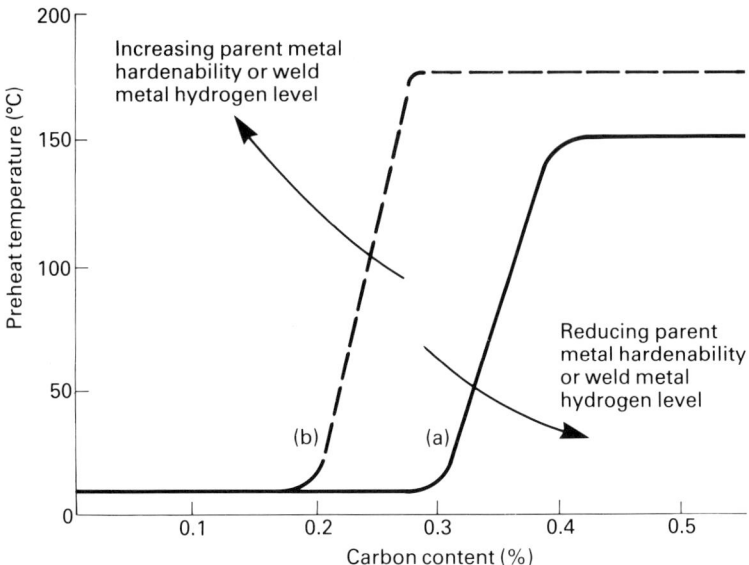

5.14 Guide to preheat temperatures when using austenitic stainless steel manual electrodes with heat inputs of 1–2 kJ/mm. (a) low restraint (e.g. material thickness <30 mm), (b) high restraint (e.g. material thickness >30 mm).

Both materials leave hard HAZs after welding and the welds are very difficult to examine non-destructively because of the different crystal structures of face-centred cubic austenite or nickel, and body-centred cubic ferrite; the only options are visual and dye penetrant examination.

Weld surfacing with austenitic stainless steels is a well-known use of this type of consumable for welding. It has been known to give cracking (often known as **underclad cracking**) at certain specific locations in relation to the weld bead pattern, particularly when welding thick-sectioned alloy steels. Although it is generally thought that this cracking is a form of reheat cracking (Chapter 6), it is possible that some is a type of hydrogen cracking brought about, in part, by hydrogen in the original (thick-sectioned) steel. Cracking of this type can often remain undetected until after PWHT and thus be confused with reheat cracking.

Finally, safe welding procedures have been developed over a period of time, so many previously unsuspected factors are now known. An example of this is shown in Fig. 5.15, where the bulk of the weld on to a mild steel (approximately 40 mm thick) was made with relatively high heat input and the final layer of 'cosmetic' beads with a lower heat input. Normally the final layer followed while the steel was still warm from

Hydrogen cracking 155

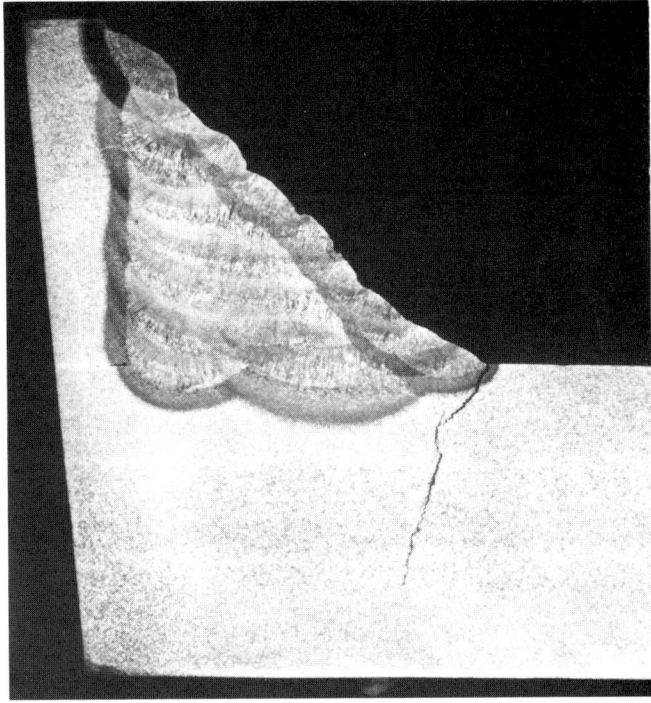

5.15 Hydrogen crack in HAZ of first run of final layer of a compensating plate/shell weld in mild steel ~40 mm thick. This has developed into a brittle fracture which arrested before the section was severed.

depositing the bulk of the weld. In the example illustrated, a delay of several days had resulted in the steel cooling to ambient temperature so that a crack formed in the HAZ of the first run of the final layer to be deposited. Unfortunately this crack initiated a brittle fracture when the steel cooled out and was then subject to an ambient temperature approaching zero. The brittle fracture fortunately arrested before the section was completely fractured.

Weld metal hydrogen cracking

The techniques for avoiding weld metal hydrogen cracking are less well developed than those for avoiding HAZ cracking, partly because cracking in the weld metal is less common than in the HAZ and partly because it has only recently been realised that control of hydrogen diffusion often plays a more important part than microstructural control. Weld metal hardness is much less important than HAZ hardness, because the possible

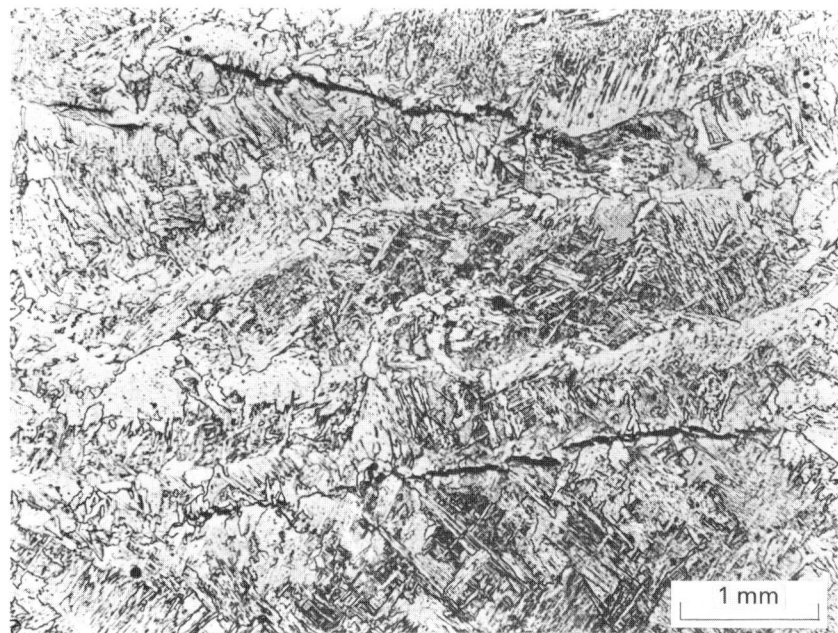

5.16 Microcracks in weld metal deposited from rutile manual electrodes.

range of variation is generally less, due to weld metal carbon contents usually being lower. However, with high hydrogen levels, cracking is possible in weld metals with a maximum hardness just below 200 HV, whereas under similar circumstances, the critical hardness for HAZ cracking would be about 350 HV.

There are three distinct regimes where hydrogen cracking is possible. The first of these is when welding heavy sectioned C and C:Mn steels without preheat and with consumables giving high hydrogen levels, i.e. with rutile or cellulosic electrodes; cracking often takes the form of fine microcracking (Fig. 5.16). Such cracking is rare today because of the widespread use of basic electrodes for welding thick sections, particularly if precautions are taken using welding diagrams, such as those in ref. 1.

The second likely cracking regime is with steels where a high degree of alloying is required in the weld metal, to achieve adequate resistance to creep, high temperature oxidation or hydrogen attack, or to achieve particular strength levels. In the last case, the development of lean alloyed steels of low carbon levels (often below 0.05% C, as in the so-called HSLA steels) to avoid the risk of HAZ cracking has made weld metal cracking a major problem, particularly as the weld metals needed to achieve high

Hydrogen cracking

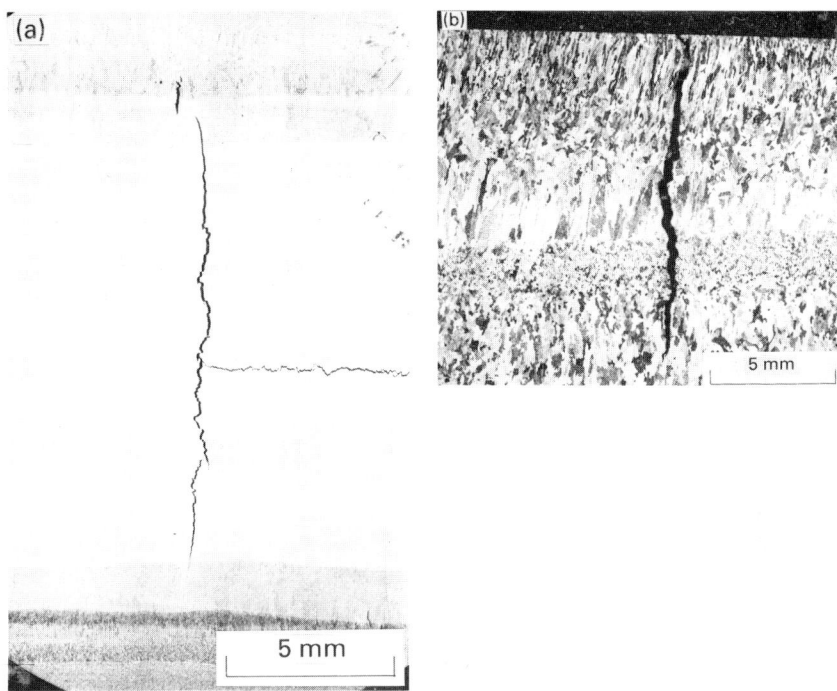

5.17 Weld metal hydrogen cracking in low alloy steel weld metal: (a) plan section (lightly etched) of 2.25% Cr : 1Mo weld bead showing transverse and longitudinal cracking; (b) longitudinal section, showing cracking to be perpendicular to weld surface.

strength will often have CEs appreciably higher than the parent steels. With the weldable high temperature steels, the disparities are not so great, as such steels do not usually have CEs below those of their weld metals.

In all these cases, any weld metal cracking is likely to be transverse to the weld length and perpendicular to the surface (Fig. 5.17), although longitudinal cracking may also occur (as in Fig. 5.17(a)) in some circumstances. Many of these steels have mandatory preheat temperatures in their applications standards, which should take care of hydrogen cracking in the weld metal, *provided* appropriate low hydrogen levels are used and well controlled.

The third type of weld metal hydrogen cracking is the so-called 'chevron cracking', which is transverse to the weld line, but at 45° to the surface (Fig. 5.18). Such cracking is most common in welds in relatively thick-sectioned C:Mn plate, welded with high heat input processes, such as submerged arc. Although 45° cracking has been seen in two-pass welds (in 40 mm thick plate), it is more usual in multipass welds. An important

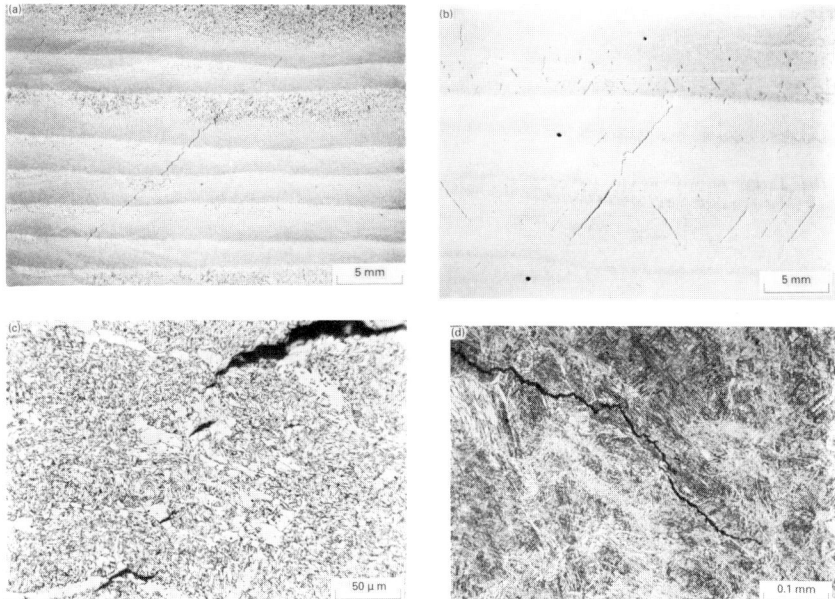

5.18 Transverse weld metal cracking at 45° to weld surface (chevron cracking) in longitudinal section: (a) multipass submerged arc weld with crack (intermittent in nature) passing through several weld runs below weld surface; (b) different crack showing dispersed and intermittent nature; (c) tip of crack showing how initial cracks tend to originate in coarser primary ferrite; (d) cracking in self-shielded arc weld.

feature of 45° transverse cracking is that with the large weld bead sizes resulting from high heat inputs, the diffusion distances for hydrogen can be so large that insufficient time is available to allow enough hydrogen to escape before the weld is cool enough for it to crack. Nevertheless, such cracking is not confined to high heat input welds and Fig. 5.18(d) illustrates an example in a self-shielded cored wire weld.

It can be shown that, although increasing heat input slows weld cooling and allows more time for hydrogen diffusion, this beneficial effect is less than the harmful effect of the greater diffusion distances resulting from the larger weld bead size. Transverse 45° cracking in C:Mn weld metals can largely be avoided by using very low hydrogen levels, i.e. $\leqslant 5$ ml/100 g deposited metal; further guidance can be found in ref. 1.

Detection and identification

Detection

Hydrogen cracks are normally very fine and difficult to detect, even when they break the surface, which they often do not. The consequences of not detecting hydrogen cracks can be disastrous. The example shown in Fig. 5.15 led to an arrested brittle fracture. For cracks which are present at the surface, magnetic particle inspection (MPI) is the preferred technique, although dye penetrant should be used if a non-ferritic filler (e.g. an austenitic stainless steel or a nickel alloy) has been used. The MPI technique has the advantage that, when using d.c., cracks just below the surface can be detected.

For buried cracks, ultrasonic examination is greatly preferred to radiography. Radiography is only capable of detecting those few cracks that are relatively wide and are almost parallel to the incident beam. However, even the use of ultrasonics is not always successful. If a crack has formed in the root region of a two-sided multipass butt weld, the residual stress pattern in the root on completion of welding (Fig. 1.16) will usually be such that the root region is in compression. This will tend to force together the faces of any root cracks, making them very difficult to detect until the weld has been give a PWHT, thus relaxing the stresses.

If it is suspected that transverse 45° weld metal cracks may be present (Fig. 5.18), it is necessary to carry out an ultrasonic examination using a 45° probe in both directions along the top of the weld seam. The usual 90° probe is not capable of detecting such cracks.

A further common reason for not detecting cracks is when inadequate time is left for cracks to grow to a detectable size before examination. It is still uncertain how long an incubation period is required for hydrogen cracks to start, and for how long they can continue growing. Several periods have been proposed, varying from overnight (16 hours) to 3 days (72 hours). There is some evidence that when a plentiful supply of hydrogen is available, cracks form soon after the weldment has cooled to near ambient temperature and grow to a detectable size within a few hours. That, however, is not the problem area. The greatest delay probably occurs when the welding conditions are close to a crack/no-crack boundary, so that appreciable time is needed for hydrogen to diffuse to the sites of cracking in harmful quantities. Also, in very cold conditions, down to 0 °C and lower, still longer times will be necessary.

The difficulties of detecting buried cracks in as-welded joints make the use of safe welding procedures particularly important. The difficulties can also lead to incorrect identification of the type of crack that has been detected: cracks found after PWHT tend to be classified as reheat cracks (Chapter 6), even if the steel is known to be not susceptible to this type of cracking.

160   Weldability of ferritic steels

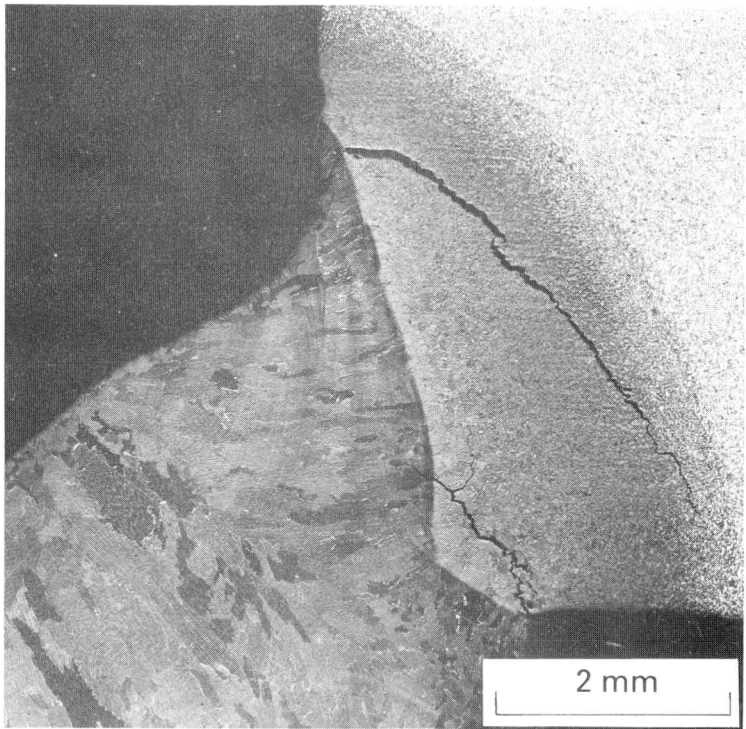

5.19 Heat-affected zone hydrogen crack which has extended into the weld metal as well as the fine-grained HAZ.

Identification

The identification of hydrogen cracks is not simple, as they can occur in a variety of locations and orientations, and can be transgranular or intergranular. The latter are more likely if the steel or weld metal is of the alloyed type or is exceptionally hard. In the HAZ, hydrogen cracks are usually parallel to the weld (unless they are extensions of transverse weld metal cracks) and usually have a portion close to the fusion boundary. However, the cracks may divert into the fine-grained HAZ, as in Fig. 5.7(a), or also into the weld metal (Fig. 5.19).

Cracks will almost never extend beyond the visible HAZ unless they have been heavily stressed after welding, or unless the parent steel itself contains much hydrogen. Without post-weld stressing, cracks extending beyond the visible HAZ are likely to be lamellar tears.

When employing fractographic techniques for crack identification, crack surfaces in C and C:Mn steel HAZs are usually transgranular,

Hydrogen cracking 161

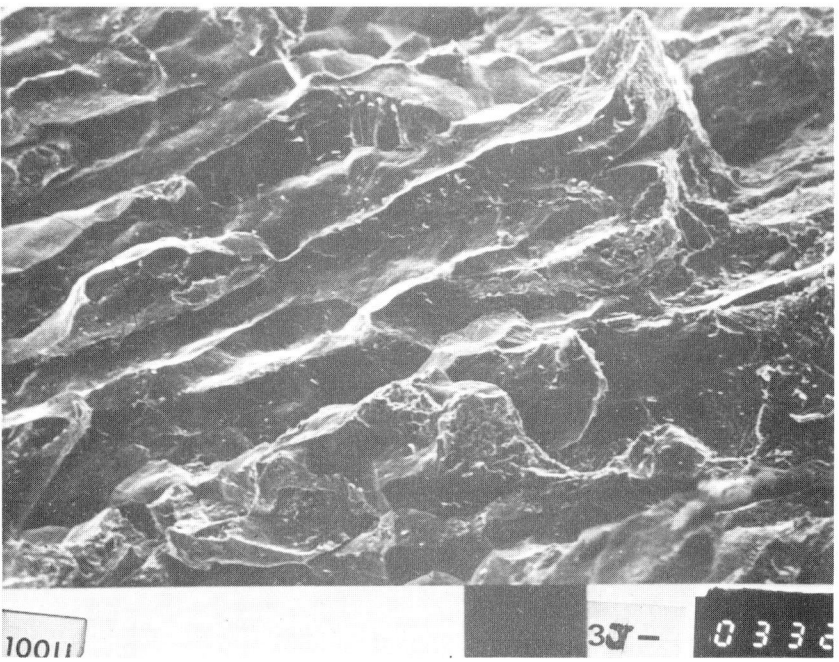

5.20 Predominantly intergranular crack surface of low alloy weld metal in SEM.

being of a quasi-cleavage type, although some ductile tearing (microvoid coalescence) may also be present. In alloyed steel HAZs, intergranular cracking is the common type. Unlike reheat cracks, hydrogen cracks will *never* show cavitation along the prior austenite grain boundaries beyond the crack tips.

Similar comments are applicable to weld metal hydrogen cracks (e.g. Fig. 5.20), with the further observations that cracking in weld metals is frequently transverse to the weld and that in weld metals of the acicular ferrite type, cracking is common in the *softer* primary ferrite (if present) at the prior austenite grain boundaries, as in Fig. 5.18(c). This is because, when the weld metal starts to crack, deformation is accommodated in the lower yield stress (softer) microstructural constituent, whose ductility becomes exhausted before the harder acicular ferrite is stressed beyond its yield point. In relatively hard, alloyed weld metals (e.g. in the Cr:Mo steels) any transverse weld metal cracking is usually perpendicular to the weld surface (Fig. 5.17b), but in C:Mn weld metals the chevron type of cracking (Fig. 5.18) at about 45° to the surface is common. An unusual type of 'weld metal cracking' is when a weld metal of relatively high carbon

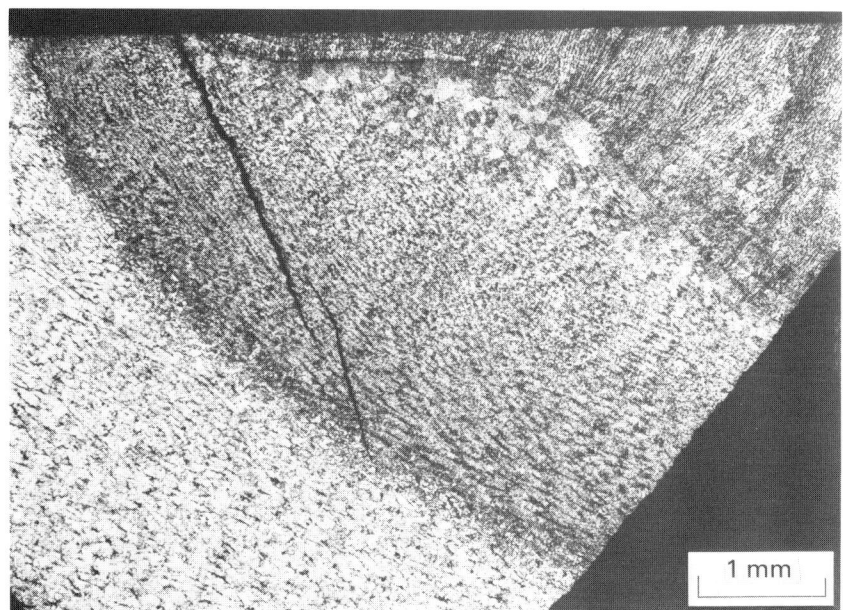

5.21 Cracking in an electroslag weld metal which has become the HAZ of a later arc weld; section is of a boat sample.

content or equivalent is welded over and becomes itself a HAZ within which cracking occurs (Fig. 5.21).

Other aspects of hydrogen in ferritic steels

Several aspects of the behaviour of hydrogen in steels are discussed below, largely to show how such behaviour relates to hydrogen cracking during welding. The most dangerous aspect is that an undetected hydrogen crack can readily initiate either a brittle fracture, as in Fig. 5.15, or a fatigue crack. More detail of some of these problems is given in Chapter 10.

Fisheyes are a most unusual manifestation of hydrogen embrittlement and are confined to as-welded weld metal fracture surfaces. As can be seen from the cover illustration and (in more detail) in Fig. 5.22, each fisheye consists of a central region (the pupil of the eye) and the surrounding region (the iris). The central region is a small imperfection – either a pore or an inclusion – which was locally rich in hydrogen when the fisheye was formed.

When weld metal is stressed beyond its yield point at a relatively slow strain rate and necking starts, the hydrogen in the pore (or concentrated around the inclusion) locally embrittles the surrounding steel so that fur-

Hydrogen cracking

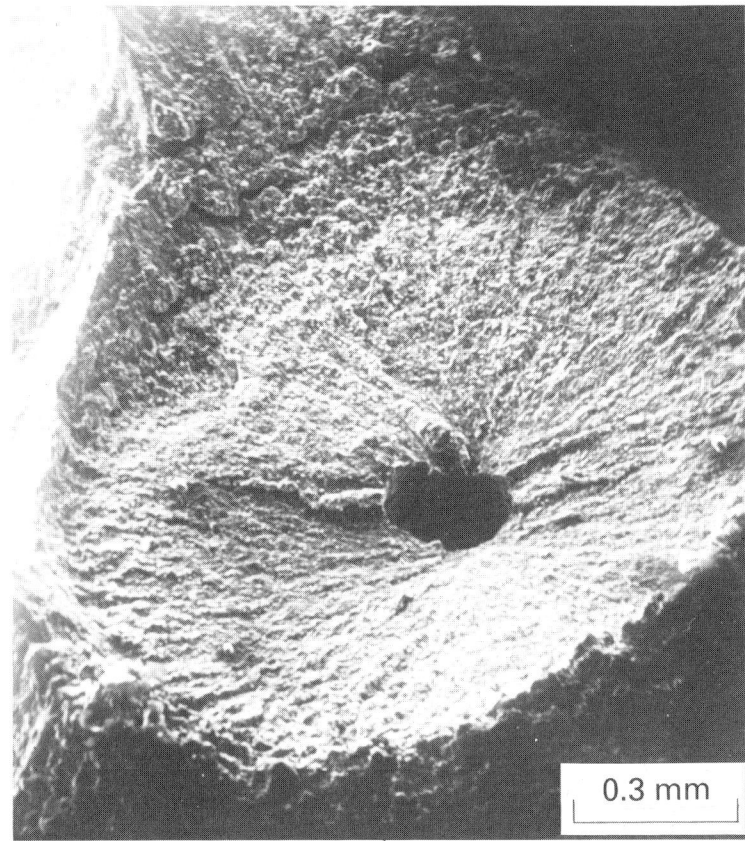

5.22 Fisheye on fracture surface as seen in SEM showing central pore and surrounding brittle fracture.

ther straining leads to a hydrogen-assisted brittle fracture – usually of the quasi-cleavage type. This fracture proceeds for a short distance and then stops for two reasons. Firstly, the plastic deformation heats the steel and, secondly, the local strain rate at the crack tip increases*. Both of these effects *reduce* the ability of hydrogen to embrittle the steel to such an extent that the brittle (hydrogen-assisted) fracture stops. Further straining leads to failure by a ductile process (microvoid coalescence) in the normal way. It will be appreciated that the only 'defect' at a fisheye is the central 'pupil', which is normally not much more than a millimetre or so across. It is *not* the whole fisheye.

* The heating produced can be felt by (cautiously) touching the neck region of a tensile test specimen immediately after fracture. The increase in strain rate at a hydrogen crack tip causing the crack to stop growing is why hydrogen cracks grow so slowly.

Fisheyes are not normally seen for two reasons: firstly, tests on all-weld metal samples – which are carried out by manufacturers of consumables for quality control – are always carried out on samples that have been heat treated (usually for about 16 hours at 250 °C) in order to remove hydrogen;[4] secondly, because ferritic steel weld metal usually overmatches the parent steel in strength, the fracture in a cross-weld tensile test normally occurs in the parent steel. Also, tests on weld metal that has been given a PWHT will not show fisheyes, because the heat treatment will remove sufficient hydrogen to avoid the problem.

The only time that fisheyes are likely is when the weld metal fails to overmatch the parent steel. This can happen if the parent steel is stronger (or the weld metal is weaker) than it should be. Fisheyes are also possible if tests are being made to establish the strength of a soft weld metal used to butter a high strength steel (even though the weld metal may be of very low hydrogen level).

In events such as those in the previous paragraph, it is not permitted to apply a hydrogen removal heat treatment. If the test specimen has given adequate strength (and, if called for, ductility) values, the presence of fisheyes is *not* a cause for rejection, unless the pores or particles at the **pupils** of the fisheyes are unacceptable. If ductility levels are adversely affected, the best advice is to extract further test specimens well before they are to be tested and let them stand in a warm part of the test house for as long as practicable, so that as much hydrogen as possible can escape. However, it is not unknown for specimens extracted from a thick-sectioned weld several months after deposition to exhibit fisheyes when tested.

Hydrogen flaking can occur in heavy forging billets when they have been cooled from casting temperatures with too high a hydrogen content. The flakes are small cracks within the steel billet caused by the residual stresses developed on cooling acting on steel embrittled by hydrogen. They are, therefore, very similar to the hydrogen cracks which occur after welding. Normally, steel for such heavy forgings should be of low hydrogen content; if this is not achieved, the only way of preventing cracking or flaking is to give the steel a prolonged heat treatment within the tempering temperature range (say, 600–700 °C) to reduce its hydrogen content before it is allowed to cool to ambient temperature.

Steel that is hot rolled to plate and sections is not normally thick enough for sufficient hydrogen to be retained to give a problem, although in the early days of continuous casting, cracking was occasionally found within centre-line segregates, which were sufficiently enriched in alloying elements to be easily embrittled.

The problem is a reminder that steel, particularly if it is alloyed and of medium or high carbon content, can contain sufficient hydrogen to give

Hydrogen cracking

welding problems. Such an example is that of a heavy forging bored out and welded inside the bore (without any prior tempering treatment) with a very low hydrogen welding process and consequently a low preheat level. The preferred solution is to rough machine out the bore, to temper the steel and only then to carry out welding.

High pressure hydrogen is normally stored in high strength steel cylinders at ambient temperatures, despite the well known embrittling effect of hydrogen on high strength steels. Although there are limits on the strength of steels that can be used for such service, the major factor in preventing cracking is that molecular hydrogen cannot enter steel that has even a thin oxide coating (rust) on its surface. If the steel vessel containing hydrogen is stressed beyond its elastic limit, the oxide is liable to crack and hydrogen can then enter the atomically clean steel surfaces thus presented. With very high strength steels, the oxide can crack before the steel reaches its elastic limit; hydrogen can then enter the steel, embrittle it and cause cracking without prior plastic deformation.

Similarly, when steel contains molecular hydrogen trapped in voids, this hydrogen cannot re-enter the steel lattice below about 250 °C; such trapped hydrogen is responsible for some of the residual hydrogen measured in weld metal hydrogen analyses, and also for the reduced hydrogen diffusion rates measured near ambient temperature (Fig. 5.3).

Hydrogen attack or **hydrogen damage** occurs when steel containing hydrogen is exposed to high temperatures. The attack occurs because the hydrogen combines with carbon in the carbides in the steel (Fig. 5.23) to form methane. This destruction of the carbides reduces the strength of the steel and the methane can form internal blisters. Attack is most likely when steel is used (as in petrochemical plant) to contain gases which contain high proportions of hydrogen at high temperatures and pressures.

The presence of chromium in a steel increases its resistance to hydrogen attack, and curves (the 'Nelson diagram') are available to show which steels can be safely used at different temperatures and partial pressures of hydrogen.[5]

Steel that has been in such hydrogen service above about 200 °C will contain high levels of hydrogen and plant is normally cooled down slowly to allow sufficient of this hydrogen to diffuse out to avoid cracking when the vessel has cooled to ambient temperature. However, it may still contain enough hydrogen to give cracking problems if the vessel has to be welded for repair or modification; reheating to diffuse out some of this hydrogen or other precautions may be required in such cases.

Sour service is a description of the use of pipework and the like to contain liquids (often slightly acidic) which contain hydrogen sulphide (H_2S) in solution. Its significance is that molecular hydrogen cannot enter steels at or near ambient temperature because the oxide on the surface prevents

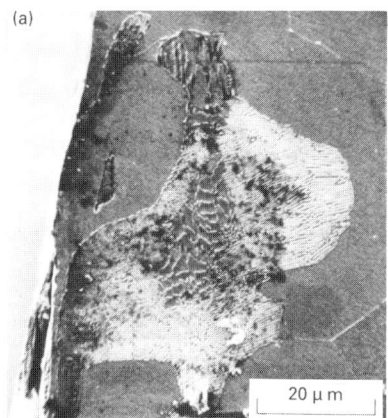
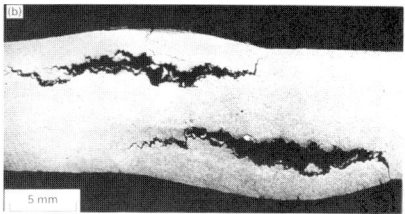

5.23 Hydrogen attack: (a) near surface of C : Mo steel. SEM micrograph, white regions are carbide lamellae in pearlite, black areas are cavities where carbides have been destroyed; (b) blistering in mild steel, optical micrograph.

the iron reacting with molecular hydrogen to give atomic hydrogen. The presence of H_2S allows atomic hydrogen to be formed and to enter the steel lattice in concentrations considerably exceeding the equilibrium solubility of hydrogen in steel at 1 bar pressure. This hydrogen can give severe cracking and blistering problems, which are described in the next few paragraphs.

Hydrogen blistering occurs in two circumstances. One (described above) is when steels are used in hydrogen service above the safe limits for such service. The hydrogen taken into the steel reacts with the carbon in the steel to form methane, which can form quite large bubbles, particularly in mild steel (as in Fig. 5.23(b)). Blistering, similar in appearance can also occur at and near ambient temperature if steel with unfavourable types of inclusions (usually rolled out MnS and manganese silicates) is put into sour service. Globular or compact inclusions (e.g. those produced by inclusion shape control with calcium or rare earth metals) or the use of steels of very low sulphur contents (i.e. <0.002% S) avoid the problem.

Hydrogen stress corrosion cracking and **HIC** (hydrogen-induced cracking) both occur in sour service, HIC being associated with similar conditions to those giving blistering in sour service, where blistering and cracking are associated with inclusion stringers and, in severe cases as in Fig. 5.24, can propagate from one inclusion plane to the next, rather like lamellar tears (Chapter 4).

Hydrogen cracking

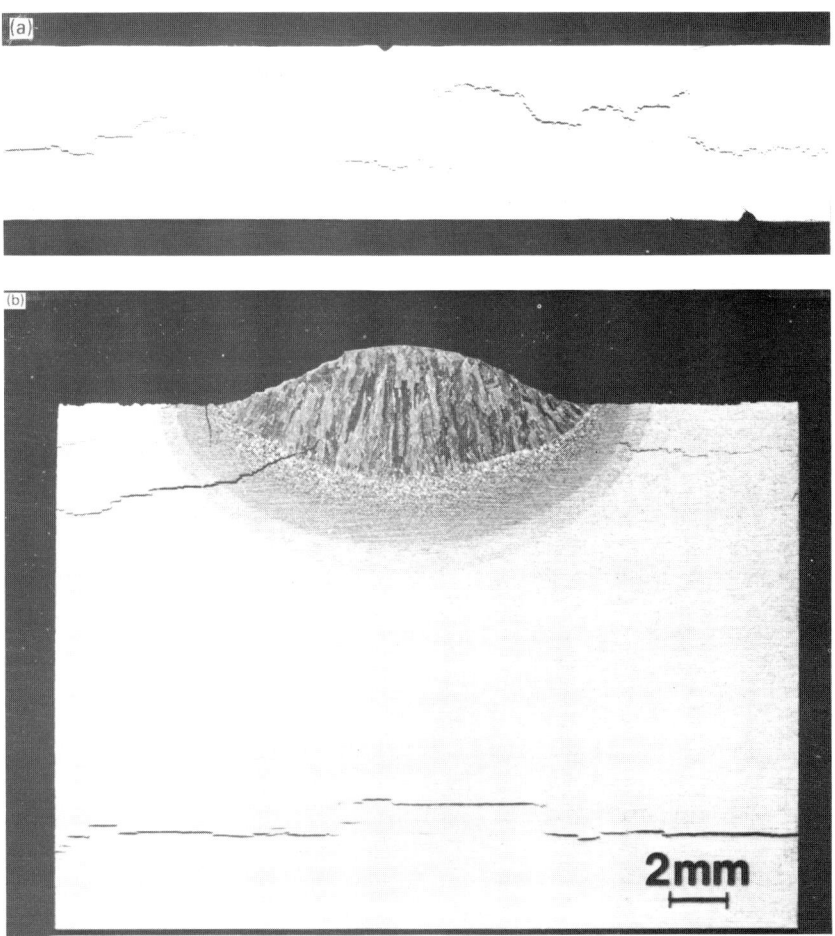

5.24 Hydrogen-induced cracking (HIC) after tests in a solution containing H_2S: (a) susceptible 5 mm thick linepipe steel; (b) bead-on-plate deposit on susceptible steel.

This cracking is a type of stress corrosion cracking (SCC) which is most likely at regions of high tensile stress, and particularly with the residual stresses associated with welds. The cracks resemble hydrogen cracks occurring during welding. Where sour service conditions are severe, the recommendations of NACE[6] should be followed. These include a maximum Rockwell hardness of 22 HRC (roughly equivalent to 248 HV) and mandatory PWHT for welds in alloyed steels. Beside their application to petrochemical plant, these requirements are often imposed for service in pipelines and associated valves and other equipment in oil and gas fields if the field is likely to become sour later in its existence.

Hydrogen in heat treatment is often added to the furnace atmosphere in some form (e.g. exothermic gas with ~15% H_2 or cracked ammonia, with up to 75% H_2) to protect the steel surfaces from oxidation. When the treatment involves heating within the austenite range, i.e. during annealing, normalising or hardening, the steel will pick up hydrogen in proportion to its partial pressure in the furnace atmosphere, about 4 ppm (~4 ml/100 g) being picked up by heating in pure hydrogen at about 800 °C. With heat treatments involving slow cooling, most of this hydrogen has time to diffuse out during the cooling and gives no further problems. If the treatment involves quenching, however, some hydrogen will be retained and, as well as being a probable cause of quench cracks, could give problems should the steel be welded without tempering, particularly in the circumstances outlined for welding in the bores of heavy forgings outlined in the section on hydrogen flaking, above.

Hydrogen in welding has other effects than in producing cracking. During welding it can be a source of porosity (Chapter 7), although its identification is difficult because it can react, while still hot, with carbon in the steel to leave methane, rather than hydrogen, in the pores.

Hydrogen in the arc atmosphere also increases penetration of the welding arc. This is most pronounced when welding with cellulosic electrodes (whose cellulose decomposes in the arc to give hydrogen) and during wet underwater welding. Its effects can, however, be noticed with more normal hydrogen levels, particularly if a procedure involving two-sided welding that requires inter-run penetration is subsequently used for consumables dried to lower moisture levels. It is always important, therefore, to check this aspect when procedures are modified to involve consumables of lower hydrogen potential. Lamellar tearing can be aggravated by using consumables of higher hydrogen level, as discussed in Chapter 4.

References

1. Bailey, N., Coe, F.R., Gooch, T.G., Hart, P.H.M., Jenkins, N. and Pargeter, R.J., *Welding Steels Without Hydrogen Cracking*, Abington Publishing, Cambridge, 1993.
2. British Standard BS5135: 1984, 'Arc Welding of C and C–Mn steels', BSI, London, 1984.
3. Bailey, N. (ed.), *Hardenability of Steels, a Select Conference*, TWI, Cambridge, 1990.
4. BS639: 1986, 'Covered C and C–Mn steel electrodes for MMA welding', BSI, London, 1986.
5. API 941, *Steels for Hydrogen Service at Elevated Temperatures and Pressures in Petrochemical Refineries and Petrochemical Plant*, American Petroleum Institute Publication 941, 4th ed., API, Washington D.C., Apr. 1990.

6 NACE Standard MR0175-92, 'Standard recommended practice for sulfide stress corrosion resistant-metallic materials for oilfield equipment', NACE, Houston, TX, 1992.

6 Reheat cracking

Description

Reheat or heat-treatment cracking is one of the rarest types of cracking that occur during the welding operation. It is only possible in welds that are given PWHT; it appears to be restricted to steels or weld metals containing at least two of the elements Cr, Mo, V and B, and it is mainly found in relatively heavy sections. Two versions exist, structural weld cracking and underclad cracking. In both types, the cracking usually occurs in the HAZ, and is usually confined to the coarse-grained regions. However, reheat cracking is also possible in coarse grained weld metal and parent steel.

Cracking normally occurs as the weldment is being heated to the PWHT temperature, probably above about 565 °C, when the creep strength is still high and ductility in the HAZ is soon exhausted. Cracking depends on the grain boundary regions at elevated temperature being less ductile and/or slightly weaker than the grain interiors. A weaker grain boundary region means that any local deformation which accompanies the relief of residual stresses occurs predominantly along the grain boundary regions. A coarse-grained HAZ (or other region) has fewer grain boundaries to share this deformation, so that ductility is more rapidly exhausted in a coarse- than in a fine-grained region. Low ductility of grain boundaries by impurities still further reduces the capacity for grain boundary deformation, so that such impurities can make an otherwise heat-treatable steel liable to crack. A coarse grain size is again harmful because it provides less grain boundary region over which such impurities can be distributed.

The relationship between the strengths of HAZ and weld metal at elevated temperature is also important. If the HAZ is appreciably stronger (particularly in creep) than the weld metal, the weld metal will start deforming first and the HAZ will not have to strain plastically. If the weld metal is

Reheat cracking

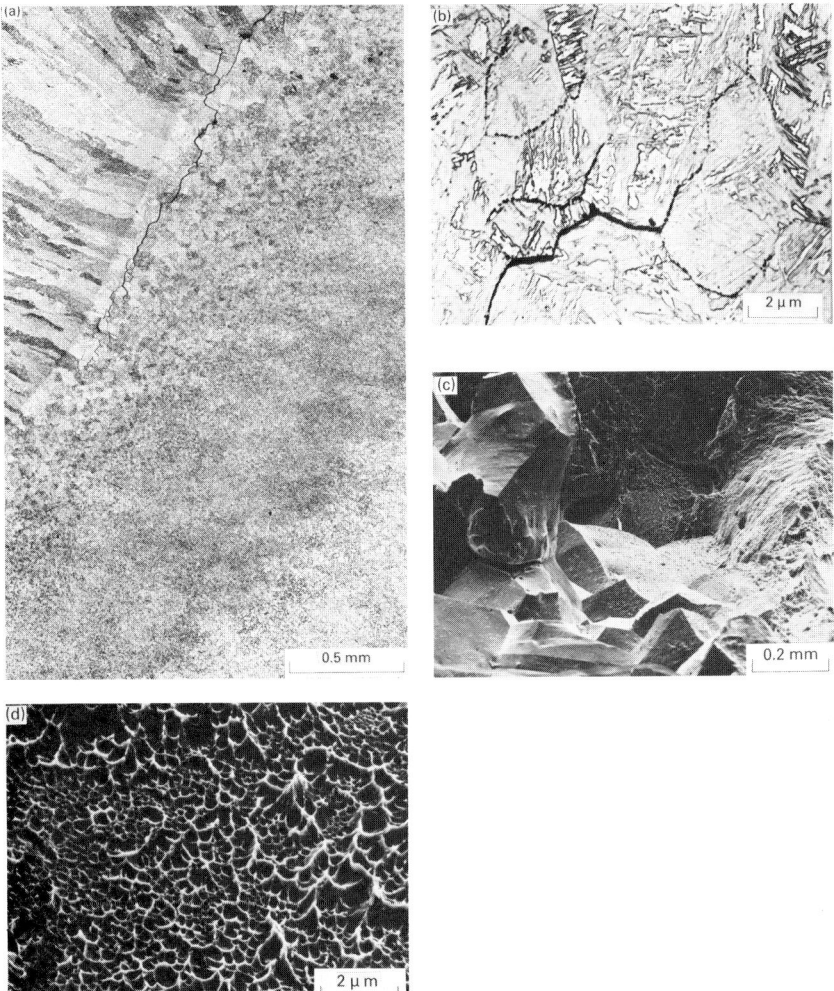

6.1 Typical appearance of reheat cracking: (a) macrosection showing cracking to be intergranular and in coarse-grained HAZ; (b) microsection near tips of cracking showing cavitation along prior austenite grain boundaries; (c) SEM fractograph showing intergranular cracking; (d) showing cavitation on grain facets.

stronger than the HAZ, plastic deformation is forced on the HAZ, which is then more likely to crack than if the strengths of HAZ and weld metal were more nearly equal. The latter situation is particularly likely when a ferritic steel is clad with a stainless steel (underclad cracking) because stainless steels are stronger at high temperatures than ferritic steels, especially when the ferritic steel is not particularly creep-resistant.

Reheat cracking takes place by the normal creep failure mechanism of cavities opening up along grain boundaries and then coalescing to form a continuous crack. Sectioning through a failure, or fractographic examination in an SEM, will often show evidence of this creep cavitation (Fig. 6.1). Although not always present on the crack surfaces, creep cavitation is strong evidence that cracking is of the reheat cracking type, provided the weld has not been in creep service, when such cavitation can also occur.

For cracking to occur, as stated above, at least two of the elements Cr, Mo, V and B are necessary. In the Cr:Mo steel types, if vanadium is absent, at least 0.5% Mo is needed to give cracking; with vanadium present, only 0.2% Mo is needed. The elements Sb, As, Sn and P are also likely to increase the risk of cracking; these are also the elements which confer temper embrittlement (Chapter 10) on the same types of steels. Finally, carbon is generally thought to be harmful.

Nakamura et al.[1] have rated the susceptibility to reheat cracking by the parameter ΔG, Eq. [1.22], and this was modified[2] to include a term for carbon, Eq. [1.23]. In both expressions values above zero denoted a risk of cracking. The ΔG formula is of limited use, as are Eq. [1.22] and [1.23] because, although they can indicate which types of steel are likely to be susceptible to reheat cracking, they do not allow the effect of the temper-embrittling impurities or (in two cases) that of carbon to be assessed. For the impurities, it could be useful to use one of the temper embrittlement parameters (Eq. [1.27]–[1.29]), discussed in Chapter 10, to be used. However, a metal composition factor (MCF) has been developed[3] for the 2.25% Cr:1Mo steel which includes factors for impurities:

$$\text{MCF} = \text{Si} + 2\text{Cu} + 2\text{P} + 10\text{As} + 15\text{Sn} + 20\text{Sb} \qquad [6.1]$$

It indicates that silicon and copper, as well as the impurities associated with temper embrittlement, increase the risk of cracking. In the tests to establish this formula, increasing heat input was noted as increasing grain boundary embrittlement.

If the steel contains segregate-rich bands, any cracking is more likely in these regions, because they are likely to contain more of the alloying, impurity elements and carbon than the bulk of the steel.

Role of composition

The relative effects of the elements Cr, Mo and V can be assessed from the modified Nakamura formula,[2] Eq. [1.23]. The effect of boron is less well understood, although it is likely to be needed in an 'active' form. When present as such, very small amounts (possibly just in excess of 0.0005%) are likely be detrimental. Carbon is also important (Eq. [1.23]);

certainly in the 0.1Cr:0.5Mo:0.25V high temperature steel, the carbon level in the British Standard Specification BS 1501: Part 2: 1988: Grade 660[4] is kept below 0.13% to minimise problems. In the same steel the vanadium content is kept below 0.28% for the same reason. Nevertheless, steels with lower vanadium contents (and containing no chromium and with molybdenum less than 0.5%) have been found to be susceptible.[5]

The steels in which reheat cracking has been found include 0.5Cr:0.5Mo:0.25V, 0.5Mo:B, 2.25Cr:1Mo, 5Cr:1Mo in descending order of susceptibility.[6] However, the risk of cracking in the 2.25Cr:1Mo steel is so low as to be of little practical importance. In addition, cracking has been identified or claimed in 2Cr:1Mo weld metal, in some of the high strength structural steels covered by ASTM A517-1990 (Grades B, E and F, i.e. of the Cr:Mo type with microalloying additions of boron and titanium), Mn:Ni:Mo pressure vessel steels in the ASTM A508 and A533 specifications when they contain a few tenths of a percent of chromium and about 0.05% or more of vanadium. Cracking has also been seen in a high strength structural steel with approximately 0.1C:1.5Mn:0.25Mo: 0.07V.[5]

The relationship for reheat cracking in 2.25Cr:1Mo weld metal (Eq. [6.1]) is similar in many respects to those developed for temper embrittlement (such as J, $\bar{X}$ and P_E, Eq. [1.27]–[1.29]). The same deleterious impurities, Sb, As, Sn and P, are important, although Si and Cu, rather than Si and Mn are also harmful. Following the finding of reheat cracking in the Mn:Mo:V steel, it would appear that the detrimental effect of Mn in regard to temper embrittlement is also relevant to reheat cracking in this type of steel.

Other influences

Of the other influences on reheat cracking, the most important is the *prior austenite grain size*. This is because the coarser the grain size, the fewer are the grain boundary regions available both for deformation during stress relief and also to which any embrittling impurities can segregate. Hence, cracking is most likely to occur at regions with a coarse prior austenite grain size. The most likely of these is the coarse-grained, high temperature HAZ. An exceptional case is known where a parent Cr:Mo:V steel of satisfactory composition was incorrectly normalised, giving it a much coarser prior austenite grain size than its coarse HAZ; reheat cracking occurred in the parent steel but not in the HAZ.

A consequence of this dependence on grain size is that, if a welding technique is used that avoids the presence of coarse-grained regions in the final weld, the risk of cracking can be considerably reduced and cracking itself avoided. The use of such techniques is described in the next section.

The presence of **notches**, particularly at weld toes, can in marginal circumstances promote cracking, so that their elimination by grinding weld toes can be helpful.

Such weld toes occur at the weld surface. If the surface is the original surface of, for example, a normalised plate, then the **decarburised layer** at such a surface is less likely to suffer reheat cracking than the unaltered surface of a forging which has been machined all over. It is believed that this effect is largely a result of the lower carbon content in the decarburised layer.

Other factors have less influence on cracking. The most important of these is the **heating rate** to the PWHT temperature, and the **PWHT temperature** itself. To reduce the risk of reheat cracking, the heating rate should be as fast as possible in order to heat the steel rapidly through the temperature range within which it is most brittle and can crack. However, the heating rate must be strictly controlled, not only because it is required by the relevant specification, but in order to avoid unequal heating of different parts of the fabrication, particularly if they are of different thicknesses. Such uneven heating can lead to distortion and/or higher residual stress levels after PWHT than are acceptable. A relatively high PWHT is also beneficial; certainly for the Cr:Mo steels this should exceed 650 °C.

The effect of *heat input* during welding is minor, largely because its effect on HAZ grain size is relatively small. Similarly the hardness of the HAZ, which admittedly cannot be varied by a large amount in the more highly alloyed steels under consideration, has been found to be without effect.

Control techniques

Once details of the composition of a suspect steel have been worked out to allow it to be safely welded and heat treated without reheat cracking, then it is important to maintain a watch on mill sheets to ensure that the compositional limits are maintained. One such example is the 0.5Cr:0.5Mo:0.25V steel, in which the limits of 0.08–0.13% C and 0.22–0.28% V in BS1501: Grade 660[4] are most important. Exceeding those limits gives a serious risk of reheat cracking; falling below the minimum vanadium and carbon contents will give inadequate creep strength and a risk of premature failure in service.

When cladding a suspect steel with an austenitic stainless steel or a nickel alloy weld metal, the risk of cracking is reduced if a two layer cladding technique is used, so that the second layer refines the coarse-

Reheat cracking 175

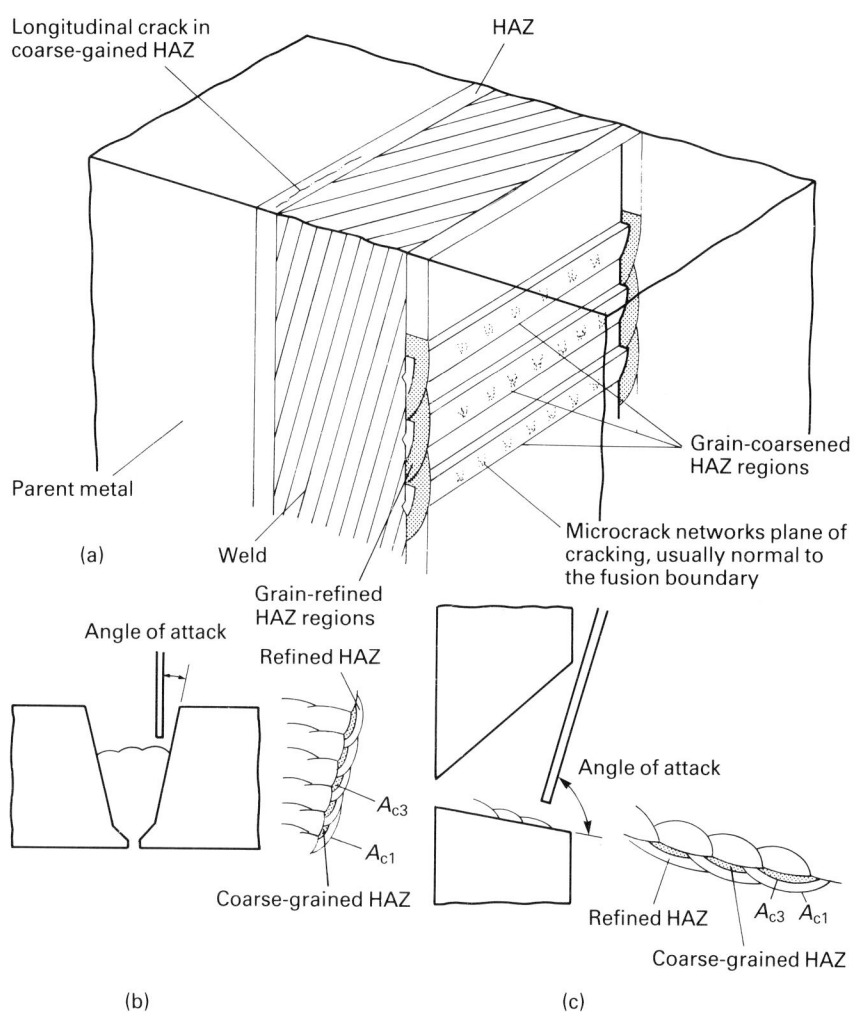

6.2 Angle of attack and HAZ refinement: (a) location of cracks in coarse-grained regions; (b) low angle of attack and high overlap increase refinement of HAZ; (c) high angle of attack and low overlap decrease refinement of HAZ; (d) macrosection of weld with high degree of HAZ refinement; (e) macrosection of weld with low degree of HAZ refinement ((d) and (e) overleaf).

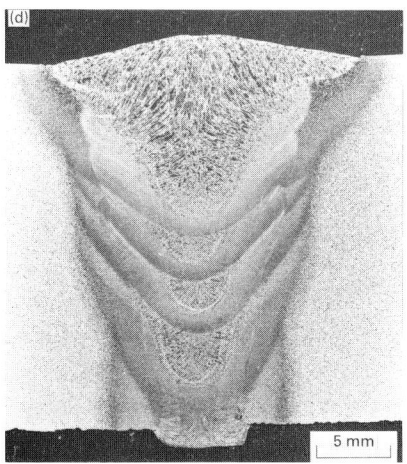 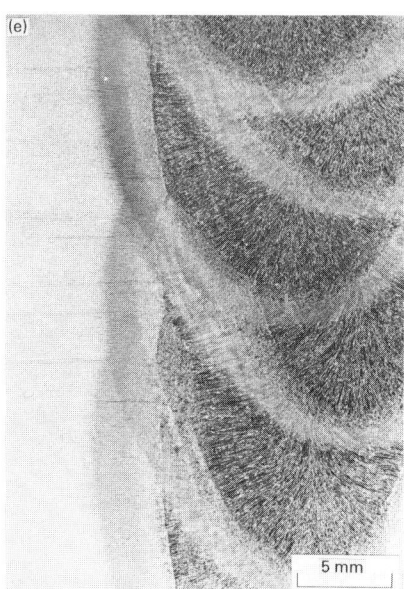

6.2 cont'd

grained, high temperature HAZ of the first. Its own high temperature HAZ should be restricted to the first layer of cladding, where it is harmless.

When depositing structural welds, two possibilities are available for ensuring a fine-grained HAZ. The first, which is most suitable for automatic welding, consists of using a steep-sided preparation, as in Fig. 6.2. The coarse-grained region, in which cracking is most likely, is that which has been heated above about 1000 °C. When reheated by subsequent passes to temperatures between about 800 and 1000 °C this coarse region is refined and, thus, made less susceptible to reheat cracking. When welding a multipass weld using constant welding conditions, the degree of refinement is greatest with a small angle of attack, as when using a steep-sided preparation (Fig. 6.2(b) and (d)). With such a technique, however, conditions must be held constant, because they are on the verge of giving defects such as lack-of-sidewall fusion. With a high angle of attack, as when carrying out the horizontal portions of horizontal-vertical welds (Fig. 6.2(b)) with constant welding conditions, a much greater proportion of coarse grained HAZ results.

The second choice, known as the two-layer technique, was developed for MMA welding,[7,8] although it can be used with semi-automatic and automatic processes. It is suitable for welding in the horizontal position, as well as positionally, but requires different welding conditions for first

Reheat cracking 177

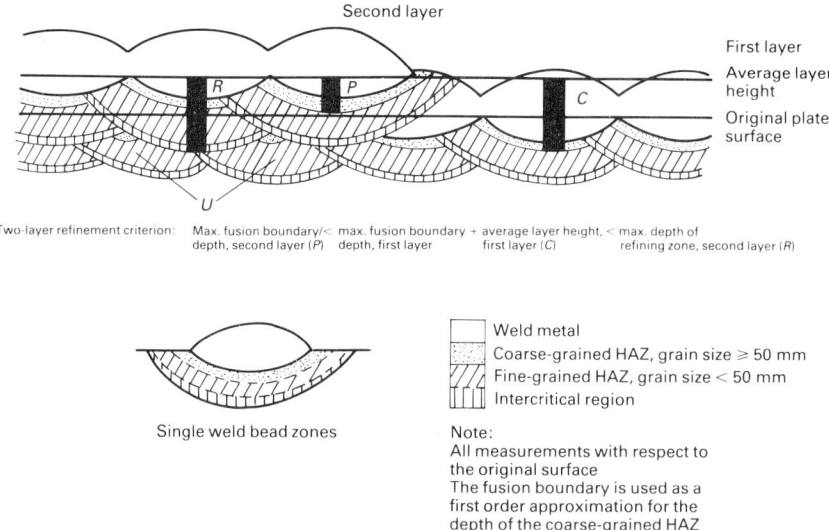

6.3 Sketch showing two-layer technique. Note that the degree of overlap is less than ideal (i.e. ~50% overlap obtained by aiming arc at toe of previous weld bead). 'U' indicates unrefined HAZ regions.

and second layers, as well as close control of weld bead overlap. The two-layer technique, although originally developed for avoiding reheat cracking in the HAZs of susceptible steels, has since been adapted to improve the as-welded HAZ toughness of weld repairs which cannot be given PWHT (Chapter 11).

The principle of the method is illustrated in Fig. 6.3. The weld metal selected should not normally be subject to reheat cracking (e.g. 2.25Cr:1Mo weld metal is used to weld the 0.5Cr:0.5Mo:0.25V steel). A larger bead is used for the second layer than for the first, so that temperatures sufficient to refine the coarse HAZ of the first layer are provided by the second layer, but excessively high (grain coarsening) temperatures do not penetrate beyond the first layer of weld metal into the HAZ of the parent steel. In addition, a controlled weld bead overlap of about 50% is necessary for both first and second layers; this can be achieved in manual welding by pointing the electrode at the toe of the previous bead. It is also important to ensure that the angle of attack is close to 90° for both layers. This necessitates a wider weld preparation than would normally be employed for economical welding.

When using MMA welding, the first layer is usually deposited with 2.5 or 3.25 mm diameter electrodes and the second with 4 mm. Tight control of welding beyond the second layer is not critical to achieve the HAZ refinement needed to avoid reheat cracking, but should be consistent with good welding practice and the level of properties needed. This so-called two-layer technique has been adapted to TIG and, with more difficulty, to MAG welding. A high level of HAZ refinement is more easily achieved in the flat position than with vertical or overhead welding. The degree of preheat required for susceptible steels (normally about 200 °C minimum) is helpful in ensuring a high degree of refinement.

Normalising the weld or cladding deposit, either by full-scale furnace normalising or by a local treatment, will also refine the HAZ grain size and reduce the likelihood of reheat cracking. Full-scale normalising is usually impracticable, but, if carried out, a check should be made that the treatment is not likely to damage the properties of any ferritic weld metal or embrittle any stainless steel cladding by precipitation of the sigma phase. Local normalising is probably most easily carried out on cladding by a TIG remelting of the cladding surface, with the welding parameters selected so that the coarse grained HAZ on the ferritic steel is refined and not re-coarsened.

If less onerous measures are required to reduce the risk of reheat cracking, the simplest is to grind the weld toes to remove any undercut and provide a smooth profile. This treatment is only useful on butt welds where the weld toes at both surfaces can be ground. If done while the steel is still warm from welding, toe grinding can also reduce the risk of hydrogen cracking at weld toes. Peening of the weld toe region may also be helpful, but must be carried out with care to avoid damaging the material.

It has been claimed that an intermediate PWHT for 2 hours at 500 °C can be helpful in some cases. Another option is to heat to the PWHT temperature in two stages with a hold at about 500 °C, sufficient to allow the temperature to equalise, and then heat to the final temperature at the maximum rate permitted by the appropriate specification. This reduces the time at the critical temperature where cracking can occur.

Detection and identification

The detection of reheat cracks is not easy in structural welds and is even more difficult under cladding, because of the different physical properties of the two materials involved. Furthermore, many reheat cracks are relatively short and, near their tips, discontinuous. For both types of reheat crack, appropriate ultrasonic examination is the only reliable test method.

Where reheat cracks come to the surface at the weld toe, MPI or (for cladding) dye penetrant can provide suitable techniques, although grinding is advisable to confirm that the cracks exist. Similarly, to confirm the existence of suspected underclad cracking detected ultrasonically, it is sometimes recommended that the cladding be ground off and the newly exposed surface of the underlying steel examined by MPI or by microscope after suitable preparation.

Unless cavitation is present near the extremities of cracks, differentiation between reheat cracks and hydrogen cracks is almost impossible. In the types of steel which are susceptible to reheat cracking, both hydrogen cracks (see Chapter 5) and reheat cracks are intergranular with respect to the prior austenite grain structure. The presence of cavitation, in either conventional microspecimens – Fig. 6.1(b) – or on fracture surfaces examined in an SEM – Fig. 6.1(d) – will confirm the presence of reheat cracking, although its absence does not rule it out. Cracking in the fine grained regions of the HAZ would usually support a diagnosis of hydrogen cracking. Careful metallographic or fractographic examination should reveal whether any liquation cracking (Chapter 3) is present, rather than (or even acting as an initiation site for) reheat cracking.

Both hydrogen and reheat cracking can be present in the same specimen, as hydrogen cracks can provide suitable notches from which reheat cracks can grow during PWHT. Furthermore, the failure to have detected cracking by NDE before PWHT does not exclude the possibility that some or all of the cracking was hydrogen-induced but could not be detected until residual stresses had been relaxed sufficiently by PWHT (Chapter 5).

References

1. Nakamura, H. et al., *Proceedings of the 1st International Conference on Fracture*, 2, Sendai, Japan, 1965, p. 863.
2. Haure, J. and Bocquet, P., 'Underclad cracking of pressure vessel components', Convention CECA 6210-75-3/303, Creusot-Loire, final report, Sept. 1975.
3. Boniszewski, T. and Hunter, A.N.R., 'Reheat cracking in 2.25%Cr:1%Mo sub merged arc weld metal. Technical note', *Metal Construction*, 1982, **14**, 495–496.
4. British Standard BS1501, 'Steels for pressure vessels: plates: Part 2: Specification for alloy steels: 1988', BSI, London, 1988.
5. Clark, J.N. and Meyers, J., 'Stress relief cracking in high strength structural steel BS4360 grade 55F', *Mat. Sci. Technol.*, 1988, **4**, 610–612. (Despite its title, the steel covered by this paper is not to the Standard quoted, but is a modification which is outside both the compositional and the strength limits of Grade 55F.)
6. Dolby, R.E., Dhooge, A., Sebille, J., Steinmetz, R. and Vinckier, A.G., 'Status on reheat cracking in nuclear vessel steels 1976', ERIW Report 3531/3/77 to CEC, April 1977.
7. Alberry, P.J., Myers, J. and Chew, B., 'An improved welding technique for HAZ refinement', *Weld. Metal Fabrication*, 1977, **45**, 549–553.
8. Alberry, P.J. and Jones, K.E., 'Two layer refinement techniques for pipe welding', 2nd International Conference on Pipe Welding, TWI, London, Nov. 1979.

7 Faults of welding

Faults and defects other than cracks in welds are often considered to be solely under the control of the welder and independent of metallurgical considerations. This is by no means the case and it is sometimes important to understand the metallurgical aspects of the problem, to realise that the solution to a cracking problem might unexpectedly introduce a fault, or that the presence of a welding fault can sometimes cause cracking. Many welding faults, although relatively harmless by themselves, can make adequate inspection for other serious imperfections and defects difficult and sometimes impossible.

This section is not meant to be a comprehensive review of welding faults, defects and imperfections; reference should be made to texts on welding engineering, and particularly on the use of different welding processes. For a brief guide to defects, a TWI slide collection with notes[1] is useful for giving illustrations of defects, their usual causes and suggestions for prevention and rectification.

Welding faults can be divided into three major types: faults of shape, inclusions of solid material and voids.

Faults of shape

Faults of shape are serious if they cause a stress concentration. Besides reducing the fatigue resistance of a joint, they can also provide sufficiently high local stresses to induce cracking – particularly hydrogen and reheat cracking – where local regions of HAZ or weld metal are relatively brittle at some stage. Surface pocking (usually confined to submerged arc welds) and root concavity are relatively harmless in this respect, and excess metal and excess penetration are benign unless the excess metal produces angles sufficiently sharp to increase the stress concentration markedly. However, excess root penetration can also give unwanted and confusing reflections when ultrasonically testing joints.

Faults of welding

7.1 Hydrogen cracking that originated at root of single-sided butt weld in 8.4 mm thick pipe as a result of linear misalignment.

Undercut, lack of fusion and incomplete penetration are all likely to give stress concentrations which can cause cracking. The degree of penetration can be decreased if consumables are dried to very low hydrogen levels to avoid hydrogen cracking; this can sometimes give rise to incomplete root penetration, i.e. a lack of fusion between the weld runs made from opposite sides, or other root defects.

Overlap can either be confused with cracking during surface inspection or can conceal a HAZ crack which would be open to the surface and readily detectable in the absence of this fault.

Linear misalignment, which can produce stress concentrations sufficient to give cracking either during welding (Fig. 7.1), in service, or both, is easy to detect. Angular misalignment is more difficult in this respect, but can be equally harmful both where fatigue resistance is required and where there is a risk of hydrogen cracking (see Fig. 7.2).[2]

Stray arcing is a fault of shape only in that the arc has been struck in the wrong place and has locally heated (and possibly melted) a small amount of the parent steel. Because the time of arcing is usually very short, the effective heat input is small and it can be assumed that the heated steel will have been cooled so quickly that the resultant microstructure will be martensitic, hard and probably brittle. If not already cracked, an arc strike is likely to provide a region where cracking can readily occur in subsequent service. When grinding out, care must be

182 Weldability of ferritic steels

7.2 Pipe joint broken open to show extent of fatigue crack (dark) which had originated from hydrogen crack at root of weld (very dark at bottom) and grown into both 9.4 mm thick pipes. The pipe ends had to be pulled in to make the joint, resulting in severe angular misalignment.

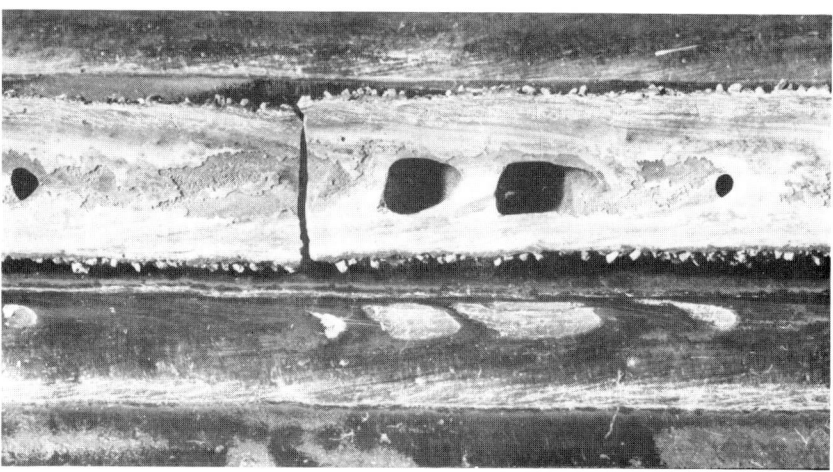

7.3 Surface pocking on submerged arc weld bead showing cavities in slag, which has been turned over to show matching cavities on its underside adjacent to the weld surface.

taken to remove any heat-affected material by going a little way below the original surface – particularly if the component is not to be given PWHT.

When welding steels which have become heavily magnetised, a phenomenon called **arc blow** can occur. This directs the arc away from where it is aimed and even, in extreme cases, extinguishes it periodically. The tendency to arc blow increases with the degree of alloying of the steel being welded, particularly with steels containing nickel. The more prone a steel is to arc blow, the more difficult it is to demagnetise it (the normal cure for severe arc blow).

A problem peculiar to submerged arc welds is that of **surface pocking** (sometimes termed **gas flats**), shown in Fig. 7.3. It is not a defect, as the normal rounded surface of the weld bead is merely flattened as a result of gas pressure in the molten slag burden. The gas is thought to be carbon monoxide given off from the weld metal late in its solidification. The flattening is rarely below the level of the adjoining plate surfaces, whilst sectioning just below the surface never reveals faults within the weld metal. Three factors have been associated with surface pocking: easily reduced oxides (e.g. FeO and MnO) in the flux, the absence of deoxidants from the flux; and the use of fluxes of high viscosity, which makes the escape of trapped gases into the atmosphere more difficult. Certainly the problem has usually been found when using basic fluxes, which give slags of high viscosity, but do not always contain deoxidants, and rarely FeO.

Inclusions

The term 'inclusion' covers any solid material within a weld which is of a different composition and structure from the parent or weld metal involved in the joint. Usually the term refers to welding slag which has been unable to separate out of the weld pool with the rest of the slag. Although inclusions are often referred to as 'continuous' or 'intermittent', their basic causes are sometimes similar, in that they are often associated with lack of fusion faults due to incorrect welding techniques or conditions. Occasionally isolated slag inclusions may result from slag dropping from damaged electrode coverings, or from welding over millscale or oxide which has not been removed from the weld preparation.

If it is required to determine whether an apparent inclusion is of welding slag, or a corrosion product resulting from service, it is possible to use the micro-analytical facilities available on an SEM to determine between these possibilities. However, slag inclusions in metallographic sections often show a dendritic structure, which demonstrates that they have been molten and cannot be corrosion products.

Weld metals and steels normally contain non-metallic inclusions resulting from their sulphur contents and the oxidation and deoxidation reactions which they have undergone. Such inclusions are normally much smaller than welding slag inclusions – those in weld metals are roughly spherical and typically less than 10 μm (<0.010 mm) in diameter. By making certain assumptions about the order in which reactions take place and the specific gravity of the inclusions, it is possible to estimate their volume fraction in a weld metal.[3] Such inclusions are a normal and necessary constituent of steel weld metals. Inclusions close to the fusion boundary may be slightly larger than average, but this is also normal.

It is also possible for metallic particles to become embedded in weld metal or HAZs. In TIG welding, small particles from the tungsten electrode can be included in the weld metal. Although harmless, these show up very strongly under radiography and can interfere with detection of other faults.

Another type of metallic inclusion can occur when metallic particles (either deoxidants or alloying elements) do not fully dissolve in the weld pool. These usually originate from MMA electrode coverings or sometimes from alloying submerged fluxes. Such particles may be harmless, as in the example of a partially dissolved ferro-molybdenum particle, shown in Fig. 1.8 and intended for a high strength weld metal. On the other hand, the area shown in Fig. 1.9 is of a manganese, or ferro-manganese particle which *has* dissolved, but incompletely mixed in the weld metal leaving a harder region, which in sour service could provide a site from which stress corrosion cracking could initiate. Both types of particles result from a weld pool that had solidified too quickly for complete dissolution and homogenisation. They can usually be avoided by increasing the weld pool temperature or its solidification time by such measures as increasing the welding current, reducing travel speed or reducing the width of weaving.

With automatic and semi-automatic welding, it is possible for copper from the contact tip to touch the weld preparation and fall into the weld pool. Unless they interfere with welding, such particles are fairly harmless if they can dissolve in the weld pool (most automatic weld metals already contain some copper from the coating on the welding wire). In some circumstances, however, fine particles abraded from the contact tip appear to be able to work their way through submerged arc slags on to the hot metal surface. Copper that comes into contact with the HAZ or solid weld metal that is hotter than the melting point of copper (1083 °C) can melt and penetrate along the grain boundaries (by a process usually known as liquid metal penetration), solidify there, and effectively embrittle the metal, often causing cracking (Fig. 3.30 and Chapter 3).

Welding over zinc-coated steel can, in some rather unusual circumstances, also give liquid metal penetration and embrittlement, as was mentioned in the last section of Chapter 2. The full conditions for this have not been fully identified, although trapping the zinc so that it cannot escape, as in a twin fillet weld, and welding on to high strength alloy steels appear to be two contributory factors.

A final form of unwanted metal, although not strictly in the form of inclusions, is weld metal spatter. This is unsightly but is not usually harmful, as the spatter is not usually hot enough to heat the underlying metal to a harmful degree.

Cavities

Cavities in ferritic steel welds are usually a result of gas porosity. Shrinkage porosity usually manifests itself as solidification cracking (Chapter 3), except sometimes near the surface in deep, high speed submerged arc welds (Fig. 7.4), in which shrinkage porosity may sometimes be found near the centre-line at or near the weld face. Crater pipes result from a mixture of shrinkage and gas being given off during the final stages of solidification.

Porosity, by itself, normally has a negligible effect on mechanical properties (including toughness) and can readily be detected by radiography, or by eye if it breaks surface. It is, however, undesirable for a number of reasons. Firstly, porosity can mask the presence of more serious defects during NDE. Secondly, its cause might also be a cause of other faults in the weld. For example, porosity due to nitrogen indicates that the weld metal has a high nitrogen content and is likely to have poor toughness (Chapter 9).

Gas causing porosity comes from a variety of sources. Wormhole porosity usually results from contamination of the surfaces being welded, particularly if these are held so close together so that the gases produced by the decomposition of oils, grease and/or rust cannot easily escape except into the weld pool, so that they are trapped by the solidifying grains (or dendrites) of weld metal. In TIG welding, the absence of deoxidants, as when autogenously welding unkilled steels, can lead to porosity.

Lack of shielding is a common cause of both start porosity and uniform porosity. Likely causes are found at the start of an MMA weld if the incorrect technique is used, or if gas shielding is temporarily blown away by wind or a draught or is impeded by spatter on the shielding gas nozzle.

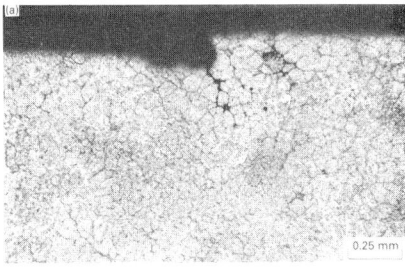 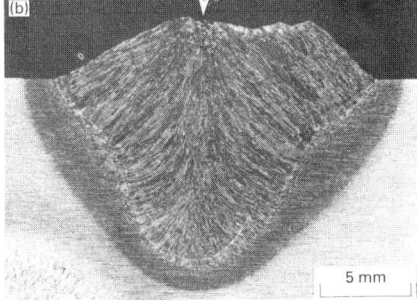

7.4 Fine shrinkage porosity at surface of submerged arc weld bead: (a) detail of porosity; (b) location (arrowed).

Uniform porosity may also be caused by consumables or the workpiece being damp.

Gas porosity is usually a result of the solubility of a gas dissolved in the weld metal falling rapidly as it solidifies, as was illustrated in the case of hydrogen in Chapter 5 (Fig. 5.2). Although hydrogen can give rise to porosity in ferritic steel weld metals, it is not the commonest cause, probably because its solubility on solidification is fairly high in relation to the amount likely to be present in the liquid. Nitrogen, picked up from the air when shielding is inadequate, is a far more usual cause. Nitrogen in excess of about 250–300 ppm invariably gives a porous weld, unless the welding has been carried out under high pressure (hyperbaric welding), when larger quantities can be retained in the solid weld metal. Porosity can also result from a chemical reaction in the molten weld metal between oxygen and carbon to give carbon monoxide:

$$C + O = CO \tag{7.1}$$

This reaction occurs when the oxygen content of the weld metal is high as a result of inadequate deoxidation.

Shielding gases can sometimes be physically trapped and unable to escape from the solidifying weld pool. Argon may be trapped during MIG welding and laser welds are frequently porous, probably because of their rapid solidification.

It is possible, using a micro-analytical technique developed at TWI, to analyse the gases in weld pores as an aid to determining the cause of porosity. The results are unambiguous in some cases (e.g. when Ar is detected from the shielding gas), but if the porosity is a result of excessive hydrogen, the gas found is likely to be methane, resulting from hydrogen reacting with carbon in the steel:

$$C + 2H_2 = CH_4 \tag{7.2}$$

Similar reactions are likely when carbon monoxide or dioxide is initially present in pores.

References

1. Bailey, N. (ed.), 'Faults in fusion welds in constructional steels', TWI Slide Set No. 2, TWI, Cambridge,1986.
2. Boulton, C.F. and Bailey, N. 'The influence of geometric weld defects on structural integrity', in *Proceedings of the Conference on 'Fabrication and Reliability of Process Plant'*, London, 1976, TWI, Cambridge, 1977, 217–226.
3. Bailey, N. and Pargeter, R.J. 'The influence of flux type on the strength and toughness of submerged arc weld metal', TWI Report Series, TWI, Cambridge, 1988.

8 Inspection for defects

Details and techniques of inspection[1-3] are not a subject of this book, but it is necessary to highlight areas where there are interactions between metallurgical behaviour and the ability to detect and identify defects and faults – particularly cracks. Because cracks are forbidden by many standards, there is a tendency to overreact when linear defects are discovered by NDE. Because many fabrications are made to strict deadlines, there is also a tendency to gouge out and repair any 'cracks' without pausing to check what they are, whether they are cracks and, if so, of what type. Only when this is known can the correct remedial action can be taken. It is for this reason that each section of the book dealing with a type of cracking includes brief sections on detection and identification.

Added to the tendency to overreact when cracks are discovered is the fact that NDE techniques are constantly improving, so that smaller and smaller faults and cracks can be detected. However, techniques of fracture mechanics analysis have been developed[4] to assess the safety of allowing structures containing cracks and other faults to go into, or to remain in, service. Such analysis can be used to determine whether possibly expensive and time-consuming remedial measures are needed at all. Fracture mechanics can also be used at an early stage in manufacture to determine what sizes of defects can be allowed without impairing the safety of the structure. This type of analysis also helps to assess the standards of inspection necessary. For fracture mechanics analysis, it is necessary to know the maximum applied and residual stresses and the minimum levels of toughness at various locations in the structure.

It is usually safer to investigate any NDE indications thoroughly and to assess whether it is possible to accept whatever is present – or possibly to reduce its harmfulness by superficial grinding. This is because weld repairs are always liable to introduce further defects – particularly if the repair is carried out hastily under the pressure of impending deadlines or delivery dates.

Inspection techniques

Visual examination, even if aided by magnifying lenses, low power microscopes, MPI or dye penetrant (DP) techniques, is only capable of detecting faults at the surface. The use of MPI with d.c. current gives a slight sub-surface capability, whilst careful visual surface examination may occasionally hint that something is amiss just below the surface. However, many faults and defects lie well below the surface and the most common techniques for their detection are ultrasonic examination and radiography. Several other techniques, such as stress wave (acoustic) emission, have been and are being developed, but their use is at present specialised and they are not common in the fabrication industry.

Visual examination is chiefly used for detecting defects of shape: undercut, under- or over-fill and misalignment. Surface porosity and open cracks (particularly solidification cracks) can also be detected. The use of MPI and DP allows fine cracks of other types at the surface to be detected. One risk, particularly with MPI, is that under certain conditions it is possible to obtain a marked response from weld fusion boundaries, or even the ripples on a weld surface, without anything being amiss. The fusion boundary effect is particularly troublesome, because it mainly results from the inevitable undercut at the weld toe (which admittedly should be controlled), but which could be mistaken for a hydrogen or reheat crack.

Radiography allows bulk faults to be detected. These include slag inclusions, porosity and wide solidification cracks. Fine cracks cannot be detected, except under very favourable circumstances when the plane of the crack is aligned with the X-ray or gamma-ray beam. Radiography will also detect many of the defects and faults found by visual examination, with the benefit that inaccessible areas, such as the root of a fillet weld, can be examined.

Ultrasonic examination is more suited to the detection of linear faults, such as cracks and lack of fusion. Care is needed to use the correct technique, not to identify linear indication as cracks too readily, and not to use too high a sensitivity – which can produce 'defects' out of slight, harmless metallurgical variations in structure.

Inspectability

Not all steels and weld metals can be non-destructively examined with equal facility, particularly when the technique used is ultrasonics or radiography. Without special techniques, welds on ferritic steels made with austenitic stainless steel or nickel alloy consumables are virtually impossible to examine. The fusion boundary acts like a defect (at the very place where defects are likely!), because it separates two materials having widely different physical properties. The only techniques easily applicable to such welds are visual examination and surface examination using DP. Although difficulties with radiography might be expected to be minor, the normal coarse grain size of non-ferritic weld metals causes diffraction effects, which can mask many faults and defects.

Coarse-grained material of any sort will give difficulties with radiography, and inhomogeneity is liable to give problems when using ultrasonics or radiography, particularly when they are used at their most sensitive. A high inclusion content can give problems, particularly with ultrasonics; in fact, ultrasonic techniques are sometimes used to assess the suitability of plate to resist lamellar tearing (Chapter 4) resulting from a high population of inclusions. Further care is also needed because inclusions that have decohered from the matrix give a much greater ultrasonic response than from the same inclusions before they decohered, as they may have been at an earlier examination.

Consequences of metallurgical features

Apart from the difficulties experienced when examining welds made with non-ferritic fillers discussed in the previous section, the major effects to be considered result from hydrogen cracking and residual stresses.

Hydrogen cracks grow very slowly, and it is necessary to allow sufficient time for them to develop before carrying out NDE for this type of cracking. Although the usual delay period is 48 hours, this time requirement is not well based, and recent evidence suggests that the nearer the welding conditions are to the crack/no crack boundary, the more slowly do hydrogen cracks develop. It is, therefore, most unsuitable to suggest that the time can be reduced by ensuring that, for example, a post-heating treatment is given to diffuse out hydrogen. This is because the post-heating is itself part of the procedure that needs to be monitored to ensure that it has not led to hydrogen cracking. Only if PWHT is used, or if hydrogen cracking is known to be *impossible* in the joint, steel and consumables under consideration, can this rule be relaxed.

The residual stress pattern in a completed multipass weld is such that some part of the weld will contain compressive residual stresses (Fig. 1.16), even in regions which, directly after they had been deposited, contained tensile residual stresses and in which cracks could have formed. The final compressive stresses push together the faces of any cracks formed earlier and make them more difficult to detect by any NDE method. The application of PWHT relaxes these stresses and makes cracking easier to discover. In addition, the surfaces of any cracks with access to air during PWHT will become oxidised, thus making the cracks even more detectable. This improvement of crack detectability after PWHT is why specifications always call for NDE after PWHT. It can, however, lead to the incorrect identification of hydrogen cracks as reheat cracks (Chapters 5 and 6).

References

1 Halmshaw, R., *Introduction to the Non-destructive Testing of Welded Joints*, Abington Publishing, Cambridge, 1988.
2 Anon., *Non-destructive Testing Aspects of the Significance of Weld Defects*, Abington Publishing, Cambridge, 1971.
3 Anon., *Procedures and Recommendations for the Ultrasonic Testing of Butt Welds*, Abington Publishing, Cambridge, 1971.
4 PD 6493: 1980, 'Guidance on some methods for the derivation of acceptance levels for defects in fusion welded joints', BSI, London, 1980.

9 Joint integrity

By a broad definition, adequate weldability includes the ability of a weld to serve its purpose as well as whether it can be made free from cracks and defects. These broader aspects of weldability are discussed below.

Strength and ductility

Unlike most other metals, it is easy to deposit ferritic steel weld metals that do not undermatch the strength of the parent metal. In fact, on the few occasions where undermatching is required, it is difficult to do this with all but high strength steels.

There are two reasons for this. Firstly, a ferritic steel weld metal has a very dense network of dislocations in its microstructure.[1] These minute imperfections in the otherwise perfect crystal lattice serve to hamper plastic deformation and thus strengthen the weld metal. The dense dislocation network is a result of the weld cooling rapidly under the restraint of neighbouring cold metal and thus undergoing plastic deformation (warm and cold working). The network is resistant to heating up to PWHT temperatures, but is destroyed at normalising temperatures, so that a normalising heat treatment considerably reduces the strength of weld metals to the levels expected for parent steels of similar composition, grain size and heat treatment.

The second factor in the strength of ferritic weld metals is that their microstructures are relatively fine grained as a result of their transformation from austenite. It is well known that a fine grain size confers increased strength, and a particular microstructure peculiar to ferritic steel weld metals – acicular ferrite – is exceptionally fine-grained without any need for a refining heat treatment.

As in other metal systems, the strength of ferritic weld metal increases as the content of alloying elements is increased. The most important strengthening element is carbon but, because too high a carbon level is often undesirable in a weld metal, other elements are usually added to confer increased strength. The use of the carbon equivalent formula, Eq. [1.2] (which is used for the parent steel in connection with hydrogen cracking, Chapter 5) is a useful way of estimating weld metal strength. For submerged arc weld metal it has been shown[2] that yield and tensile strength both increase linearly with the CE of the weld metal, as in Eq. [1.9]–[1.11].

It is useful for the strength of a weld metal to overmatch that of a parent metal. Overmatching ensures that if a structure is overloaded, the parent metal deforms before the weld. It is, therefore, more important for the yield strength to overmatch than for the tensile strength. Again this is a common situation, because weld metals tend to have ratios of yield strength to tensile strength which are higher than for most wrought steels. Some degree of overmatching is advantageous because in most cases the parent metal (particularly in the wrought condition) is less likely to contain defects sufficient to cause premature failure than weld metal.

However, it is not always easy in high strength steels to ensure overmatching and several instances are known where some undermatching, coupled with a high level of weld quality, has been shown to be acceptable. Deliberate undermatching when welding other types of ferritic steel is only employed when welding the 9% Ni steel for cryogenic applications, because the only weld metals capable of giving adequate toughness at the very low temperatures involved (i.e. down to −196 °C) are nickel alloys of lower strength than 9% Ni steel.

Despite its advantages, overmatching of weld metal strength has some drawbacks. The first of these is that the highest tensile residual stress in an as-welded joint is approximately equal to the yield strength of the weld metal itself. This means that the stronger the weld metal, the higher will be the maximum residual stress, with increased risks of cracking during welding and lower defect tolerances in service. The second disadvantage is that as the strength of weld metal is increased, the risk of its suffering weld metal hydrogen cracking is also increased. A further disadvantage is confined to service in environments where too hard a weld metal may result either in stress corrosion cracking or in hydrogen cracking in service.

Weld metal strength generally increases as the temperature falls and decreases as the temperature is increased until it is vanishingly small as the weld metal starts to melt. In the as-welded condition, the smooth fall in strength with increasing temperature from ambient is interrupted by the effects of strain ageing between about 100 °C and 200 °C, where tensile strength increases with temperature. Above this temperature range, the actual hot strength properties of a weld metal follow those of a parent steel of similar composition, and determine the design stress values, until the temperature is reached where creep becomes important.

Post-weld heat treatment generally reduces weld metal, HAZ and often parent metal strength,[3,4] so that repeated PWHT can reduce strength levels below specification values. However, this is not a marked effect, and most applications standards are sufficiently flexible for some low values to be acceptable.

Ductility is normally associated with tensile properties, but in the case of welds more significance is attached to the ductility in the presence of a sharp notch, i.e. to fracture toughness. In sound weld metals, normal tensile ductility at ambient temperature is not very variable, although it falls slightly as the strength of a weld metal is increased. It is difficult to show any embrittlement by hydrogen in a plane tensile test (although it is marked in a notched test; Fig. 5.8). However, in tests at elevated temperatures, ductility in the as-welded condition is reduced between about 150 °C and 300 °C because of strain ageing.

At much higher temperatures, when the weld metal is austenitic, there is a so-called ductility dip range, which can sometimes give problems during severe hot-forming operations (Chapter 3), particularly if weld and parent metal exhibit ductility dips over different temperature ranges.

Finally, close to the melting temperature, ductility falls to zero. If the ductility is lost at a temperature appreciably lower than that at which strength is lost, the weld metal is liable to liquation cracking (Chapter 3), because of the presence of the liquid films responsible for liquation cracking at the grain boundaries. Such liquid films have measurable strength but no ductility.

Toughness

As important as adequate strength, and harder to accomplish, is the achievement of satisfactory toughness in both HAZ and weld metal. Toughness is conventionally measured in two ways, impact tests on notched specimens – usually the Charpy test on 10 mm square specimens – and by slow strain rate tests using full-thickness notched and pre-cracked test specimens – fracture toughness tests. The former are much cheaper and simpler to carry out but give less meaningful results than fracture toughness tests, particularly as most welded structures are strained at slow strain rates. Both tests exhibit a transition from ductile to brittle behaviour as the test temperature is reduced.

In the case of full-thickness fracture toughness specimens, fracture toughness can be directly related to the behaviour of the structure in service, enabling sizes of allowable and unacceptable defects to be calculated. Although correlations between Charpy and fracture toughness test results exist, they are only applicable to the type of test conditions for which they were devised. That is to say, a correlation for HAZ results will not be applicable to weld metal, nor can a correlation for basic MMA weld metal necessarily be applied to cellulosic electrode welds.

A further disadvantage of impact tests is that they are not sensitive to the presence of hydrogen. Even with fracture toughness tests, special precautions (testing at a very slow straining rate, minimising heating during specimen extraction and machining, and testing as quickly as possible after welding, but with a slow straining rate) are required.

Two aspects of toughness are important: the factors controlling each are not all the same. The more important aspect is resistance to brittle fracture and, unless stated otherwise, it can be assumed that in the present text good toughness refers to a good resistance to brittle fracture, i.e. a low transition temperature between ductile and brittle fracture. Brittle fracture in welds is particularly dangerous because it can lead to failure apparently below the yield stress of the parent steel and without prior deformation, which could have given warning of failure. Fracture toughness, in the form of crack tip opening displacement (CTOD) testing is particularly suitable to this regime, as it is concerned with fracture initiation. The other aspect of toughness is resistance to ductile fracture, or tearing resistance, for which fracture toughness testing techniques, such as J R-curve tests, also are available.

As a gross simplification, brittle fracture is least likely in low strength steels (the term steels including weld metals and HAZ material), in steels having fine grain sizes, in steels that have not been embrittled by hydrogen, strain ageing, secondary hardening and temper embrittlement, and in steels containing nickel. In addition, microstructural features other than grain size can also influence toughness. For example, in low carbon, low alloy steel HAZs a martensitic microstructure is preferable to bainite (Fig. 1.4(d)), although the latter will be softer (i.e. less strong). This effect has led to controls on maximum heat input when welding many strong, tough low alloy steels in order to ensure the formation of martensitic HAZs.

Of the other factors controlling toughness, the grain size is controllable in multipass welding by ensuring reheating to refine the prior austenite grain size, particularly in the HAZ. The value of this is that, on transformation, a fine prior austenite grain size will lead to a fine grain size in the transformation product and better toughness.

HAZ toughness

Before the almost universal use of killed steels, necessary for continuous casting, the major toughness problem in welding was the toughness of the outer, sub-critical regions of the HAZ which were embrittled during welding by strain ageing, the effects of which were removable only by PWHT. This is still a problem when older steels have to be welded for repair or modification, but is otherwise not an important consideration.

Joint integrity

Where good toughness is currently required, steels of low carbon content and designed to give good HAZ toughness are normally employed, so that the major area of concern is the weld metal. Nevertheless HAZ toughness is still an important concern and rules have been formulated[5] to ensure both the selection of steel compositions to give good as-welded HAZ toughness and to ensure that this toughness is maintained by good welding practice. The first few of these rules are applicable generally to C, C : Mn, microalloyed and low alloy steels:

1. Select a low carbon content and, in the case of C and C:Mn steels containing 0.12% C or more, a low CE.
2. Select a fine-grained steel, i.e. Al-treated, as this ensures a fine grain size and good toughness in the lower temperature regions of the HAZ.
3. Ensure that both the total and interstitial (free) nitrogen contents are low.
4. Ensure that the steel has inherently good toughness.
5. Select a clean steel with a low content of non-metallic inclusions or, if this not possible, one where the inclusion shapes are controlled.
6. Use PWHT. Even if the measured toughness is reduced, the overall fracture resistance of the joint will be better because of lower residual stresses.
7. For steels containing ⩾0.12% C, avoid Nb and/or V if high heat inputs (above about 3.5 kJ/mm) are to be used.
8. Microalloyed steels with <0.12% C are more tolerant of microalloying elements at high heat inputs. Steels with a deliberate titanium addition to minimise HAZ grain growth are helpful at high heat inputs. Alloying intended to reduce the transformation temperature from austenite is helpful, unless very fast cooling times ($t_{8-5} \leqslant 5$ s) are expected.
9. For low alloy steels, the alloy content should be sufficient to ensure a low carbon martensitic, rather than a bainitic, HAZ microstructure. Additions of nickel are beneficial.

The following recommendations[5] relate to good welding practice, although some may conflict with the requirements to avoid cracking:

10. Reduce heat input and, if possible, increase welding (travel) speed in order to reduce HAZ width.
11. Avoid very low heat inputs, particularly at weld toes, as they can give fully hardened martensitic HAZs.
12. Adopt welding techniques that maximise refinement of multipass HAZs. These include allowing sufficient overlap of adjacent weld beads, steep-sided preparations (consistent with good fusion) when

welding in the flat and overhead positions or the use of the two-layer technique.[6]
13. Raise preheat/interpass temperatures (but not excessively) to increase the proportion of refined HAZ in the microstructure.
14. Temper beads to refine and temper the HAZ of the final weld run should be adopted with caution, as they require very accurate positioning of the weld toe with respect to the toe of the previous weld run (Fig. 5.11).

The final recommendations[5] concern the fracture toughness testing of welds:

15. Match the test conditions, including the degree of restraint, to the service conditions. This may require the adoption of shallow notches if it is accepted that the most likely defect is a shallow crack and not a through-thickness defect.
16. Although slots, or machined notches, are easier to machine and position accurately than fatigue cracks, the results of such tests will require factoring in order to obtain a test result that can be used to assess the safety of welded structures likely to contain cracks.
17. Check the actual position of the crack tip after testing by appropriate fractographic examination and/or metallographic examination of sections through the fracture.
18. Simulate long term nitrogen ageing (strain-ageing) by heating test specimens for 2 hours at 100 °C before testing, or by using a standardised long time between welding and testing.
19. Consider whether the tests should be made with or without hydrogen resulting from welding. In the former case, the heat treatment to produce nitrogen ageing will inevitably remove some hydrogen and it may, therefore, be prudent to omit it. In the latter case the hydrogen removal heat treatment will also serve as a strain-ageing treatment.

The recommendations about steel composition are aimed at obtaining a soft HAZ or, in a low alloy steel, one that has a low carbon martensitic microstructure rather than bainite. In addition, the presence of non-metallic inclusions has two detrimental effects. Firstly, inclusions lower the resistance of a steel to ductile fracture by aiding the tearing process. This normally proceeds by microvoids forming in the plastic zone ahead of the crack tip and then tearing back to the crack itself – hence the expression 'microvoid coalescence' for this type of failure. Such microvoids form preferentially at inclusions or other brittle features. Secondly, the inclusions provide internal notches at which brittle fractures can start. The larger the inclusion, the easier the initiation, hence few small inclusions are the best to have but, if that is not possible, small rounded inclusions produced by inclusion shape control are less harmful than the long stringers of manganese sulphides and silicates that would otherwise be present.

The welding procedural recommendations include keeping heat input low (but not too low) in order to minimise the width of the HAZ. The rationale for this is to keep the HAZ too small to develop a crack of a critical size, as well as to prevent any growing crack reaching an unsupportable size before it reaches a tougher material (i.e. the parent steel or weld metal). If the heat input is allowed to fall too far (except with low carbon, low alloy steels), there is a likelihood that hard martensite will form and remain untempered, thus leading to a region susceptible to hydrogen cracking, brittle fracture and failure in service through stress corrosion. A check should be made as to whether any mandatory heat input requirements relate to heat input or arc energy. In the latter case, the values should be different for processes having different arc efficiencies, as discussed in Chapter 1.

The recommendations on welding concern the refinement of the HAZ by a number of techniques. Any steel heated to the lower part of the austenitising temperature range will transform to a fine-grained austenite on heating. If the grain size is not allowed to grow, the transformation products on cooling this fine austenite will also be fine-grained and therefore relatively tough. As the times at austenitising temperatures are very short during welding, temperature is the only factor allowing the austenite grains to grow. However, the temperature in the HAZ at the fusion boundary reaches the melting temperature of the steel itself and the presence of the usual grain refiner – aluminium – will not restrict the maximum HAZ grain size at such temperatures, although it does slightly restrict the width of the coarse grained HAZ at the lower temperature end.

A controlled addition of titanium to the steel can prevent HAZ grain growth altogether, but the conditions needed to control this titanium are so difficult to achieve that the process is only available for a limited range of steel compositions and thicknesses.

In a multipass weld, a region of the HAZ of an underlying weld run is always heated by the next run (either adjacent or succeeding) to the lower part of the austenitising temperature range and is thus refined. The extent of this refined region is controlled by the factors that determine the details of how the heat input is distributed within the steel and its HAZ. The factors of importance are the degree of weld bead overlap, the angle of attack of the electrode, the weld preparation and the interpass temperature. The last acts by reducing the temperature gradients within the steel, particularly at lower temperatures, thus widening the region which is refined relative to the coarse HAZ region.

Techniques for maximising the degree of refinement were discussed in

Chapter 6, because a refined HAZ is also required in some steels to achieve a satisfactory resistance to reheat cracking. The two-layer technique (Fig. 6.3) has been developed at TWI as a means of carrying out as-welded repairs to C, C:Mn and Mn:Ni:Mo steels[6] in thick sections when PWHT is impracticable, and is currently being examined for Cr:Mo steels. An alternative repair technique, known as the half-bead technique, involves grinding away half of the first layer of weld beads before depositing the second layer. However, it appears to be a more cumbersome way of achieving a fine-grained tough HAZ than the two-layer technique.

However, the two-layer technique is not the only way of achieving a high degree of HAZ refinement, as steep-sided weld preparations (as used in narrow gap welding) coupled with a shallow angle of attack to the side of the preparation (short of leading to lack of fusion defects) will also gave a highly refined HAZ, as shown in Fig. 6.2.

If it is called for, PWHT gives two advantages. Firstly, it lowers residual stresses, so that less toughness is required for a given acceptable or tolerable defect size. Secondly, the softening of the HAZ of most steels accompanying PWHT improves toughness. Unfortunately, steels containing some of the microalloying elements (particularly Nb and V when present together), as well as Cr:Mo and similar steels, give secondary hardening at the usual PWHT temperatures for C:Mn steels (normally ~600 °C). In such cases, the PWHT temperature can either be reduced to 550 °C or less (with considerably less relief of residual stresses), or it can be increased to 650–675 °C (with some loss of strength) for the microalloyed steels, or appreciably higher for the Cr:Mo type. With alloy steels containing Ni as well as Cr and Mo, care is needed that the PWHT temperature does not exceed the lower transformation temperature for the steel – the A_{c1} – which is lowered by the presence of nickel.

A further threat to the toughness of the HAZs of multipass welds is posed by the presence of local brittle zones (sometimes termed LBZs). These have not yet been fully investigated, although it is known that some are regions of the HAZ which have been heated by a succeeding pass to a temperature just above the A_{c1} (Fig. 1.5). Such heating causes the parts of the steel richest in carbon to transform to austenite. On cooling, these transform back either to martensite or to pearlite, both of relatively high carbon content. Both these transformation products are harmful because of their low resistance to brittle fracture and their presence can give low test results (and the possibility of service failures) in fracture toughness tests. They are unlikely to be detected by conventional Charpy impact tests, which combine fracture initiation and its propagation over a larger volume of metal. Application of PWHT should normally temper, and thus remove the harmful effects of, martensitic LBZs, but is less likely to improve toughness when pearlitic regions are present.

Weld metal toughness

As with other systems, ferritic steel weld metals normally have their toughness reduced as their strength is increased. However, there is a major exception to this rule. If conditions (as discussed in Chapter 1) are suitable for the development of acicular ferrite, the toughness of as-deposited ferritic steel weld metal will improve if the amount of alloying elements is increased until coarse primary ferrite has been replaced by fine-grained acicular ferrite. Within this range, strength and toughness both increase. However, with further alloying, toughness begins to fall owing to the introduction of bainite or martensite when the microstructure is completely, or almost completely, one of acicular ferrite.

The benefits of having a fine acicular ferrite microstructure survive PWHT, but not normalising, either as a separate heat treatment or by succeeding weld passes. However, the fine ferrite produced by normalising by succeeding weld passes (Fig. 1.11f) ensures the retention of good toughness.

When it is necessary for a weld metal to have a yield strength in excess of about 650 N/mm^2, it is not possible to maintain an acicular ferrite microstructure, because the degree of alloying required is such that bainitic or martensitic microstructures result. Except in the Cr:Mo weld metals (particularly 2.25Cr:1Mo) which are bainitic, it is advantageous to aim for a martensitic microstructure of low carbon content. In a fully martensitic weld metal, nickel is the only element known to be capable of improving toughness. However, other factors, such as a high degree of refinement in multipass welds and low inclusion contents are also known to be helpful.

The only way to increase the strength of fully martensitic weld metals in the as-welded condition is to increase the carbon content – and this poses severe problems in regard to avoiding cracking, particularly hydrogen cracking. If heat treatment is to be carried out, then weld metal strength can be increased in some low carbon weld metals by ageing treatments, such as are applied to maraging steel welds.

Strain-ageing

Strain-ageing[7] is a problem to which virtually all weld metals in the as-welded condition are prone, because all contain sufficient nitrogen (1–2 ppm, i.e. 0.0001–0.0002% N) and do not usually contain enough metallic aluminium (i.e. aluminium *not* combined as the oxide Al_2O_3) to form aluminium nitride (AlN) and remove the free nitrogen that causes the problem.

Strain-ageing is an interaction between free nitrogen (and free carbon) in the steel and dislocations in the atomic lattice. When it occurs, it increases strength and reduces ductility and toughness. Heat treatment at a sufficiently high temperature (i.e. PWHT) removes its effects by giving time for the free nitrogen to combine with elements such as iron and manganese, which form nitrides slowly.

When weld metal cools, it is constrained to deform plastically to accommodate contraction and, in cooling through the strain-ageing temperature range (from about 300 °C down to 150 °C or below), it undergoes strain-ageing. This ageing is most noticeable in the weld root, which is not only likely to pick up more nitrogen from the atmosphere than the rest of the weld but is also likely to be more heavily strained. Ageing of a steel (or weld metal) that has been previously strained is known as static strain-ageing.

When a completed weld is strained within the strain-ageing range, it undergoes dynamic strain ageing. Weld metal (and susceptible steel generally) tested within the strain-ageing temperature range shows higher tensile strength and lower ductility than at ambient temperature. The maximum strength does not, however, coincide with the minimum ductility – probably because there is an interaction between strain rate and strain-ageing, so that the faster the strain rate, the higher is the temperature at which strain-ageing has its greatest effect. Various properties measured in the tensile test result from different stages during the test which, because of necking (which concentrates straining within a small length of the specimen and increases the local strain rate within that region), occur at different rates. The maximum load (and tensile strength) occurs before the onset of necking when the straining rate is uniform and relatively slow. Reduction of area occurs during necking, when the local straining rate is increased; hence, the maximum effect of strain-ageing on reduction of area occurs at a higher temperature than the maximum effect on tensile strength. Tensile tests carried out in the strain-ageing range up to the temperature giving the maximum tensile strength show marked serrations (Fig. 9.1) in the stress/strain curve as a result of discontinuous yielding, after the initial yielding.

Similar behaviour is shown by fracture toughness tests. These also exhibit serrated loading curves, with a minimum in (ductile) fracture toughness at about 225 °C. Charpy tests are not normally used to reveal dynamic strain-ageing, because the temperature of maximum effect is a few hundred degrees Celsius higher than in slow strain rate tests. Charpy tests are, however, often used to exhibit static strain-ageing by testing specimens that have been pre-compressed (by typically 5%) and then heated for a short time (usually at 250 °C) before testing in the usual way.

Because of the risk of poor toughness in the weld root caused by strain-ageing, it is customary, where a high level of toughness is required, to gouge or grind out the root region of multipass welds and re-weld with sound weld metal. If this is not practicable (for example, in one-sided welding), particular care should be taken in the selection of consumables

Joint integrity

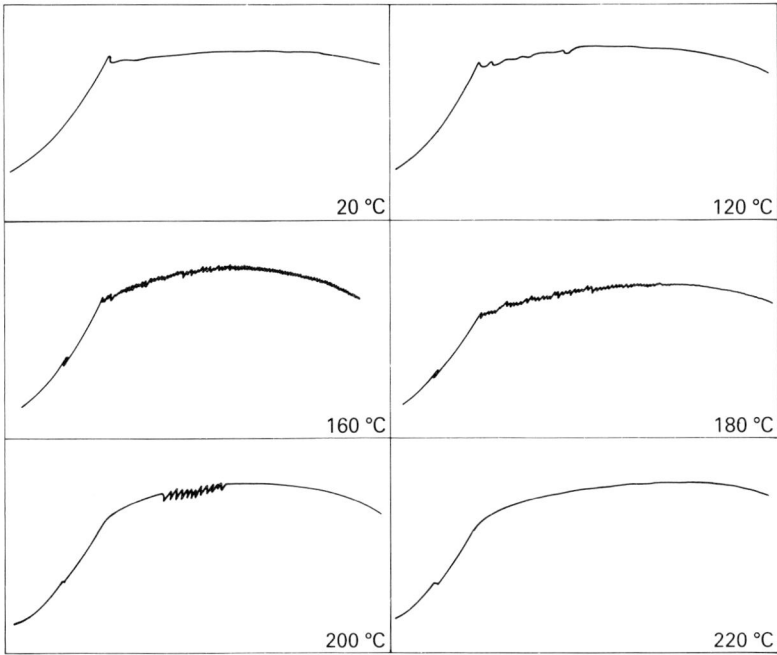

9.1 Stress/strain curves at slow strain rate for Mn:Ni:Mo weld metal showing serrated appearance due to discontinuous yielding from 160 to 200 °C.

for root runs, possibly using a tougher weld metal for the root than elsewhere. One such approach has been the use of submerged arc weld metals containing boron, as 'free' boron can combine with nitrogen and possibly mitigate the effects of strain-ageing, as well as prove an inherently tougher weld metal which will thus retain a higher level of toughness after strain-ageing than a conventional C:Mn composition.

As stated earlier, strain-ageing is not a serious problem with the HAZs of modern steels, which are usually normalised (or given a similar heat treatment) and aluminium-treated if high toughness is required. However it is possible that steels that are of the normalised rolled type may suffer strain-ageing if they do not contain a deliberate addition of titanium. This is because the normalising treatment in such steels can be too short to allow the combination of aluminium with nitrogen; titanium nitride forms much more readily.

Post-weld heat treatment
The application of PWHT is generally beneficial to the toughness of weldments because of the reduction of the residual stress level, regardless

of whether the measured toughness is altered. For a useful reduction in residual stress levels within an acceptable time of PWHT, a minimum nominal temperature of 600 °C should be used with conventional C and C : Mn steels. This temperature of 600 °C (at the rate of 1 hour per 25 mm thickness) reduces residual stress levels to about 30% of the parent plate yield strength.

PWHT at 600 °C is likely to improve the measured toughness of HAZs, unless they contain elements such as Cr, Mo, Nb and V, which are likely to give secondary hardening. For such steels, higher temperatures, generally at 650 °C or higher, are recommended and normally specified by applications standards.

Although a single PWHT is normally sufficient for any fabrication, there are some circumstances where two or even more heat treatments are necessary. This can happen when the fabrication is too large to fit inside a furnace, so that the component requires heat treating in sections, and the overlapping sections receive two treatments. Also, if repair welding is required to rectify defects found after the original PWHT, a subsequent treatment will be required for the repair welding. It is not unknown, indeed, for several repairs and heat treatments to be carried in isolated cases. Each PWHT will reduce strength slightly, and guidance on the likely change in properties can be found in ref. 3 and 4. Standards such as the British Pressure Vessel Standard, BS5500,[8] can make allowance for such occurrences.

Where two different steels require different PWHT temperatures, the temperature should be selected for the more important component and biased towards the less important, as discussed in Chapter 2. In cases of extreme incompatibility, it is possible to butter the steels requiring the higher temperature with two layers of weld metal, PWHT at the appropriate higher temperature, complete the joint and give the completed joint a PWHT at the lower temperature. This lower temperature PWHT will not harm the higher temperature steel.

For weld metals, PWHT does not always improve measured toughness. Deterioration in toughness may be a result of residual austenite in the weld metal transforming on PWHT to martensite or carbides, or to secondary hardening by alloying or microalloying elements, possibly diluted into the weld metal. However, the removal of static strain-ageing damage, coupled with the reduction of residual stresses, is likely to improve the overall joint toughness and the PWHT will also prevent dynamic strain-ageing in service at slightly elevated temperatures.

A recently discovered problem associated with PWHT is the finding of poor toughness in weld metals and HAZs accompanied by intergranular fracture. This appears to be associated with phosphorus segregation, and thus to be a type of temper embrittlement (see next section and Chapter

Joint integrity

10). However, the steels and weld metals involved are not usually of the type associated with temper embrittlement, and some further ingredient appears to be necessary. This may be that one or more microalloying elements (for example, titanium) could be acting to remove soluble carbon and thus allow the segregation of temper embrittling elements to the prior austenite grain boundaries.

Temper embrittlement

This form of embrittlement more often results from service at elevated temperatures than from PWHT, and is dealt with in Chapter 10. However, some steels, particularly alloy steels such as the Ni:Cr:Mo, and 3–9% Ni types, with relatively high manganese and silicon contents (as are sometimes met with in castings) and also their weld metals and other (C:Mn) types – probably with moderately high titanium contents – can be susceptible to temper embrittlement. Embrittlement can be so marked that heavy sectioned components in these steels can suffer embrittlement during PWHT unless the rate of cooling from the heat treatment temperature is relatively fast. Examples of such problems are in heavy-sectioned components in 9% Ni steel, on the rare occasions when it requires PWHT, and in some high strength Ni:Cr:Mo castings, where quenching may be necessary, even though the residual stress due to quenching may be almost as harmful as those left from welding.

Hydrogen embrittlement

Hydrogen embrittlement reduces the ductility of most ferritic steel HAZs and weld metals to some degree after welding. For a given weld hydrogen level, embrittlement is most marked in the as-welded condition and in thick sections, where the time for diffusion of all the hydrogen from the centre of a section may take many years.[9] The hydrogen content, particularly near free surfaces, is considerably reduced by PWHT, although an appreciable proportion of that left after welding may still be present in the centre of welds thicker than about 50 mm.

When assessing the toughness of as-welded structures by fracture mechanics testing, i.e. when using slow strain rate tests (as opposed to Charpy impact tests, in which the effect of hydrogen is negligible), consideration should be given to whether the structure sees its full loading as soon as welding has been completed (when most hydrogen still remains), or whether the structure will be in an unloaded condition for several weeks or months, when much hydrogen will be lost. Extreme examples of these two cases are cantilever bridge construction, when the loads may be higher during construction than in service, and submarine building, when the hull of the boat may be on the slipway with preheating mats switched on for several months and with subsequent fitting out taking several months more before seagoing trials are undertaken.

In the first case, any fracture toughness tests should be carried out as soon after welding as is possible in order to retain the maximum amount of hydrogen. In the second case, some delay between welding and testing is advantageous but, if this is not possible, low temperature heating (below about 300 °C) may be used to remove some hydrogen from the weld before testing. Reference 9 should be consulted to estimate suitable times and temperatures for diffusing hydrogen out of welds.

Hardness

The hardness of weld metals and HAZs parallel their yield and tensile strengths. Equations [1.12] and [1.13] have been developed for estimating these properties for weld metals and HAZs, respectively. For the tensile strength of parent steel:

$$TS_{steel} = 3.05 \, HV - 221 \qquad [9.1]$$

where HV represents Vicker's hardness numbers and TS the tensile strength in N/mm^2. It is likely that Eq. [9.1] is applicable to weld metals and HAZs as well as to parent steels. No simple relationship can be developed between the yield strength and hardness of parent steels because there are several relationships, depending on the thermomechanical history of the steel.

Although HAZ hardness has been related to the risk of hydrogen cracking, its use for post-weld checking of multipass welds for this purpose is not recommended, because the so-called 'critical hardness' varies with too many factors (Chapter 5 and ref. 9) and the hardness of any HAZ is usually reduced by the tempering it receives from the subsequent weld runs. However, weldment hardness is important when the weld is to undergo service in various corrosive environments (particularly solutions containing H_2S) and when used in high pressure hydrogen service. This is further discussed later and in Chapter 10.

Creep resistance

Creep behaviour is a specialised subject, which is only lightly touched on in the present text. In ferritic steels, creep resistance is conferred by fine particles of complex carbides of chromium, molybdenum and sometimes vanadium. For optimum creep resistance, appropriate heat treatment is necessary and in welded structures this usually requires a conventional PWHT at a temperature around 700 °C.

Creep-resistant steels are usually welded with fillers giving weld metals of roughly matching composition, although the 0.5%Cr:0.5%Mo:0.25V steel is normally welded with 2.25%Cr:1%Mo fillers because Cr:Mo:V steel and weld metal are susceptible to reheat cracking (Chapter 6). The creep-resistant steels for welding are usually of relatively low carbon content (usually <0.15%) and, although the weld metals are normally lower in carbon content, the difference is not great because weld metals need at least 0.05% C to achieve adequate creep resistance.

Creep failure in service may occur in a number of locations, but the most damaging is in the lower temperature HAZ and adjacent parent steel (so-called Type IV failure) because, even if the creep damage is cut out and re-welded, the HAZ and nearby parent steel will have suffered so much creep damage that failure may be expected after a relatively short life after repairing such damage by welding.

At high temperatures, austenitic stainless steels and some nickel alloys have superior creep resistance to ferritic steels, so that in complex high temperature constructions such as power station plants, there will be a need for transition joints between the ferritic steels and these other metals. These need careful planning and design because of the differences in the coefficients of thermal expansion between the different metals; also the risk of carbon migration across the fusion line between the two materials gives risks of local precipitation and embrittlement. Fillers are usually of the nickel alloy type, preferably with expansion characteristics intermediate between ferritic and austenitic steels.

Corrosion resistance

Two aspects of corrosion require some consideration, and both are further discussed in Chapter 10. One is simple corrosion, the other is corrosion involving cracking, i.e. stress corrosion.

Simple corrosion occurs in gases and in liquids, although in the absence of moisture, significant oxidation (i.e. gaseous corrosion) only occurs at high temperatures. Such oxidation is best resisted by additions of chromium to the steel, with matching additions in the weld metal. No particular precautions are needed with regard to welding, other than the usual ones when welding chromium steels.

Corrosion in moist air – away from salt-laden marine atmospheres – has led to the development of special weldable weather-resistant steels containing small additions of elements such as chromium, copper and sometimes phosphorus. It should be emphasised that such steels have no better resistance to corrosion in marine atmospheres or in salt water than conventional C:Mn steels.

Because of the alloying additions, the weldability and strength of weathering steels is similar to the medium strength C:Mn structural

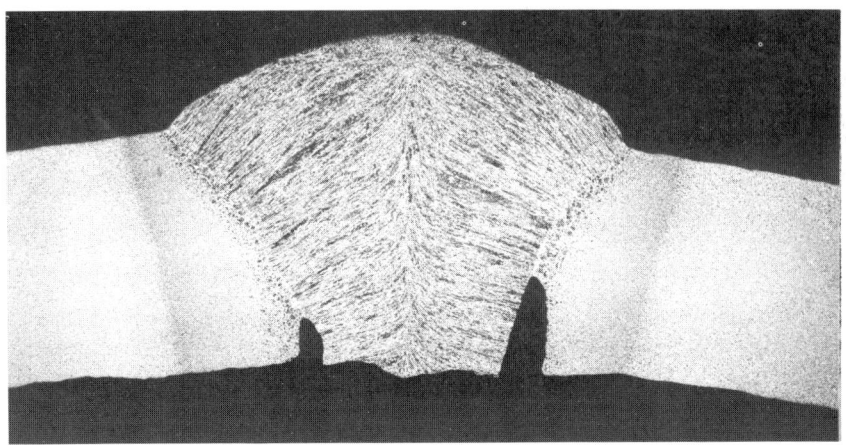

9.2 Selective 'tramline' corrosion in coarse-grained HAZ of a single-pass weld in 8 mm thick pipework.

steels. They gradually build up an oxide coating which is adherent and resists further corrosion so that they may be used unpainted. If painted, the paint should last longer than it would with conventional steels before re-painting is needed. When welded, any welds exposed to the atmosphere should be made with appropriately alloyed fillers, otherwise the welds will not only corrode perceptibly faster but their location will always be apparent from the different colour of their oxide coating.

Aqueous corrosion involves electrolytes in the water, so that anodes and cathodes are always present. A small anodic area in electrical contact with a larger cathode will suffer accelerated corrosion, whereas a large anode will protect a smaller cathode from corrosion without having its corrosion rate unduly increased. Thus, small zinc anodes will be severely corroded in protecting the hull of a ship from corrosion, whereas a steel hull of a ship will not corrode perceptibly faster if it is in electrical contact with a bronze propeller, which it is helping to protect from corrosion. Even in the absence of a second metal, the mill scale on wrought steels is cathodic to the steel itself and can give rise to problems in welded fabrications. If the mill scale is present on the bulk of the steel, it will be absent from the HAZ and weld metal, which will consequently corrode faster in water.

Anode/cathode reactions can also take place at welds because of small differences in composition between weld and parent metals. These can work both ways; for example, certain acidic mine waters can give rise to marked preferential attack ('tramline corrosion') in the HAZ (Fig. 9.2), whereas in other circumstances preferential corrosion of the weld metal

Joint integrity

9.3 Selective corrosion of a weld metal.

(Fig. 9.3) can occur in seawater. Even in the absence of compositional effects, preferential corrosion of the weld region can occur in as-welded joints. This is because the higher tensile residual stresses allow corrosion to proceed slightly faster than in the less highly stressed parent steel.

As implied earlier, cathodic protection (either using anodes or impressed currents) can be used to avoid serious corrosion in seawater. One reason for this is that the galvanic currents deposit a chalky, protective film on the steel surface. However, this chalking only occurs in water more than a few degrees above freezing; in very cold Arctic waters, the film is not deposited and some protection is lost.

A special type of corrosion occurs in de-aerator vessels which are subject to corrosion by water at high temperatures. If the vessel has not been given PWHT, a type of pitting corrosion can occur. This is serious because the pits can develop into multiple cracks which can penetrate the vessel wall.

Corrosion in non-aqueous liquids such as molten metals is a specialised topic, largely outside the scope of this book. In molten lead or zinc a steel needs to be essentially free from silicon, and suitable low silicon consumables are available for this application.

In addition to normal corrosion, several types of stress corrosion can affect welded ferritic steels. The oldest to be discovered is known as caustic cracking (Fig. 9.4), which can occur in boilers and other vessels producing water at high temperatures together with free alkalis (usually caustic soda, NaOH). Problems are likely above temperatures ranging from 50 to 80 °C in the as-welded condition, as the content of alkali in the water is increased. Post-weld heat treatment allows temperatures about 25 °C higher than these to be used, as residual stresses add to the service stresses needed to produce cracking, so that cracks tend to occur at weld areas and in as-welded vessels. It should be noted that local evaporation of water can lead to quite high concentrations of alkali, particularly in crevices.

Currently, the most worrying type of stress corrosion cracking is that due to hydrogen, usually picked up from sulphide solutions, although it may also occur in cathodically protected structures if the level of protection is too high. Hydrogen sulphide (H_2S) poisons the oxide surface and allows hydrogen to diffuse into the steel itself. This can give rise either to cracking at relatively low hardness levels if the inclusion population is of the wrong type (HIC cracking) or, with hardness levels exceeding 22 HRC (approximately 248 HV) and an acidified H_2S solution, to the cracking

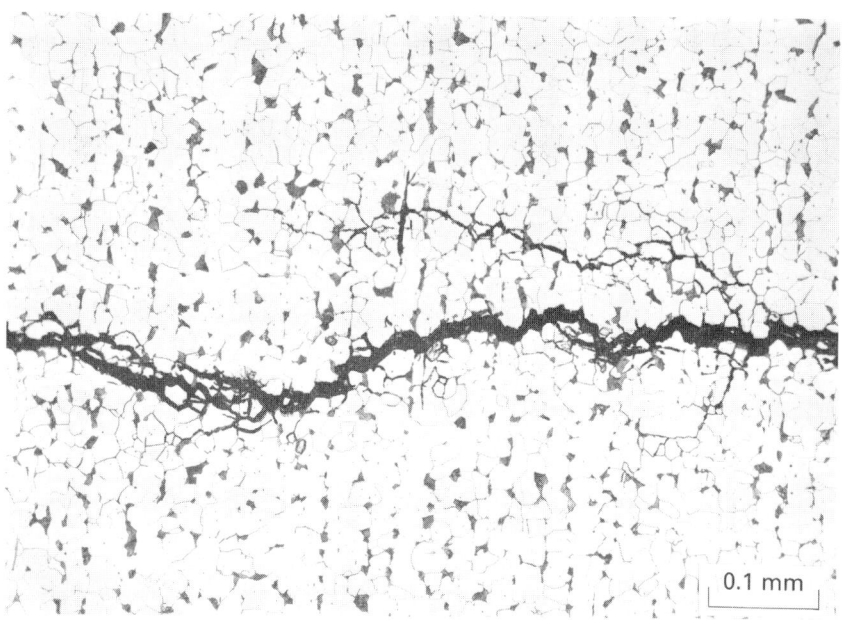

9.4 Typical caustic cracking in parent steel; note transgranular nature of cracking.

following a similar pattern to that encountered with hydrogen cracking during welding, as described in Chapter 5.

Conditions liable to give cracking are frequently met in oilfield systems when the oil goes sour. As high strength steels (e.g. up 700 N/mm² minimum yield stress) are used for valves and fittings, the need to meet the severe maximum hardness requirement makes PWHT mandatory when welding such steels,[10] and even then the hardness may be difficult to control.

Another form of stress corrosion can occur in liquid ammonia, when a trace of water (10–500 ppm, containing low oxygen levels of about 15–100 ppm) is present. Although cracking can be prevented by deliberate additions of water (>0.1%), problems can occur in ammonia storage vessels (very large spheres, which are not amenable to PWHT), when ammonia evaporates and re-condenses within the sphere – but without sufficient water to inhibit cracking.

Other types of stress corrosion cracking may occur, but they are of very specialised application, and are not dealt with here. Any type of stress corrosion cracking which is sensitive to hardness can be initiated at local hard spots, such as were described on p. 21, 22 in Chapter 1.

References

1 Wheatley, J.M. and Baker, R.G., 'Factors governing the yield strength of a mild steel weld metal deposited by the metal arc process', *Brit. Weld. J.*, 1963, **10**, 23–28.
2 Bailey, N. and Pargeter, R.J., 'The influence of flux type on the strength and toughness of submerged arc weld metal', TWI Report Series, TWI, Cambridge, 1988.
3 Gulvin, T.F., Scott, D., Haddrill, D.M. and Glen, J., 'The influence of stress relief on the properties of C and C–Mn pressure vessel steel plates', *J. West Scotland Iron Steel Inst.*, 1972-3, **80**, 149–175.
4 Lochhead, J.C. and Speirs, A., 'The effects of heat treatment on pressure vessel steels', *J. West Scotland Iron Steel Inst.*, 1972-3, **80**, 188–219.
5 Dolby, R.E., 'HAZ toughness of structural and pressure vessel steels – improvement and prediction', *Weld. J.*, 1979, **59**, 225s–238s.
6 Jones, R.L., 'Development of two-layer deposition techniques for the MMA repair welding of thick C:Mn steel plate without PWHT', TWI Res. Report 335/1987, TWI, Cambridge, April 1987.
7 Baird, J.D., 'The effects of strain ageing due to interstitial solutes on the mechanical properties of metals', Metallurgical Review No. 149, *Mat. Metals*, 1971, **5**(2), 1–18.
8 British Standard BS5500: 1991, 'Unfired fusion welded pressure vessels', BSI, London, 1991.
9 Bailey, N., Coe, F.R., Gooch, T.G., Hart, P.H.M., Jenkins, N. and Pargeter, R.J., *Welding Steels Without Hydrogen Cracking*, Abington Publishing, Cambridge, 1993.
10 NACE Standard MR0175-92, 'Standard recommended practice for sulfide stress corrosion resistant-metallic materials for oilfield equipment', NACE, Houston, Tx, 1992.

10 Service problems

This chapter considers how service problems impact on welding requirements; it is not aimed at giving fully detailed descriptions of the problems involved, although sufficient information is given to appreciate when and why precautions are necessary.

Fatigue and corrosion fatigue

Fatigue is still the most common mechanism of failure of welded joints.[1] This is because the stresses needed to produce fatigue failure are much lower than for simple tensile failures and also because fatigue is very sensitive to stress concentrations – which most welds contain in the form of both shape changes and welding faults (e.g. Fig. 5.7). Fatigue of welds is so much a result of built-in stress intensifiers that metallurgical considerations are unimportant and the long-life fatigue strength of weldments in high strength steels is no better than of those in mild steel. Only if short fatigue lives can be accepted, are high strength steels of advantage. The fatigue strengths of welded joints depend essentially on the joint type, as exemplified by standards such as the British Code of Practice for fatigue in the design of bridges.[2]

Good fatigue resistance is obtained by avoiding defects and blending the weld profile to a smooth contour. This is possible for many butt welds but not for fillet welds and those single-sided butt welds whose undersides are inaccessible. Various grinding and re-melting techniques are available for improving fatigue resistance, but when applied to fillet welds they merely move the site of cracking from the weld toe, where detection of cracks is easy, to the weld root, where it is not.

Resistance to corrosion fatigue is lower than for fatigue in air, but follows similar rules. Cathodic protection can restore fatigue strength levels to the values in air, although care must be taken not to over-protect, because of the risk of introducing hydrogen into the steel (particularly at welded joints) and thus causing embrittlement.

Service problems

Corrosion

Several aspects of corrosion were dealt with in the previous chapter. Corrosion is a major cost to industry – particularly as weldments tend to corrode faster than parent steels. Some of this is due to the residual stresses locked up in welds and some is due to inevitable differences in composition between weld and parent metals leading to galvanic effects.

Once a material has been chosen, ways of minimising corrosion include protective coatings, galvanic protection and avoiding water traps when components are subjected to occasional wetting. It is not appropriate to detail the benefits of different schemes – only to point out any complications and precautions that need to be taken when welding.

If a steel is pre-coated, either primed or fully painted, the coating will be damaged by welding and will need to be made good for some distance from the weld. Advice should be taken as to whether the coating needs to be removed *before* welding, because welding over a thick paint film (as opposed to a weld-through primer) may lead to hydrogen cracking and/or porosity. Welding on to galvanised or other zinc coatings is normally acceptable, Chapter 2), apart from any extra precautions which may be necessary to deal with the fumes, unless the zinc can build up in a confined space, such as between two opposing fillet welds, or if the steel is of very high strength. Welding on to carburised or nitrided coatings is similarly acceptable, although local hardened areas on the HAZ can be expected.

Stress corrosion

As discussed in the previous chapter, several types of stress corrosion are possible in ferritic steels. The most widespread of these is when sulphides are present, particularly in acidic solutions, and hydrogen can enter the steel and cause a type of cracking. The risk of cracking in service increases markedly with the maximum hardness, so that less and less H_2S is needed to give cracking as the hardness is increased. An instance is known where an alloy steel plate was flame cut and stood outside in a damp, marine climate. Some days later, the flame-cut region was found to have completely cracked off, owing to its exceptionally high hardness of around 600–700 HV, rather than to the presence of H_2S.

When the conditions are acid and the liquid saturated with H_2S, then the maximum acceptable hardness in the parent steel or weldment is 22 HRC (approximately 248 HV). In addition, alloy steels need to be given a PWHT at a temperature exceeding 620 °C.[3] There have been moves to relax this requirement when conditions are less severe, but such relaxations have not yet been universally accepted. Examples where some relaxation may be sought are under biological fouling in seawater and other environments where the acidity and/or sulphide concentration are less severe than assumed for the NACE requirements.[3] Conditions can also be relaxed on the outside of pipework carrying H_2S solutions, where the hydrogen content will be lower near the outside surface than near the inside. In the latter case, some specifications (e.g. BS4515)[4] allow the maximum hardness to be increased from 250 HV on the inside of pipework to 275 HV on the outside (with a possible increase in the future to 300 HV).[5]

In other cases where stress corrosion cracking is likely, the application of PWHT to give tempering of any particularly hard HAZ regions (or weld metal hard spots), coupled with appreciable reduction of residual stresses, is usually adequate.

Loss of toughness in service

Strain-ageing

Strain-ageing is usually a more serious problem during welding itself than in subsequent service at moderately elevated temperatures, where it can only cause concern in the as-welded condition. At the temperatures involved (150–300 °C), the parent steel and all regions of weldments are operating at temperatures where the likely mode of fracture is ductile and not brittle. Nevertheless, at the most sensitive temperature (about 225 °C), the tearing resistance of a susceptible (as-welded) material can be roughly half of its ambient temperature value.

Although steels can be made immune to strain-ageing by immobilising free nitrogen and free carbon with aluminium, suitable alloying additions and/or the correct heat treatment, this is not possible with most weld metals in the as-welded condition because they do not contain sufficient aluminium (if any); if they did, sufficient time is not usually available in the welding cycle for the formation of AlN.[6] Strain-ageing in service can be avoided by the application of PWHT, if practicable, otherwise it is advisable to select consumables and welding conditions that will give a sufficient margin to allow for the lower ductile toughness at elevated temperatures.

Temper embrittlement

Temper embrittlement is a serious form of embrittlement suffered by alloyed steels (i.e. steels that do not contain *free* carbon) and containing certain impurities when they are exposed to temperatures within the range 350–600 °C. The impurity elements P, Sb, Sn and As are most damaging if the alloying elements Mn and Si are relatively high in content. The impurity elements segregate to the prior austenite boundaries at temperatures below about 600 °C and thus embrittle the steel, raising its transition temperature – below which it fractures intergranularly. The embrittlement is fully reversible if the steel is reheat-treated at a temperature above about 600 °C.

The amount of embrittlement is usually assessed by measuring the shift in Charpy transition temperature (usually the 50% shear FATT – fracture appearance transition temperature) before and after an embrittling heat treatment, such as the **step ageing** heat treatment, which consists of longer times at successively lower temperatures. One such treatment[7] consists of the following stages:

593 °C, hold for 1 hour, furnace cool to:
538 °C, hold for 15 hours, furnace cool to:
523 °C, hold for 24 hours, furnace cool to:
495 °C, hold for 48 hours, furnace cool to:
468 °C, hold for 72 hours, furnace cool to:
415 °C, remove from furnace.

Other authorities favour longer times (60 and 125 hours, respectively) at the two lowest temperatures. Charpy specimens are then prepared and tested either to determine the transition temperatures with and without step ageing or to ascertain whether the aged steel meets some arbitrary requirement, such as 55 J at 10 °C.[7]

In comparing the results of such tests with the chemical composition of the steel, two compositional parameters have been found useful. One is the sum of the manganese and silicon contents, (Mn + Si), in mass %. The other is the factored sum of the impurities, the factor varying slightly from one steel to another. The factor for the 2.25Cr:1Mo steel is termed $\overline{X}$ and is given in Eq. [1.29a].[7]

Factors for other steels are similar but vary in the weight given to the elements Sb, Sn and P (the least embrittling element is always As). Figure 10.1[8] shows the degree of temper embrittlement on a plot of (Mn + Si) against the impurity parameter. Other workers multiply (Mn + Si) with a simplified version of the impurity parameter to give the J-factor[8] as in Eq. [1.27], and recommend a maximum value, such as 200, for a particular application.

Because the $\overline{X}$ and J parameters are specific to particular steels, a more general parameter, P_E, has been developed for both weld metals and parent steels,[9] Eq. [1.28]. This study also claims that copper (a frequent constituent of weld metals originating from the copper coating on automatic welding wires) has no effect on temper embrittlement. In multipass weld metals, embrittlement is confined to the columnar and coarse reheated regions, worsening with the coarseness of the grain size.[9]

The steels most at risk of temper embrittlement are the 2.25Cr:1Mo

214 Weldability of ferritic steels

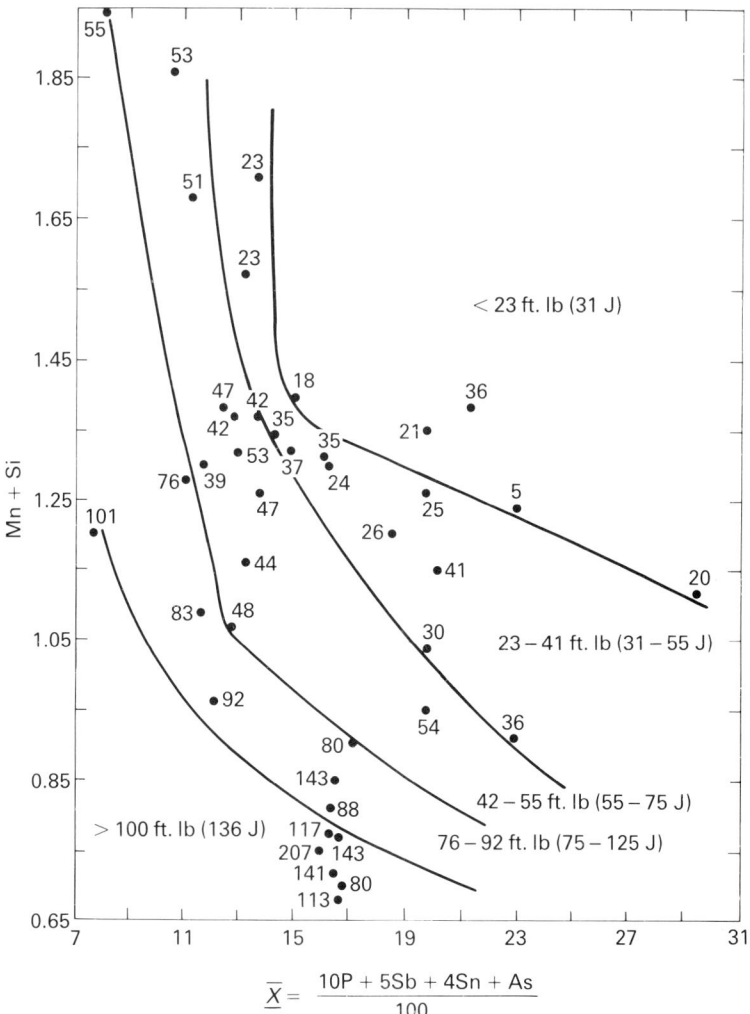

10.1 Temper embrittlement of 2.25Cr : 1Mo weld metal from ref. 7. Toughness (in ft lb; 1 ft lb = 1.357 J) of different weld metals after step ageing in a plot of (Mn+Si) against X.

and 3Cr:Mo steels, nickel steels containing 3–9% Ni and many of the Ni:Cr:Mo high strength steels. Of these, only the Cr:Mo type are regularly used in high temperature service, although turbine rotor steels in some Ni and Ni:Cr:Mo steels can also be used at high temperatures and may sometimes require welding. It has been shown that simple C and C:Mn steels with no carbon (or, strictly speaking, no *free* carbon) can also be prone to temper embrittlement.[10,11] There is a remote possibility that such steels could lose their free carbon in service at high temperatures (e.g. by graphitisation – see a later section) and thus become liable to temper embrittlement.

If a component has been embrittled in service, precautions are necessary to prevent failure, particularly if welding is needed for modification or repair. This is usually effected by maintaining the item at a moderately elevated temperature (e.g. above 150 °C) during and after welding and until PWHT can be applied. Particular care is needed if other embrittling or weakening mechanisms (such as hydrogen embrittlement or creep damage) are also likely to be present. Application of PWHT will remove the temper embrittlement – although it will occur again in service if conditions are unchanged.

Hydrogen embrittlement
Hydrogen embrittlement during welding was discussed in Chapter 5 and, in more detail, in ref. 12; precautions to be taken when testing the toughness of steels and weld metals in the presence of hydrogen were outlined in Chapter 9; stress corrosion effects due to pick-up of hydrogen processes are also described in Chapter 9 and earlier in this chapter, and the phenomenon of hydrogen attack (or hydrogen damage) in a later section of this chapter. The present section discusses how hydrogen embrittles welded steels when the steel is used to contain gaseous hydrogen at high pressure. Two cases are considered: ambient and high temperature behaviour. The two differ because hydrogen from gaseous, molecular hydrogen (H_2) cannot enter dry oxidised steel at temperatures below about 250 °C but can at higher temperatures.

Hydrogen gas can, therefore, be contained in steels at about ambient temperature – provided the stresses are sufficiently low. If the steel is stressed too highly, the oxide film can crack and hydrogen can enter the steel through the exposed unoxidised steel surface. For steels of low and medium strength levels, this is only likely to occur when the steel starts to deform plastically – which should not happen in service. With very high strength steels, differences in elastic properties of the steel and the oxide mean that the strain required to cause cracking of the oxide can occur before the yield strength of the steel is reached, which is an unsafe situation. Such conditions are most likely at welds because of residual stresses and the inevitable stress concentrations.

At elevated temperatures, steels can readily absorb hydrogen from the molecular state. The amount of hydrogen absorbed increases with the partial pressure of hydrogen and the temperature (Fig. 10.2). Provided these parameters are not so high as to lead to hydrogen attack, there are no problems at elevated temperature, because steel is not embrittled by hydrogen at temperatures of 250 °C and higher. However, when the steel

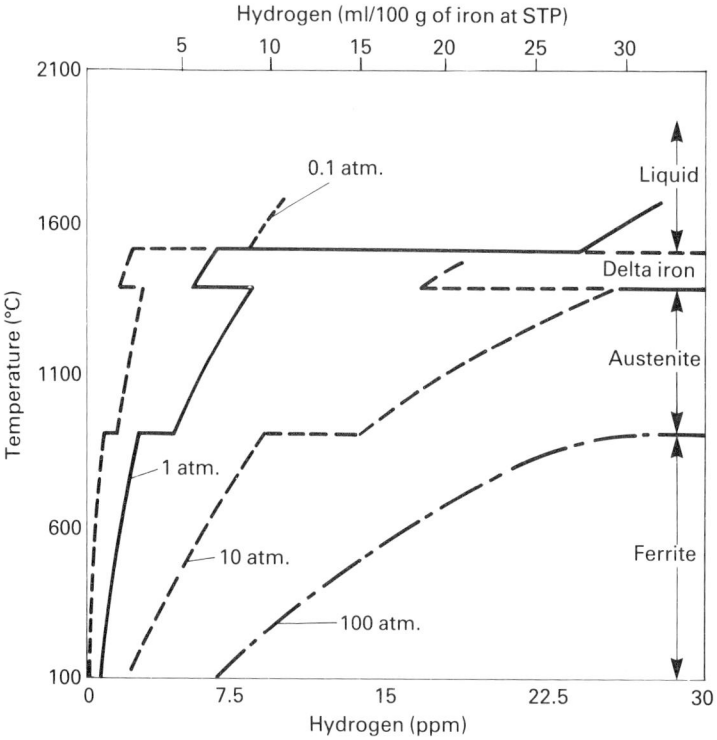

10.2 Influence of temperature on the solubility of hydrogen in iron at different pressures, after ref. 14.

cools, much of the hydrogen will remain in the steel unless sufficient time is allowed for its removal during the cooling cycle. This hydrogen embrittles the steels progressively as the temperature falls below about 250 °C; the embrittlement can lead to cracking or failure on stressing the component cold (e.g. by hydrotesting).

The precautions necessary to avoid these problems are to replace hydrogen in the vessel by an inert gas, such as nitrogen, and then hold at temperature or cool the component sufficiently slowly to allow most of the hydrogen to diffuse harmlessly away. It is also important to ensure that regions of general or locally high hardness are not present, by taking precautions if any welding is required and avoiding the use of steel of too high a strength. Some guidance on the tolerable strength levels for Cr:Mo pressure vessels is given in Fig. 10.3.[13] Hardness values have been added to the original plot (using Eq. [9.1]) so that the integrity of any hard weld zone regions can be assessed. Although the restriction on strength or hardness may seem unwarranted, it is a necessary precaution as it may

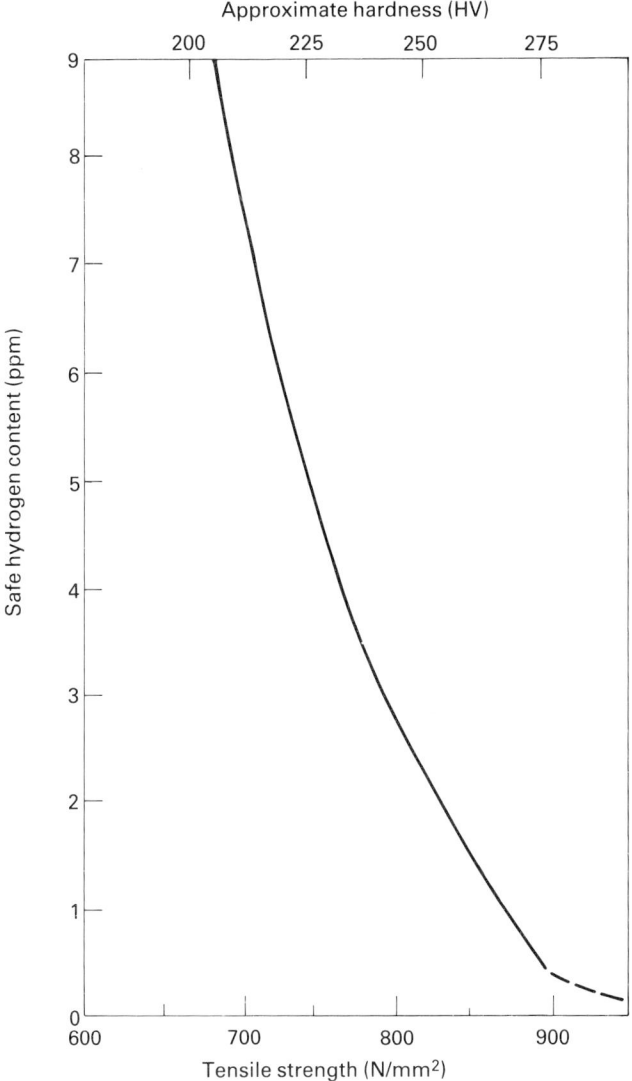

10.3 Safe hydrogen concentrations to avoid hydrogen crack growth below 150 °C in relation to strength and hardness of parent steel, assuming the presence of a 25 mm deep crack in the vessel wall and an applied stress of 1/3 yield (after ref. 13).

not always be possible to follow the slow cooling requirements in an emergency.

The type of plant containing high pressure hydrogen at elevated temperatures occurs in refineries and chemical plant – hydrotreating, reforming and hydrocracking units and ammonia converters being typical examples of such plant. Normally on shutdown, the contents of such vessels are replaced with an inert gas and the plant cooled over several days. However, in an emergency, such precautions may not always be possible and the plant may cool naturally much faster.

High pressure hydrogen plant is unlikely to be operated at such a high temperature as to suffer creep damage, but temper embrittlement is always possible, particularly in older plant, and the presence of two simultaneous embrittling mechanisms calls for special precautions whenever the steel is near ambient temperature.

Welding of steel containing hydrogen for repair or modification should be carried out with care. Normally, the steel is heated for long enough to remove most of the hydrogen before welding starts. However, this may not always be necessary, particularly if the preheating temperature is sufficiently high to be above the hydrogen embrittlement temperature range. Post-heating – or direct heating to the PWHT temperature after welding – may then be advisable.

High temperature service problems

Several of the problems likely when welded fabrications are used at high temperatures are discussed elsewhere. Oxidation and creep resistance are outlined in Chapter 9, strain-ageing and temper embrittlement in Chapter 10 and in previous sections of this chapter and some of the problems of high temperature hydrogen in the previous section.

Hydrogen attack

Hydrogen attack (or hydrogen damage) is distinct from hydrogen embrittlement and occurs at temperatures above about 220 °C. For any steel, the safe temperature decreases as the partial pressure of hydrogen is increased. Of the steel types likely to be used in hydrogen service, only those containing the element chromium appear to have enhanced resistance to hydrogen attack. Thus, as the required temperature and pressure are increased, the chromium content of the steel employed should also be increased. The requirements for safe hydrogen service are shown in the Nelson diagram (Fig. 10.4);[14,15] the latest version of ref. 15 specified by the API should always be consulted.

Although the Nelson diagram has been in existence for nearly a quarter of a century,[14] it has been built up mainly from records of service failures, as

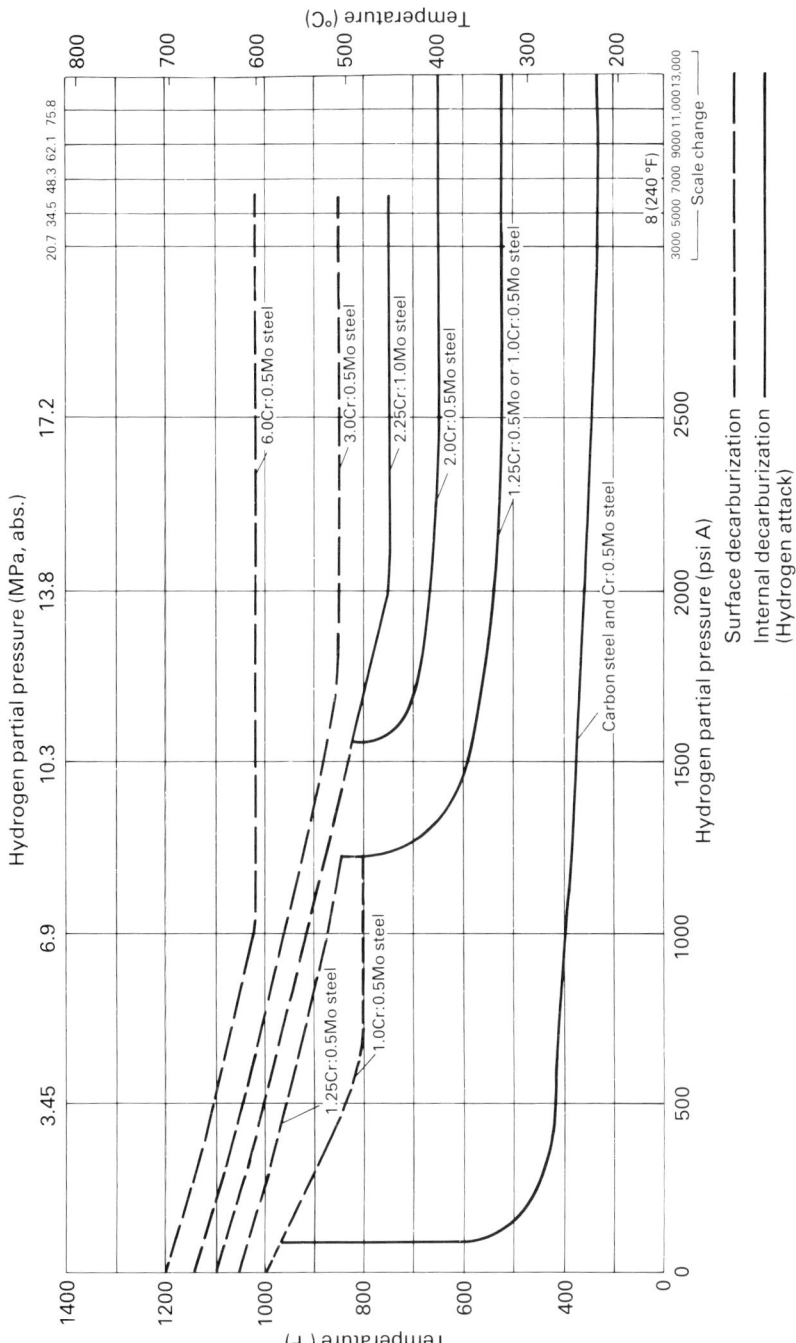

10.4 Nelson diagram, showing regions safe from hydrogen attack for different steel types at different temperatures and pressures, after ref. 15.

no accepted laboratory test method for simulating hydrogen attack has been found – partly because the incubation time can be very long (over 10 000 hours, 13 months). This means that it has not been possible to study systematically the conditions leading to attack, and new evidence is still accumulating. It was originally thought that the C:0.5%Mo steel resisted hydrogen attack better than simple C or C:Mn steels, but this is no longer regarded as the case.[15] Hence, revisions to the diagram are expected as knowledge grows.

Hydrogen attack occurs because hydrogen diffuses into steel (and weld metal) and reacts with the carbon in the steel to form methane (Eq. [7.2]). At high temperatures, above about 560 °C and at pressures below ~13 bar, surface decarburisation is the usual form of attack. This attack weakens and lowers the ductility of the steel locally. With higher pressures, attack is not confined to the surface layers but occurs throughout the thickness of the steel.

Microscopic examination will reveal that the carbides in the pearlite (Fig. 5.23(a)) or elsewhere in the steel have been attacked and that small voids or microcracks have formed. The destruction of the carbides and the simultaneous formation of cracks reduces the strength, ductility and toughness of the steel. Hydrogen attack can also lead to blistering (Fig. 5.23(b)) if the steel has a population of stringers of inclusions of manganese sulphide or mixed sulphide and silicates. Creep resistance is also reduced, although for most steels the safe temperatures for exposure to hydrogen are well below those at which creep occurs.

Detection can be difficult without the extraction of test samples (i.e. for bend, crush or tensile tests), although careful microscopic examination, analysis for carbon (which will be reduced from its original level) and ultrasonic examination can all be helpful. The last will usually show scattered reflections characteristic of fine cracking.

Graphitisation

Graphitisation was originally found in petrochemical plant and steamlines, and this led to a thorough investigation at the time.[16] It is not a current problem, because susceptible steel compositions are no longer allowed to be used within the dangerous temperature range. However, with current pressures to extend the life and/or efficiency of plant, it must not be ignored as a possible problem area.

In C, C:Mn and C:Mo steels, the true phase in equilibrium with iron is carbon (graphite) and not cementite (Fe_3C). The approach to equilibrium is so slow that graphite is never seen under normal circumstances. However, with some steels, graphite nodules (Fig. 10.5; similar to those seen in spheroidal graphite cast irons) can form after very long periods at ele-

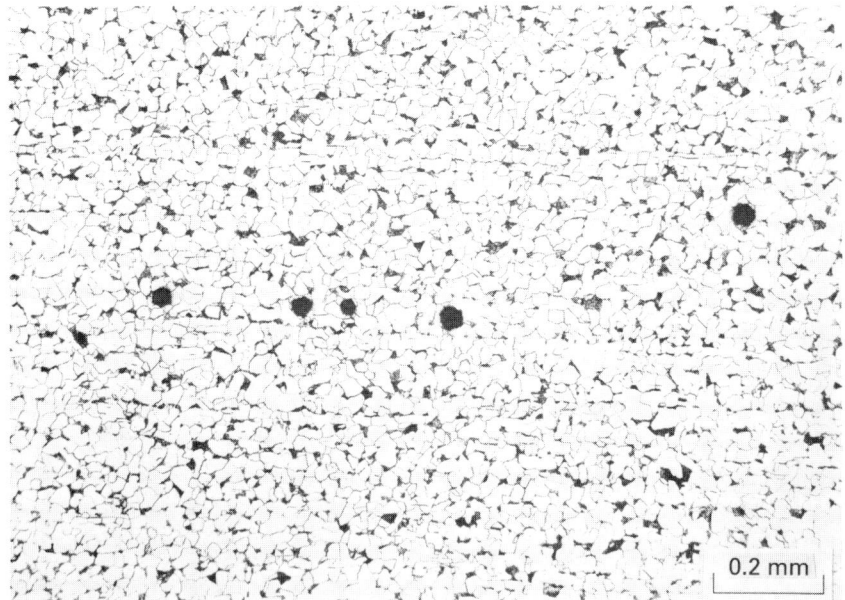

10.5 Graphitisation in mild steel.

vated temperatures. The effect of the transformation from cementite to graphite is to reduce strength and, in some cases, toughness.

Graphitisation has only been seen in certain steels subjected to temperatures in excess of 455 °C for periods of at least four years.[16] Above the A_{c1} temperature steels start transforming to austenite in which carbon is much more soluble than in ferrite. The reaction to graphite is most rapid at temperatures about 60 °C below the A_{c1}.[17] Graphitisation is accelerated by prior plastic deformation, and is said to be more rapid in martensitic microstructures than in ferrite.[17] It occurs in weld metal as well as in parent steel.[16]

Graphitisation has only been found in simple carbon and C:Mo steels.[16,17] Steels containing about 0.5% Cr or more do not suffer this type of degradation, presumably because chromium forms more stable carbides than do iron and molybdenum. Graphitisation is also retarded by the elements Mn, Mo, Ti and V. A combination of low aluminium ($\leqslant 0.003\%$) and high nitrogen ($\geqslant 0.03\%$) also appears to prevent graphitisation. Aluminium and silicon are believed to accelerate the process.

The amount of carbon remaining in solid solution in equilibrium with graphite is much less than in the original condition, when Fe_3C (cementite) is the quasi-equilibrium constituent. This may reduce the risk of carbon strain-ageing, but it could also make a plain C or C:Mo steel susceptible to temper embrittlement in the graphitised condition.

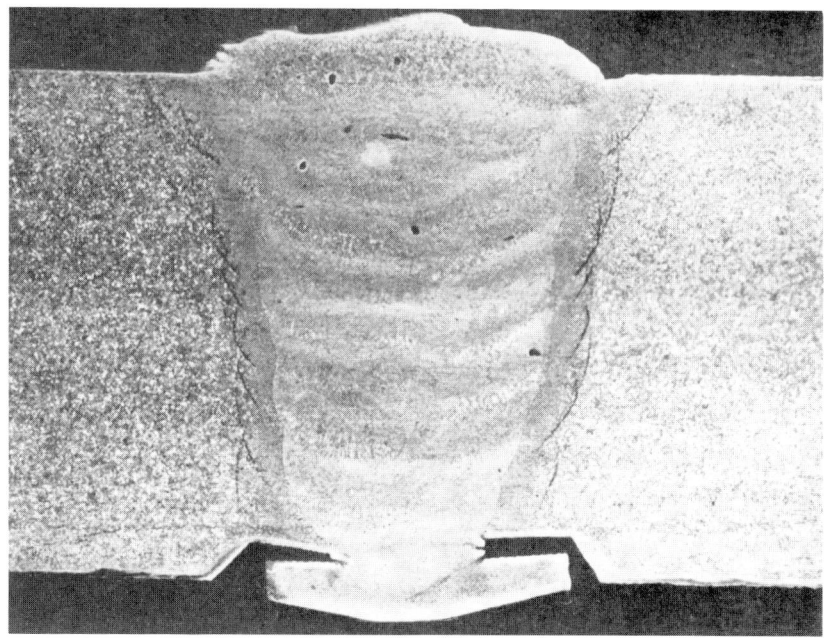

10.6 'Eyebrow' graphitisation in multipass weld in C:Mo steam piping, from ref. 16.

Although graphitisation weakens steel, its effect on strength and toughness is usually relatively small, except in the case of welded joints. In the HAZ, graphitisation is concentrated at a certain distance from the fusion boundary, at about the limit of the visible HAZ, to form chains of graphite nodules (Fig. 10.6), sometimes termed 'eyebrows'. This concentration of a line of weak graphite reduces ductility and toughness significantly.[16,17] Tests to examine the influence of graphitisation on creep properties showed no effect on parent steel; tests on welded specimens gave inconclusive results,[16] although it is to be expected that creep ductility and strength could be reduced under particular stress conditions.

References

1 Maddox, S.J., *Fatigue Strength of Welded Joints*, 2nd edn, Abington Publishing, Cambridge, 1991.
2 British Standard BS5400: 'Steel, concrete and composite bridges: Part 10: Code of practice for fatigue', BSI, London, 1980.
3 NACE Standard MR0175-92, 'Standard recommended practice for sulfide stress corrosion resistant-metallic materials for oilfield equipment', NACE, Houston, TX, 1992.

4 British Standard BS4515: 1984, 'Process of welding steel pipelines on land and offshore', BSI, London, 1984.
5 Walker, R.A. *et al*, 'The significance of local hard zones on the outside of pipeline girth welds', in Proceedings of the International Conference on 'The New Realities in Pipeline Design, Construction and Operation', IIR, London, 1992.
6 Terashima, H. and Hart, P.H.M., 'Effect of aluminium in C–Mn–Nb steel submerged arc welds', *Weld. J.*, 1984, **63**, 173s–183s.
7 Bruscato, R.M., 'Embrittlement factors for estimating temper embrittlement in 2¼Cr : 1Mo, 3.5Ni–1.75Cr–0.5Mo–0.1V and 3.5Ni steels', ASTM Conference, Miami, Florida, Nov. 1987.
8 Ishiguo, T., Murakami, Y., Ohnishi, K. and Watanabe, J., '2.25Cr–1% Mo pressure vessel steel with improved creep rupture strength', in *Proceedings of the Symposium on 'Applications of 2.25%Cr–1% Mo steel for thick-wall pressure vessels'*, Denver, Col., 1980, ASTM STP755, 1982, pp. 129–147.
9 Sugiyama, T., Hatori, N., Yamamoto, S., Yoshino, F. and Kiuchi, A. 'Temper embrittlement of Cr–Mo weld metals', IIW Doc. XII-E-6-81, IIW, 1981.
10 Erhart, A. Grabke, H.J. and Onel, K., 'Grain boundary segregation of phosphorus in iron-base alloys: effects of C, Cr and Ti', in *Proceedings of the Metals Soc Conference on 'Advances in the Physical Metallurgy and Applications of Steels'*, Liverpool, 1981, The Metals Society, London, 1982, pp. 282–285.
11 Weng Yu-Qing and McMahon, C.T., 'Interaction of P, C, Mn and Cr in intergranular embrittlement of iron', *Mat. Sci. Tech.*, 1987, **3**, 207–216.
12 Bailey, N., Coe, F.R., Gooch, T.G., Hart, P.H.M., Jenkins, N. and Pargeter, R.J., *Welding Steels Without Hydrogen Cracking*, Abington Publishing, Cambridge, 1993.
13 Erwin, W.E. and Kerr, J.G., 'The use of quenched and tempered 2¼Cr : 1Mo steel for thick wall reactor vessels in petroleum refinery processes: an interpretive review of 25 years research and application', *WRC Bull.* No. 275, Feb. 1982.
14 Nelson, G.A., 'Interpretive report on effect of hydrogen in pressure vessel steels; Part 2: action of hydrogen on steel at high temperature and pressures', *WRC Bull.*, No. 145, pp. 33–42, Oct. 1969.
15 API 941, *Steels for Hydrogen Service at Elevated Temperatures and Pressures in Petrochemical Refineries and Petrochemical Plant*, American Petroleum Institute Publication 941, 4th edn, API, Washington D.C., Apr. 1990.
16 Wilson, J.G., 'Graphitisation of steel in petroleum equipment' and 'The effect of graphitisation of steel on stress-rupture properties', *WRC Bull.*, No. 32, Jan. 1957.
17 Ternon, F., 'Bibliographical study of graphitisation in low carbon, low alloy steels', Electricité de France report No. HT/PV G 154 MAT/T 41, Oct. 1982.

11 Repair

Analysis prior to repair

Repair welding (including the modification of equipment in service by welding) is no different in principle from ordinary welding, although extra care usually needs to be taken, either because welding may have to be carried out in conditions that are far from ideal or because the repair may be to rectify a weld failure, in which case the cause of the failure should be known to avoid a repetition of that failure. Before attempting a repair, five questions must be answered:

1. Why does the repair need to be made?
2. What are the composition and properties of the steel to be repaired?
3. If a weld is involved, what was the original welding procedure?
4. Do any circumstances prevent the original welding procedure being repeated?
5. How long must the repair last?

Once these questions have been answered, a repair welding procedure can be established. The reasons for these questions are detailed below.

Need for repair

If the repair was due to damage caused by a simple overload, there should be no problems in answering this question, unless there is a call for strengthening the component. A more likely cause of failure is by fatigue, and this could be a result of defects in the weld or of under-design. If weld defects were the sole cause of failure, then a repair giving a better quality weld should avoid a future problem in this area, provided the defects were exceptionally large and are unlikely to exist in similar welds elsewhere in the component or structure.

If the defects are not large, then failure is probably due to overloading in fatigue, and there are two possibilities for a weld repair. One is to re-weld and grind or otherwise dress weld toes and any other stress

Repair

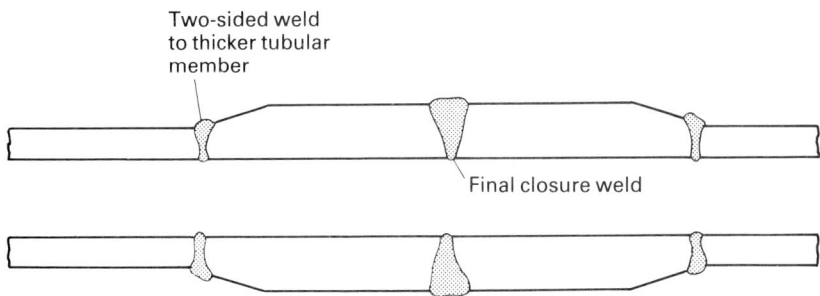

11.1 Sketch of repair to fractured tubular component when thickening is required to increase fatigue strength of single-sided closing weld.

concentrators so that the joint fatigue classification is improved.[1,2] This may not be possible, however, if fillet or inaccessible one-sided butt welds cannot be avoided. In that eventuality, it may be necessary to cut out an area including the crack and replace it with thicker steel so that the stress on the critical joint is reduced. This could involve three new welds, two to attach thicker steel to the existing members (these would need to be two-sided in order to carry out any weld dressing on the otherwise inaccessible weld faces in order to improve their fatigue strength) and the third to join together the two thicker pieces by single-sided welding, i.e. by a weld of lower fatigue strength but with lower stresses because of the increased thickness. A sketch showing such a repair to a fatigue failure in a tubular construction is given in Fig. 11.1.

A third possibility is of brittle fracture. Although this will probably have originated from a defect, the failure indicates a lack of toughness in the original steel or weldment. If the poor toughness was in weld metal, then the safest action is to gouge out the existing weld metal and re-weld with a tougher grade. In fact, partial replacement would be adequate if the problem is ascribed, for example, to the root or to the cap regions, or to an isolated run. A failure due to poor HAZ toughness would again require removal of HAZ material (unless the application of PWHT could provide the solution) and the adoption of a different welding technique (Chapter 9). This might involve lower heat inputs, better control of interpass temperatures or the use of a HAZ-refining method, such as the two-layer technique. Consideration would also need to be given to other similar welds in the structure. Was the failure due to a lapse on a single weld or are there likely to be more similar failures?

Other failure scenarios, involving such problems as stress corrosion cracking or creep failure, can be approached in a similar fashion.

Preferred repair method

For locally undersize components, or shallow defects, grinding out and blending the contour may be adequate – and much cheaper than repair welding – particularly if fatigue (rather than tensile strength) is the major consideration.

For components in brittle materials and which are not welded, repair welding should be approached with considerable caution, as the repair could lead to worse problems than the original fault.

Underwater weld repairs are particularly difficult and expensive, and repairs are frequently carried out by transferring the load path to reinforcing members, which are bolted or grouted in plastic.

Steel composition

It is important to have a good idea of the composition of the steel, and particularly of its carbon content. The best information to have is the mill sheet for the actual steel involved. Knowing the specification, and the grade within the specification, is also helpful – but these may have changed since the steel was supplied! Descriptions such as 'mild steel', 'low alloy', 'chrome-moly', 'rail steel' and the like are not helpful, and serious consideration should then be given to having the steel analysed by a reputable concern, remembering that having an accurate figure for the carbon content is of the utmost importance. For a standard spectrographic analysis, a sample weighing at least 30 grams is adequate, but establishments such as TWI are able to carry out useful analyses on much smaller samples.

Although the analysis will supply most of the assurance needed that a satisfactory repair can be effected, further tests may be advisable if the component is large, is of doubtful toughness (e.g. due to temper embrittlement) and is not known to have been previously welded. In such cases, limited Charpy testing on a piece of steel cut out from the component in an area which is redundant or can easily be replaced (e.g. taking a boat sample), might provide the necessary assurance that welding will not lead to an unexpected disaster!

Steel strength

If a matching strength repair is necessary, the strength of the steel must be known so that appropriate consumables and welding procedures can be selected. However, the higher the strength required of the repair weld metal, the more difficult will it be to avoid cracking and severe distortion, particularly if there are limitations on preheat levels and the use of PWHT. Furthermore, the stronger the steel, the less likely are sufficiently strong consumables to be available quickly or at all. It may be quite adequate to carry out repairs with an undermatching weld metal, provided a

Repair

high quality repair is made, and proper precautions are taken to avoid cracking and welding faults.

If the steel strength is not known, it will normally be possible to estimate it from hardness testing, carried out on-site if necessary, using the correlations between hardness and strength in Eq. [9.1] and [1.12].

Original welding procedure

If the original welding procedure is known, then it should be used, if possible, as a basis for the repair, unless there is evidence (from cracking or lack of toughness or other properties) that it was inadequate. Any heat input limitations should be adhered to in the repair, bearing in mind that if the welding process for the repair differs from the original, the heat input limitations should be altered to reflect the difference in arc efficiencies of the two processes. SEE P. 10.

Limitations on repair welding technique

It is not always possible to repeat the original welding procedure for a number of reasons. The component might be part of a structure that cannot be moved, thus preventing welding being carried out in the flat position or precluding the use of PWHT. The structure might be under water, thus requiring some type of underwater welding. If delicate components are fixed to the component, preheat (or PWHT) cannot be used and if the component contains liquids or gases under high pressure, hot tapping techniques will be needed.

Many of these limitations will necessitate the use of lower heat inputs than were originally used. That is not always a problem when considering the achievement of good toughness, although it is likely to increase the risk of HAZ hydrogen cracking in most steels and it will probably increase the risk of welding faults. The most common processes for repair welding are MMA and TIG, as they are manual processes which are in widespread use, can be operated in confined spaces and out of position and are well understood in terms of achieving good toughness and avoiding cracking and welding faults. Manual metal arc welding is used for repairs to the more common weldable steels, TIG for the more difficult steels, such as the low alloy engineering steels with medium to high carbon contents.

The other process of major importance for repairs is MAG welding and its variants – particularly the use of cored wires. With gas shielding, care must be taken to avoid draughts and air currents sufficient to disrupt the gas shielding. The use of self-shielded wires avoids this problem, although if good toughness is needed, single-pass welds should be avoided and refining multipass techniques employed. Other processes are rarely used but may have applications in specialised cases, such as the use of friction surfacing, submerged arc welding or surfacing for building up worn or corroded surfaces.

Expected life of a repair

This factor has a major bearing on the level of quality required from the repair. In underwater repairs, wet welding will often be chosen for a temporary repair in preference to welding in the dry. If a fatigue or creep failure led to the repair, the original welding technique can be repeated if the life required is short. Similarly, lower levels of weld metal strength, of welding quality and of inspection could be acceptable – although these steps are likely to lead to a possibly catastrophic brittle fracture if standards are allowed to fall too low.

Limitation of preheat

When preheat temperatures must be limited, or when preheat cannot be used at all, the first option is to use ferritic consumables giving low, very low or ultra-low weld hydrogen contents. These include TIG welding, MIG welding with solid wires and with certain cored wires (although probably not those of the self-shielded type) and very low hydrogen basic electrodes. These last may be used in the normal way and dried at high temperatures (350–450 °C, depending on the manufacturer's recommendations) or taken from specially wrapped vacuum packs containing small numbers of electrodes which should last for a shift under normal conditions without the need for re-drying.

It must be emphasised that strict controls of cleanliness and dryness, together with control of the issue and correct use of consumables, are essential to the successful use of low hydrogen welding procedures.

If ferritic fillers cannot be selected to give freedom from hydrogen cracking when not preheating – and this is likely with low alloy steels of most types – it should be possible to use non-ferritic electrodes, either of the austenitic stainless steel type or suitable nickel alloys, as detailed in Chapter 5. The austenitic stainless steel type is probably the easier to use – and is less liable to give problems with solidification cracking when welding steels not having low sulphur levels. However, preheat is recommended when welding steels containing above 0.2 to 0.3% C, (see Fig. 5.14). For such steels, nickel alloys are more suitable, and have been used reliably for repairs to large power station components in Cr:Mo and Cr:Mo:V steels.

Need for post-weld heat treatment

It is most unlikely that a full PWHT can be carried out on-site. Despite many claims, the efficacy of vibratory stress relief has not been satisfactorily demonstrated for welded constructions in general, or for repair welds in particular. Local stress relief may be possible, and the appropriate application code will usually give guidance on such matters as the maximum temperature gradients which can be allowed. If no appropriate guidance exists, one of the pressure vessel codes (e.g. Clause 4.4.4 of the latest edition of BS5500[3] should be consulted.

In the absence of any PWHT, it must be recognised that a high level of toughness will be required for as-welded repairs, because the level of residual stresses will be high. However, techniques are available for grain refinement, thus ensuring a reasonably high level of toughness in the HAZ. Consumables should be selected to give an acicular ferrite weld metal microstructure, or a welding technique should be used which gives a multipass refined structure. In any case, it is helpful to use the lowest strength of suitable filler, although it is recognised that this may not be compatible with achieving a tough acicular ferrite microstructure if a multipass refined structure is not possible. A fracture mechanics analysis should here be able to decide whether the best available weld metal toughness, the lowest possible weld metal yield stress (i.e. the lowest residual stress level) or somewhere in between is the best option. Where high temperature steels are to be welded and not subject to PWHT, a low carbon filler is advised, provided it can achieve adequate creep strength.

As outlined in Chapter 6, there are three ways of obtaining a refined HAZ structure. The first, which is more suitable to the original welding (preferably using a mechanised technique, because it requires a high degree of process control) than to a repair, is to use a steep-sided weld preparation and a shallow angle of attack. The second method is the half-bead technique, which was the first to be adopted in the ASME Boiler Code[4] and the third is the two-layer technique, originally developed for refining Cr:Mo:V steel HAZs in order to avoid reheat cracking[5,6] but since developed for the repair of C, C:Mn and Mn:Ni:Mo steels.[7] Both these techniques require a fairly high level of control and understanding by welding personnel of what the objectives are.

In the half-bead technique, the first layer of weld metal is deposited on to the repair cavity and the top half of this layer is ground off. Welding of the second (and subsequent layers) is continued using the same welding parameters as for the first layer.

The two-layer technique requires no inter-layer grinding (thus removing a difficult variable to control) but requires the second layer to be deposited with a higher heat input than the first layer, so that the heat to produce refinement of the first layer HAZ penetrates as far as the limits of the coarse-grained HAZ (Fig. 6.3). To assist in this refinement, the angle of attack should be steep and the overlap of adjacent weld beads controlled to about 50% by aiming the welding arc at the toe of the previous bead. As with any critical repair, low hydrogen consumables should be used to minimise any risk of hydrogen cracking, although preheat is helpful in achieving a fully refined microstructure.

Weld metal adjacent to the HAZ will probably be fully refined, but the repair process cannot be continued with ever-increasing heat inputs in order to achieve a fully refined weld metal microstructure. It is, therefore, essential to select consumables that give good as-welded toughness with normal welding techniques at relatively low heat inputs; the welding conditions for the second layer are usually continued to completion of the repair, with the adoption of a temper bead layer (Fig. 5.11) to reduce the risk of a hard HAZ from the final run. The technique can be used with several welding processes. With MMA the first layer is deposited with 2.5 or 3.2 mm diameter electrodes and the second layer with 4.0 mm. The MAG process has been used, but not fully developed. TIG can readily be adapted, and has the advantage that a refined weld metal microstructure is easier to obtain than with the other processes.

Considerable pilot plant experience has been accumulated in the use of the two-layer technique for the repair of C, C:Mn and Mn:Ni:Mo steels; current research effort at TWI and elsewhere is studying its application for as-welded repairs in Cr:Mo and similar steels, and it is regularly used for Cr:Mo:V steels where HAZ refinement is essential to avoid reheat cracking during PWHT.

Special circumstances

Presence of hydrogen

Repair welding of steels containing hydrogen is likely when the component has been in hydrogen service or in sour service in the oil industry. One way to avoid problems is to heat the steel sufficiently for damaging amounts of hydrogen to diffuse out; guidance on diffusion times can be found in ref. 8. This approach is particularly useful if circumstances make the use of high preheat temperatures during welding impracticable as, for instance, when welding must be carried out inside a component within a restricted space.

However, hydrogen will only cause cracking when the steel is below about 200 °C (and also when other circumstances are favourable – see Chapter 5). If the component is maintained above that temperature during welding and for some time after welding, sufficient hydrogen will diffuse out to avoid problems. Because the diffusion times for complete removal of hydrogen are long, work is being carried out at TWI to investigate the problem more thoroughly.

Repair

Welding components containing fluids

A special technique, known as hot tapping, has been developed for welding on to pipework and similar components containing liquids and gases under pressure. The reason for needing this technique, which is covered by standards,[9] is to avoid the expense and waste of having to empty long pipelines when making repairs, or when carrying out modifications such as adding branches. The technique consists of welding a patch on to the pipe and then, if adding a branch, using a special cutter to cut through the pipe wall and open the branch to the main fluid flow.

From the welding standpoint, consideration should be given to three factors:

1 The weld must be leakproof, and for this reason at least two passes are preferred. The use of more than one pass also allows the second pass to be deposited against the patch plate, thus tempering any hard HAZs on the wall of the pipe itself.
2 The wall of the pipe must *not* be penetrated, or even overheated, because if the temperature is too high the internal pressure can cause a bulge in the pipe with a high risk of a dangerous failure. This requirement leads to the use of relatively low heat inputs, although the fluid in the pipe helps to keep the pipe cool, particularly if the wall is thin and the fluid not hot. Guidance is available on appropriate heat inputs for different wall thicknesses, taking into account the temperature of the contents of the pipe.
3 The use of low heat inputs, possibly enhanced by the cooling effect of the contents of the pipe, increases the risk of hydrogen cracking. Particular importance is, therefore, needed to ensure the use (and maintenance of) a low hydrogen welding procedure. The welding procedure must take account of the cooling effect of the fluid in the pipe, which acts to increase the effective thickness of the pipe.

Welding under fluctuating loads

It is possible to repair structures under fluctuating loads, provided adequate precautions are taken. Examples of such repairs are repairing bridges carrying traffic and repairs to structures in the sea subject to wave action. Assuming that precautions have been taken to avoid hydrogen cracking, the main risk is of solidification cracking, with perhaps ductility dip cracking also, during the tensile cycle of the fluctuating load. Normally, welding under fluctuating loads to repair cracks and other planar defects (i.e. where welding through the complete thickness), should be restricted to gaps not exceeding about 0.8 mm. If larger gaps are present and it is not possible to stop the load fluctuating, it may be possible to wedge open the crack at its widest, in order to stop its width fluctuating, before welding. If this is not possible, welding a length of about 100 mm at each end of the crack (where the deflection is least) should reduce the deflection at the middle of the crack to an acceptable level.

Because of its better resistance to solidification cracking, MMA welding would normally be preferred to automatic or semi-automatic processes (except TIG). Electrodes have been specially developed[10] to cope with the problem. It is claimed that, besides having a good resistance to solidification cracking, such electrodes should give low contents of Ni, B and P in the weld metal to minimise risks of ductility dip cracking.

Underwater repairs

In practice, welding for repairs and modifications of structures under the water can be divided into welds that are critical and those that are not. For the former, by far the most are carried out in the dry, either by building a caisson or dam and pumping out the water, or by building a pressurised (hyperbaric) chamber around the site of the repair and displacing the water with the gas under pressure, which acts as the atmosphere for the hyperbaric habitat chamber. Less critical welds may be made in the water – wet welding – provided circumstances are favourable. Any underwater welding is expensive but (unless existing hyperbaric chambers are available without much modification) hyperbaric welding repairs are particularly costly.

Hyperbaric welding

Welding can be carried out at depths down to about 550 m, and experimental welding has been carried out at twice that depth. The atmospheres of hyperbaric chambers must be at a pressure to support the depth of water (the floors of the chambers are open) so that the pressure (in bars) is approximately one tenth of the depth in metres. The atmosphere is usually selected to be breathable by the welders and can be of compressed air down to about 20 m, but for greater depths helium–oxygen mixtures (sometimes with nitrogen) are used. The oxygen content is aimed to be close to 0.2 bar partial pressure (as at the surface) so the percentage of oxygen can be low.

Because of the open floors of the chambers, the atmosphere is very humid and, although deep-sea water is cold, the temperature inside can be high due to the heat generated by welding, preheating and any other electrical equipment in the chamber. Consequently, the absolute humidity is high and needs to be allowed for when devising welding procedures.

Welding itself is affected by the high pressure, which constricts the arc. Also, for shallow repairs the welding sets may be at the surface, in which case heavy voltage drops will occur along the leads. The usual processes for repair welding can be carried out hyperbarically, namely TIG, MAG (including the use of cored wires with gas shielding) and MMA. Special consumables are usually required because the pressure in hyperbaric chambers alters element transfer during welding, so that welds tend to have higher carbon contents than when deposited at surface pressure. With gas-shielded processes, there is a preference for using shielding gases based on helium, rather than argon, because large quantities of the latter are not suitable for breathing. A further factor that needs to be taken into account is that the ability of welder-divers decreases as the depth increases, so that everything needs to be as foolproof as possible. This has led to the development of fully automatic welding techniques, which can be employed for some repairs, particularly when they consist of cutting out a length of pipe and replacing it in a manner similar to that used for the original tie-in.

Wet welding
Although wet welding is sometimes regarded as a last desperate method of patching up a component until it can be dried out and repaired 'properly', it can be, and is, used as a satisfactory method of fixing 'secondary' structures, i.e. those where failure of one component is tolerable. Wet welding is regularly used to replace mild steel anode holders with new ones by welding to the mild steel brackets at depths down to at least 85 m in the North Sea.

Welding by MMA is by far the most common method, as it obviates the need for wire feeders, a supply of shielding gas and other complications that are required for MAG welding. The use of MAG, with a shielding gas curtain to displace the water, has been examined experimentally. However, most underwater wet welds for repair are of the patch type and the step at the edge of a patch could give difficulties with the gas curtain.

Of the other welding processes, friction welding has been used successfully to weld studs at considerable depths[11] and is being experimented with at TWI to assess the feasibility of making 'stitch welds' to repair cracks.[12] With friction welding, some form of plastic shroud is used to delay cooling of the joint and so avoid the hard martensitic structures that would otherwise form (due to the quenching effect of the water) and probably reduce toughness and stress corrosion resistance or even give cracking problems during welding. Tests on flux cored, self-shielded wires have also given promising results in the Ukraine.

The major problem with wet welding is that of hydrogen cracking due to the very high hydrogen levels brought about by electrolysis of the water in which welding is carried out, and exacerbated by the very fast cooling rates due to quenching by the water: 800–500 °C cooling times as fast as 1.5 seconds have been measured. Basic electrodes do not run well in water and are certainly not low hydrogen! Rutile ferritic electrodes run the best, but again are not low hydrogen; nevertheless, they are used for the bulk of wet welding – provided the steels being welded have CEs below about 0.40. One semi-experimental repair has been made using a temper bead technique on a steel with a slightly higher CE, but the temper bead technique is difficult to carry out properly in air and so is unlikely to find favour for more than a few special repairs.

Better resistance to hydrogen cracking has been claimed for the use of ferritic electrodes with oxidising coverings[13] and with nickel alloy electrodes.[13,14] However, despite their availability, the former have not found favour owing to their inferior running characteristics, whilst the latter are not yet developed to the point of being a commercial product. One type of electrode, often used for repairing difficult steels in the dry, but which is *not* recommended for use when welding ferritic steels in water is the austenitic stainless steel type, as such electrodes invariably give fusion boundary hydrogen cracking.

Wet welding is best carried out with electrodes which have been waterproofed, using d.c. positive polarity. A variety of waterproofing techniques is employed, ranging from paraffin wax to complex multilayer schemes involving synthetic varnishes, aluminium paint, etc. Although d.c. positive gives better running characteristics than electrode negative (a.c. must *not* be used on grounds of safety), it does give high penetration (which is increased by hydrogen from the water). High levels of dilution of the parent steel into the weld pool must, therefore, be taken into account – particularly when using nickel alloy electrodes which can give solidification cracking if the sulphur content of the weld metal is too high.

Provided welds are made that are free from defects, mechanical and corrosion properties are not significantly different from those of welds made on the surface.[13] Repair by underwater welding at depths down to 10 m can be regarded as routine; such depths are adequate for most repairs to ships, harbour defences and coast protection work. Welding down to at least 85 m has been carried out commercially; experiments at deeper depths have shown that severe porosity is likely to be a major problem at much deeper depths, e.g. 550 m, although experiments have been carried out at simulated depths of 1000 m.

Welding in space
Little is available on this topic in the usual welding literature, except that the major problems are the very low gravity (micro-gravity), the near-absolute vacuum in space itself and the very high visual contrast between light and dark, with no intermediate shades. The first greatly alters the balance between gravity and surface tension in forming a weld pool – and makes the term 'positional welding' redundant, although thermal cutting

is likely to provide problems. The high vacuum means that arc welding is not possible and the processes likely to be used are electron beam, laser, soldering, brazing and the solid state processes. High visual contrast can give serious problems for the operators.

References

1. Maddox, S.J., *Fatigue Strength of Welded Joints*, 2nd edn, Abington Publishing, Cambridge, 1991.
2. British Standard BS5400: 'Steel, concrete and composite bridges: Part 10: Code of practice for fatigue', BSI, London, 1980.
3. British Standard BS5500: 1991, 'Unfired fusion welded pressure vessels', BSI, London, 1991.
4. ASME Boiler and Pressure Vessel Code, Section XI, 'Rules for Inservice Inspection of Nuclear Power Plant Components', Article IWA-4000, Para. IWA-4513 'Repair welding process, -A, SMA', ASME, New York, July 1992.
5. Alberry, P.J. and Jones, K.E., 'Two layer refinement techniques for pipe welding', 2nd International Conference on Pipe Welding, TWI, London, Nov. 1979.
6. Alberry, P.J., Myers, J. and Chew, B., 'An improved welding technique for HAZ refinement', *Weld Metal Fabrication*, 1977, 45, 549–553.
7. Jones, R.L., 'Development of two-layer deposition techniques for the MMA repair welding of thick C:Mn steel plate without PWHT', TWI Res. Report 335/1987, TWI, Cambridge, Apr. 1987.
8. Bailey, N., Coe, F.R., Gooch, T.G., Hart, R.H.M., Jenkins, N. and Pargeter, R.J., *Welding Steels Without Hydrogen Cracking*, Abington Publishing for TWI, Cambridge, 1993.
9. API 1107, *Pipeline Maintenance Welding Practices*, 2nd edn, API, Washington D.C., 1991.
10. Nakanishi, Y., Nakamura, Y., Sakai, K., Kohno, T., Satoh, K., Kawai, Y., Yamaguchi, T. and Nishiyama, N., 'A study on welding in service conditions – development of electrodes for welding under pulsating stress', IIW Doc. XV-634-87, IIW, 1987.
11. Nicholas, E.D., 'Friction welding under water', in the *Proceedings of the International Conference on Underwater Welding*, Oslo, June 1983, Pergamon Press, Oxford, pp. 355–362.
12. Andrews, R.E. and Mitchell, J.S., 'Underwater repairs by friction stitch welding', *Metals Mat.*, 1990, **6**, 796–797.
13. Gooch, T.G., 'Properties of underwater welds', *Metal Construction*, 1983, **15**, pp.164-167, 206-215.
14. Bailey, N., 'Exploratory tests on Ni-based wet welding electrodes for ferritic steels', in *Proceedings of TWI Conference on Advances in Joining and Cutting Processes*, Harrogate, 1989, Abington Publishing, Cambridge, 1990, 135–150.

Further reading

In addition to the references given at the end of each chapter, the following publications from Abington Publishing, mostly authored by staff or consultants of TWI, provide useful supplementary information.

Weldability
Bailey, N., Coe, F.R., Gooch, T.G., Hart, P.H.M., Jenkins, N. and Pargeter, R.J., *Welding Steels Without Hydrogen Cracking*, 1993.
Bailey, N. (ed.), *Faults in Fusion Welds in Constructional Steels* (with slide set), 1986.
Bailey, N. (ed.), *Welding Dissimilar Metals*, 1986.
Bailey, N. and Pargeter, R.J., *Influence of Flux Type on Strength and Toughness of Submerged Arc Weld Metal*, 1988.
Cottrell, C.L.M., *Welding Cast Irons*, 1985.
Granjon, H., *Fundamentals of Welding Metallurgy*, 1991.
Lancaster, J.F., *Handbook of Structural Welding*, 1992.
Pargeter, R.J. (ed.), *Quantifying Weldability*, 1988.
TWI Conference on the Hardenability of Steels, 1990.
TWI Seminar on The Effects of Residual, Impurity and Microalloying Elements on Weldability and Weld Properties, 1984.

Welding processes and consumables
Boniszewski, T., *Self-shielded Arc Welding*, 1992.
Dawes, C.T., *Laser Welding*, 1992.
Dolby, R.E. and Kent, K.G. (ed.), *Repair and Reclamation*, IoM/TWI Conference, 1986.
Houldcroft, P.T., *Submerged-arc Welding*, 1989.
Houldcroft, P.T., *Which Process? A Guide to the Selection of Welding and Related Processes*, 1989.
IIW, *Compendium of Weld Metal Microstructures*, IIW, 1985.
Keats, D., *Professional Diver's Manual*, 1990.

Larbey, M. et al., *International Standards Index*, 1991.
Lucas, W., *TIG and Plasma Welding*, 1990.
Lucas, W. (ed.), *Process Pipe and Tube Welding*, 1991.
Muncaster, P.W., *A Practical Guide to TIG (GTA) Welding*, 1991.
Pokhodnya, I.K. (ed.), *Metallurgy of Arc Welding*, 1991.
Street, J.A., *Pulsed Arc Welding – An Introduction*, 1990.
Widgery, D.J., *Tubular Wire Welding*, to be published 1994.
Yeo, R.B.G., *Specifications for Steel Welding Consumables*, 1988.
TWI, *Standard Data for Welding*, 1975.

Engineering and design
Gurney, T., *The Fatigue Strength of Transverse Fillet Welded Joints*, 1991.
Halmshaw, R., *Introduction to the Non-destructive Testing of Welds*, 1988.
Harrison, J.D. (ed.), *Weld Failures*, TWI International Conference, 1989.
Hicks, J., *A Guide to Designing Welds*, 1989.
Maddox, S.J., *Fatigue Strength of Welded Structures*, 1991.
TWI International Conference on the Effects of Fabrication Related Stresses on Product Manufacture and Performance, 1987.
TWI International Conference on the Fatigue of Welded Structures, 1988.

Abington Publishing also supply directories (1988) of:

Covered electrodes for C and C : Mn steels.
Covered electrodes for low alloy steels.
Welding consumables' tradenames.
Welding manufacturers and suppliers.

Glossary

This glossary contains a number of terms used in the book. Those extracted from BS499, Part 1: 1991, 'Welding terms and symbols' are in italics and are reproduced by permission of the British Standards Institute, 2 Park Street, London W1A 2BS, from whom copies of the Standard may be obtained. A few terms (denoted by an asterisk, *) are based on definitions given in the NACE *Corrosion Engineers' Handbook* (ed. R.S. Treseder, NACE, Houston, TX, 1980). Terms are not necessarily fully defined, but are described in sufficient detail to understand them in the text.

A_{c1}
The temperature at which lower temperature transformation products start to transform to austenite on heating. (The A_{c3} is the temperature at which this transformation is completed; A_{r1} and A_{r3} are the corresponding temperatures on cooling; A_1 and A_3 are used when heating or cooling is not relevant.)

Acicular ferrite
A type of weld metal microstructure comprising Widmanstätten ferrite grains formed intergranularly; also isolated laths of ferrite of high aspect ratio.

API
American Petroleum Institute; a source of specifications and recommended practices, including guidance in steels suitable for high temperature hydrogen service (API 941) and hot tapping (API 1107).

A_{r1} see A_{c1}

Arc atmosphere
The atmosphere surrounding the welding arc during welding.

Glossary

Arc blow
A lengthening or deflection of the welding arc caused by asymmetric distribution of magnetic flux around the arc.

Arc efficiency
The efficiency of transfer of heat from the arc to the material being welded.

Arc energy (see also **Heat input**)
The amount of heat introduced by the welding process per unit length of weld, defined (for manual metal arc welding with covered electrodes) as:

$$\frac{VI}{w} \times 10^{-3}$$

where w *is the welding speed in mm/s.*

Arc energy can be used to estimate the cooling cycle of a weld and its HAZ in order to estimate the microstructure of the HAZ and, hence its risk of suffering hydrogen cracking. When multi-arc techniques are used, the separate arc energies should be added together, provided that a single weld pool is produced.

Arcing, stray (stray flash)
The accidental striking of an arc away from the weld and/or the damage on the parent metal resulting from the accidental striking of an arc away from the weld.

Arc strike see **Arcing, stray**

As-deposited
A weld, single or multipass, after deposition at whatever preheat and interpass temperatures were used; a weld is often considered to be in the as-deposited condition if it has been post-heated at a temperature appreciably below the normal PWHT temperature for that steel.

Austenite (gamma, γ)
The solid solution of carbon in the face-centred cubic crystal structure of iron. Austenite is only stable at high temperatures in ferritic steels but may be retained by fast cooling (*retained austenite*). In austenitic stainless steels it is stabilised at ambient and lower temperatures by the nickel in the steels.

Autogenous welding
Welding without the addition of filler; the weld metal consists of melted parent metal.

Automatic welding
Welding in which all the welding parameters are controlled. Manual adjustments may be made between welding operations but not during welding.

Back gouging
Removal of all or most of the first side root run by carbon arc gouging before depositing the root run on the second side to be welded.

Backing strip
A strip or bar of steel, usually of similar composition to the parent steel, which is held under the joint before welding to hold the weld pool. It is welded to the weldment and may be ground or chipped off or left in place.

Bainite
A transformation product of austenite formed during moderately slow cooling, and comprising a fine mixture of **ferrite** and iron carbide with a particular configuration.

Baking
Often used when MMA electrodes and submerged-arc fluxes are dried at high temperatures, e.g. above about 250 °C, to remove moisture. When fluxes are dried at temperatures above about 800 °C the term *pre-sintering* is sometimes used.

Balanced welding
Welding a two-sided weld so that one or more passes on one side are followed by one or more passes on the other side; this type of sequence is followed to the completion of the weld. Balanced welding evens up the distortion and longer range residual stresses.

Basic electrode (basicity)
A covered electrode in which the covering is based on calcium carbonate and fluoride.

A **basic** (submerged arc) **flux** or MMA cover contains a high proportion of basic oxides, and gives a weld metal of relatively low oxygen content, which is usually associated with good weld metal toughness. With MMA electrodes, the term is almost synonymous with *hydrogen controlled*; however, basic submerged arc fluxes do not necessarily give lower weld hydrogen contents than other types.

Glossary

Basic flux
A flux (usually for submerged arc welding) with a high proportion of basic minerals such as calcium fluoride.

Bay region
The wide region of a HAZ where the finger penetration (if present) joins the upper part of the weld bead.

Bead-on-plate
A single weld bead deposited on a flat surface of metal and not in a joint preparation.

Biological fouling
Growth, on steel structures in (sea) water, of marine organisms; their death and decomposition can give local concentrations of sulphides in the water which can lead to sulphide stress cracking of the steel and its weldments.

Body-centred cubic (BCC)
A type of crystal structure in which atoms occur in the crystal lattice at the corners and at the centre of a cube. Ferrite is the BCC form of iron.

Brittle fracture
Fracture by a brittle mechanism, such as cleavage or intergranular fracture, without significant plastic deformation. Such fracture is particularly dangerous in weldments, because the presence of residual stresses and stress concentrations can increase the local stress above the yield stress of the parent steel, even if the applied stresses are low, thus making such fracture easier.

Buttering
Depositing weld metal on to the weld preparation of one or both sides of a joint so that the actual weld is deposited on to weld metal. This can be a technique for reducing the risk of lamellar tearing or, if the buttered half of the joint is given a PWHT, for giving PWHT at different temperatures to the HAZs on two different parent steels.

Carbon arc gouging
Thermal cutting by melting using the heat of an arc between a carbon electrode and the metal to be cut.

A competent operator can usually detect when a defect has been removed.

Carbon equivalent (CE) value
A number, calculated from the chemical composition of the steel by means of an empirical formula, which summarises some aspect of the steel's behaviour, particularly its hardenability or its resistance to hydrogen cracking. The term may also be used for weld metals.

Cathodic protection*
Shifting the corrosion potential of a metal to a less oxidising potential by applying an external EMF (electromagnetic force; voltage) in order to reduce the rate of corrosion.

Cellulosic electrode
A covered electrode in which the covering contains a high proportion of cellulose.
   Cellulose contains a high proportion of hydrogen, which helps to confer high penetration during welding.

Cementite
A compound, Fe_3C, between iron and carbon which is in quasi-equilibrium with iron in C and C:Mn steels at temperatures below the transformation from austenite.

Combined aluminium
Aluminium that is chemically combined with oxygen.

Combined carbon
Carbon that is in the form of carbides and, therefore not able to participate in strain-ageing.

Combined nitrogen
Nitrogen that is combined with nitride-forming elements, particularly Al, Ti and B, and that cannot participate in strain-ageing.

Combined thickness
The sum of the thicknesses (in mm) of the metal through which the heat produced during welding can flow away from the joint. In texts before about 1970, the term *thermal severity number (TSN)*, measured in ¼ inch (6.35 mm) units was used in Britain.

Consumable guide
A wire guide, which may be coated or uncoated, made of material similar in composition to that being welded and progressively consumed to form part of the weld metal.
   Consumable guide welding is a variant of electroslag welding.

Continuous cooling transformation (CCT; see also **Transformation**)
The transformation of austenite to lower temperature transformation products such as martensite, bainite, pearlite and ferrite in a steel that is being steadily cooled (as in a weld HAZ) as opposed to being transformed at a constant temperature (*isothermal transformation*). Diagrams of the CCT and isothermal transformation behaviour for many types of steel are available in the literature.

Cored wire
A consumable electrode having a core of flux or other material.

It comprises an outer sheath of steel (usually mild steel) which contains flux and/or metallic ingredients. The sheath may be a fine hollow tube or a strip wrapped round the core. Cored wires are chiefly used for MAG, MIG or submerged-arc welding.

Covered electrode
An electrode having a cover of flux or other material. NOTE: the term flux in this context is used in its ordinary engineering sense. The covering is sometimes loosely described as flux, whether or not other materials are present.

Crater (Crater pipe)
A depression due to shrinkage at the end of a weld run where the source of heat was removed.

Creep
Deformation of a material at high temperatures under stress with time.

Critical hardness
The lowest maximum HAZ hardness at which hydrogen cracking is found in the HAZ under the particular test conditions used. The critical hardness varies with weld hydrogen content, welding process, preheat level, joint geometry and steel type.

Decohesion
The separation of inclusions, or other second phase particles, from the steel matrix: the first stage of lamellar tearing and microvoid coalescence.

Defect
An imperfection or fault in a metal sufficiently large to give a risk of failure on testing or in service.

Defect tolerance
A term used in fracture mechanics to denote the size of defect which can

be tolerated in a structure without giving a risk of unstable fracture under service or pre-service testing stresses.

Dendrite (dendritic)
The usual mode of solidification of ferritic weld metals is for grains to start growing rapidly along an axis, for branches to grow perpendicularly from this axis and for secondary branches to grow from the primary branches. The skeleton existing early in solidification resembles the branches and twigs of a tree – hence dendritic (tree-like).

Deoxidant
A material that combines with oxygen in liquid (weld) metal to form oxide(s) which may or may not separate completely into the slag during solidification. The most common deoxidants added to ferritic steel weld metals are Si and Mn; Al, Ti, Mg and Ca may also be used in various forms, the oxides of the last two usually separate completely into the slag.

Deposited metal
Filler metal after it has become part of a welded joint.

The increase in weight of a parent material due to the deposition of filler metal is particularly used in tests to measure weld hydrogen contents.

Depth/width ratio
The ratio between the maximum depth and the maximum width of a single weld bead. The ratio is of importance in solidification cracking.

Diffusible hydrogen
That portion of the hydrogen content of a ferritic steel which will be released by diffusion when the steel is held for several days at ambient (25 ± 5 °C) temperature.

Diffusion bonding (*Diffusion welding*)
A joining process wherein all the faces to be welded are held together by a pressure insufficient to cause readily detectable plastic flow, at a temperature below the melting point of any of the parts, the resulting solid state diffusion, with or without the formation of a liquid phase, causing welding to occur.

Dilution
The alteration of composition of the metal deposited from a filler wire or electrode due to mixing with the melted parent metal. It is usually expressed as the percentage of parent metal in the weld metal.

Dip transfer
A method of metal-arc welding in which fused particles of the electrode wire in contact with the molten weld pool are detached from the electrode in rapid succession by the short-circuit current which develops every time the wire touches the weld pool.

Dislocation
A mismatch on the atomic scale in the crystal structure of a metal. Dislocations hinder the plastic flow of a crystal and thus strengthen it. Plastic flow is hindered when dislocations interact with interstitially dissolved atoms (e.g. nitrogen and hydrogen), thus causing embrittlement.

Drying
Heating of MMA electrodes or flux to remove moisture. The term is sometimes restricted to low temperature heating up to about 250 °C, which will remove only loosely bound water (see **Baking**).

Ductile (fracture)
Fracture preceded by significant plastic deformation; however, in certain circumstances (e.g. creep, ductile grain boundary fracture and lamellar tearing) the overall deformation may by very small if it is confined to regions near the grain boundary, or if the number of inclusions is large.

Ductility dip
Face-centred cubic metals (e.g. austenite) containing certain impurities can exhibit reduced ductility (called the ductility dip) within a temperature range below the melting point (see **Hot cracking**).

Dye penetrant
A suitably coloured liquid of low surface tension applied for NDE which can enter a surface crack and which can be subsequently detected (after the excess on the surface has been removed), after it seeps out again, by staining a white 'developer', or by fluorescing under ultraviolet light.

Electron beam welding
Fusion welding in which the heat for welding is generated by the impact of a focused beam of electrons.

Electrogas welding
Arc welding using a gas-shielded consumable electrode to deposit metal into a weld pool, retained in the joint by cooled shoes which move progressively upwards as the joint is made.

Electroslag welding
Fusion welding using the combined effects of current and electrical resistance in a consumable electrode(s) and a conducting bath of molten slag, through which the electrode(s) passes into the molten pool, both the pool and the slag bath being retained in the joint by cooled shoes which move progressively upwards. After an initial arcing period, the end of the electrode is covered by the rising slag, and melting then continues [by resistive heating] *until the joint is completed.*

The process requires a high heat input, giving slow cooling and coarse microstructures, which may require normalising to achieve satisfactory toughness.

Embrittlement
The reduction in the normal toughness of a material by the addition of other materials, heat treatment, deformation processes or a combination of these.

Eutectic
In a series of alloys between two metals, a metal and a compound or two compounds, if the metals are not fully soluble in the solid state, the addition of either metal to the other can reduce the melting point of the other so that a minimum value (the eutectic) is reached, at which point both metals solidify simultaneously. In a binary system the eutectic has an invariant temperature and composition. Similar behaviour in the solid state gives a eutectoid, an example of which occurs in the Fe:C system between austenite, ferrite and cementite at 730 °C and 0.8% C.

Excess penetration bead
Excess weld metal protruding through the root of a fusion weld made from one side only.

Excess weld metal
Weld metal lying outside the plane joining the (weld) toes.

Face-centred cubic (FCC)
A type of crystal structure in which atoms occur in the crystal lattice at the corners and at the centres of the faces of a cube. Austenite is the FCC form of iron.

Fatigue
Continually variable stressing of a metal. Fatigue failure can occur at much lower stresses than the tensile strength of the metal and is very sensitive to stress concentrations.

FATT
Fracture appearance transition temperature, a term most frequently used in Charpy impact toughness testing. The transition temperature is defined as that at which the fracture surface contains a specified amount (frequently 50%) of ductile (or brittle) fracture.

Fault
An indication found by NDE which originates from a feature smaller than an unacceptable defect. Sometimes called an 'imperfection'.

Ferrite
The solid solution of carbon in the body-centred cubic crystal structure of iron. The phase is stable at low temperatures (alpha, α) and also, in some low carbon ferritic steels, at temperatures close to the melting temperature, when it is called delta-ferrite (δ).

Ferrite with aligned second phase
A type of weld metal microstructure comprising two or more parallel laths of ferrite whose aspect ratio is >4 : 1. The terms ferrite with aligned MAC (martensite, austenite and carbide), ferrite side plates, upper bainite, feathery bainite and lamellar product have also been used to describe this constituent.

Ferritic steel (weld metal)
A ferritic steel or weld metal is one containing little (usually <10% by volume) or no austenite, any austenite being of the residual type.

Filler metal
Metal added during welding, braze welding, brazing or surfacing.

Fisheye
A small bright area of cleavage fracture, caused by the presence of hydrogen, that is visible only on the fractured surface of weld metal.

The centre of the feature (the 'pupil') is a pore or inclusion, the area round it (the 'iris') is hydrogen-assisted brittle (cleavage) fracture and the remainder of the fracture is ductile.

Fitup
The gap between the two faces of a joint immediately prior to welding. The terms close fit, machined fit, normal fit and poor fit are sometimes used.

Flare angle
The angle formed between the finger penetration and the upper part of the weld bead (see **Bay region**).

Flux
Material used during welding, brazing, braze welding or surfacing to clean the surfaces of the joint chemically, to prevent atmospheric oxidation and to remove impurities. In arc welding, many other substances, which perform special functions, are added.

Although basic fluxes remove S, less basic fluxes may add certain impurities which they contain, including S and P. All fluxes give a quasi-equilibrium oxygen content to the weld metal which is appreciably higher than that of a killed steel.

Flux-cored wire
A type of **cored wire** in which most of the core is of non-metallic, fluxing ingredients, although some iron powder, deoxidants and metallic alloying additions may also be present.

Fractography
The study of fracture surfaces, which can be aided by microscopic examination, particularly in an SEM. To examine cracks in ferritic steels and their welds, it is convenient to cool the specimen to a very low temperature (e.g. by immersing in liquid nitrogen) and break it open while still cold. Any existing crack surface can usually be easily distinguished from the brittle fracture produced on opening the crack.

Fracture toughness (see also **Toughness**)
The ability of a material containing a sharp defect, such as a crack tip, to resist fracture (or crack growth) when stressed. The fracture toughness of steels (and weld metals) is reduced when the temperature is reduced and also near ambient temperature when they contain hydrogen. Parameters measured include K_c, the critical stress intensity at the crack tip, and δ_c, the critical crack tip opening displacement (CTOD).

Free-cutting steel
A steel containing additions such as sulphur and lead to improve machineability.

Free nitrogen (Al, C, etc)
The nitrogen (Al, C, etc.) which is not combined as stable compounds, such as nitrides, carbides and oxides, and which is in interstitial or normal solid solution in the ferrite matrix.

Glossary

Friction welding
A welding process in which one component is moved relative to, and in pressure contact with, the mating components to produce heat at the faying surfaces, the weld being completed by the application of a forge force during or after the cessation of relative motion.

Fused metal
That part of a welded joint which has been melted during welding. It comprises added weld metal and fused parent metal. The term is particularly used in tests to measure weld hydrogen contents and it is usually estimated from the area of fused metal in a weld cross section which has been ground and etched.

Fusion boundary
The boundary between the HAZ and the melted metal in a weld. Because grains usually grow epitaxially from the HAZ into the weld metal, the actual boundary may need careful examination to detect when examining a metallographic section. The presence of very small round inclusions is a good guide to the extent of the weld metal.

Fusion face
The portion of a surface, or of an edge, that is to be fused in making a fusion weld.

Fusion welding
Welding in which the weld is made between surfaces brought to a molten state, with or without the addition of a filler metal, and without the application of pressure.

Galvanic corrosion*
Corrosion in an ionized liquid (e.g. seawater) when two metals exhibit different electrical potentials; that which is anodic corrodes and protects the other.

Gas flat see **Surface pocking**

Grain refinement
Refinement of the grain size of a metal by heat treatment or other means. In the case of a ferritic steel, heating to a low temperature within the austenite range and cooling refines the **prior austenite grain size**. This may or may not refine the transformed ferrite **grain size**.

Grain size
The mean diameter of the grains that make up the microstructure of a material. When used of ferritic steels, the term often refers to the **Prior austenite grain size**. For the grain size of transformation products, terms such as ferrite grain size and martensite lath size are used.

Graphite (graphitisation)
Graphite, and not cementite, is the phase in equilibrium with iron in C and C : Mn steels. However, the approach to equilibrium is so slow that the formation of graphite by decomposition of cementite to graphite and ferrite (graphitisation) takes several years at temperatures between 450 °C and the A_1.

Half-bead technique
A welding technique designed to produce a refined HAZ on the parent steel in order to achieve acceptable toughness in repair and other welds which cannot be given a PWHT. The first layer of weld metal is deposited on the steel to be repaired and is ground off to half its depth. Welding is then continued with the same welding parameters. This technique should ensure that the refining temperature range from the second layer of weld metal refines the coarse-grained HAZ on the parent steel left from deposition of the first layer.

Hardenability
The tendency of a ferritic steel to produce a hard microstructure as a result of heating and rapid cooling, typically in the HAZ during welding. The hardenability of a steel increases as its content of alloying elements (and, to a smaller degree, carbon) is increased.

Hardening, secondary
After hardening by quenching, some steels (containing such elements as Cr, Mo, V and Nb) can harden further if they are tempered at temperatures around 600 °C as a result of processes (akin to precipitation hardening) leading to the formation of certain carbides.

Hardness (Hardness value)
The resistance of a material to deformation, usually measured by pressing an indenter into the surface. Of the several types of hardness measurement, the *Vicker's hardness test (HV)* is most commonly used in the context of welding, as the small size of the impression produced by the indenter is suitable for measuring local variations in hardness found in a weld HAZ, whose hardness varies over short distances. When measuring HAZ hardness, it is important to make sufficient measurements, particularly in the

region close to the fusion boundary, to obtain a realistic idea of the maximum hardness of the HAZ. In connection with hydrogen cracking during welding, the hardness of the HAZ associated with a weld bead before tempering (either by a subsequent weld pass or by PWHT) is the important parameter.

Heat-affected zone (HAZ)
That part of the parent metal that is metallurgically affected by the heat of welding or thermal cutting, but not melted.

In the context of ferritic steels, the term is often used for that portion of the HAZ (strictly, the *visible HAZ*) which is visible to the naked eye on an etched macro-section of a weld. Close to the fusion boundary, where the temperature had approached the melting point of the steel, appreciable coarsening of the **prior austenite grain size** occurs. Although the term HAZ is normally applied to a HAZ on parent steel, weld metal can also become the HAZ of a succeeding run or weld and a term such as 'HAZ on weld metal' should be used if there is any doubt.

Heat input
The amount of heat supplied to the parent metal by welding. The heat results from the energy of the welding arc (**arc energy**), reduced by the efficiency of transfer of the heat (**arc efficiency**). The terms *heat input* and *arc energy* are often used synonymously and, in Europe at least, heat input is likely to be used for both parameters.

Heat treatment
Heating of metals without melting to alter their properties. The term should not be used to describe preheating for welding.

HIC (hydrogen-induced cracking)
Hydrogen cracking in service in which the steel is immersed in an acid sulphide solution (**sour service**). Although the abbreviation is used exclusively for this type of cracking, the term 'hydrogen-induced cracking' is strictly applicable to cracking caused by hydrogen embrittlement at any stage of the life of the component, especially for cracking during welding.

Hot cracking (see also **Solidification cracking**)
A discontinuity produced by tearing of the metal while at an elevated temperature.

Although the term usually includes **solidification cracking**, there is no evidence that tearing of solid metal occurs during solidification cracking.

Hot tapping
Welding (particularly of branches) on to pipework or similar components while they contain their normal working fluids, e.g. oil or gas under pressure.

Humidity
The amount of moisture in solution in the air at any time. It can be expressed in two ways: the *relative humidity*, which is the humidity expressed as a percentage of the maximum amount of moisture it is possible for air to contain at the particular temperature and pressure of measurement; and the *absolute humidity* is expressed as the weight of moisture in a unit volume of air. The latter is probably the more significant as regards the amount of hydrogen likely to be introduced from the atmosphere into a weld pool during welding.

Hydrogen attack (damage)
When steel is used to contain hydrogen at high temperature and pressure, damage can occur if the service temperature is too high. This is caused by the hydrogen diffusing into the steel and combining with the carbon in the carbides in the steel to form the gas methane. Besides removing the carbides, which strengthen the steel, the gas produced forms cavities, either adjacent to the carbides or along chains of large inclusions in the steel to cause blistering.

Hydrogen blistering*
Sub-surface voids produced by hydrogen absorption in (normally) low strength steel plate with resultant surface bulges. The hydrogen, usually resulting from sour or high temperature hydrogen service, can collect along planes of inclusions in the plate and build a high pressure.

Hydrogen cracking
This is cracking produced in steel that has been embrittled by hydrogen, which may originate from the steel melting process, from hydrogen picked up during welding or from hydrogen picked up during service, either directly in hydrogen service or as a result of corrosion reactions. The cracking itself occurs in the temperature range within which hydrogen embrittles steel, namely within about 200 °C of ambient. In the welding context, it usually refers to cracking produced during welding.

Hydrogen embrittlement
Ferritic steels (and other metals) are embrittled to an increasing degree at temperatures around ambient when containing increasing amounts of hydrogen. Embrittlement is most severe at slow strain rates and is most readily apparent in steels with relatively poor toughness; in HAZ material, this is synonymous with high hardness.

Hydrogen level see **Weld hydrogen level**

Glossary

Hydrogen potential
The hydrogen level of a welding consumable, which may be estimated from the hydrogen content of a welding wire and/or the total moisture (free and combined) of any fluxes, including the electrode covering, involved.

Hydrogen stress (corrosion) cracking
A type of stress corrosion cracking (more usual in high strength steels) where hydrogen, picked up by the steel from the water (usually containing H_2S – see **Sulphide stress cracking**) as a result of corrosion reactions, embrittles the steel sufficiently to cause cracking under the influence of the residual and/or applied stresses present.

Hyperbaric welding
Underwater welding in a chamber pressurised to resist the external pressure of water.

Imperfection see **Fault**

Impact (notch impact) toughness
Toughness as measured in an impact test (usually as the energy expended in fracture initiation plus propagation). A specimen, usually notched, is struck by a falling hammer with a fixed energy.

Impurity (Impurity elements)
Elements usually of a non-metallic nature (see also **residual elements**) which are not intentionally added to a metal. In ferritic steels the usual ones analysed for are S and P, together with N, O and, in some cases, As, Sb and Sn. Impurities can affect the properties of steel and make it more prone to various types of embrittlement and cracking, particularly during welding.

Inclusions (non-metallic inclusions)
Small particles within the steel consisting of a mixture of deoxidation products (oxides) and sulphides. These can act as 'traps' for hydrogen, and hence reduce its diffusion rate, particularly at and below ~150 °C. Inclusions can also act as nuclei for the formation of higher temperature (and softer) transformation products when steel is cooled from high temperatures, thus reducing the hardenability of a steel.

Indication
An indication of a fault or defect during NDE, particularly when revealed by ultrasonic examination. Care should always be taken not to describe

an indication as a defect or as a crack until further examination has been carried out.

Ingot cast
Steel that has been cast into an ingot mould as opposed to a shaped mould (steel casting) or that has been continuously cast.

Intercritical
A term usually applied to the part of a HAZ in a ferritic steel which has been heated above the A_{c1} temperature but below the A_{c3}. Thus, only part of the microstructure has been transformed to an austenite of relatively high carbon content and re-transformed on cooling. It is the part of the visible HAZ furthest from the fusion boundary.

Intergranular
Use to refer to cracking (or such phenomena as precipitation) along, or in the region of, a grain boundary.

Interpass temperature
In a multirun weld, the temperature of the weld and adjacent parent metal immediately prior to the application of the next run.

In welding procedures it is possible to specify a minimum and/or a maximum interpass temperature.

Interstitial (solution)
A type of solid solution whereby small atoms of elements such as C, N and H can dissolve within the atomic lattice. Larger atoms which are in normal solid solution (e.g. Mn and Ni) replace the iron atoms from positions on the lattice itself.

Iron powder electrode
A covered electrode in which the covering contains a high proportion of iron in the form of a powder which acts as additional filler metal.

Iron powder can also be added to the weld pool during submerged-arc welding. Strictly, the term should be applied only to pure iron powders; when containing alloying and/or deoxidant additions, the term **metal powder** should be used.

Isothermal transformation see **Continuous cooling transformation)**

Joint
A connection where the individual components, suitably prepared and assembled, are joined by welding or brazing.

Killed steel
A steel to which sufficient deoxidants (usually Si and/or Al) have been added to avoid porosity when the steel solidifies as an ingot. The word 'killed' is used because such a steel will not effervesce as a result of the evolution of gases during solidification when cast.

Lack of fusion
Lack of union in a weld:

(a) *between weld metal and parent metal, or*
(b) *between parent metal and parent metal, or*
(c) *between weld metal and weld metal.*

Lamellar tearing
Fracture, by a tearing mechanism at or just above ambient temperature, of steel plate under a welded joint in which a sufficiently high population of certain types of inclusions occur.

Laser welding
Fusion welding in which the heat for fusion is produced by a coherent beam of monochromatic radiation from a laser.

Lattice diffusion
Diffusion within the crystal lattice, which obeys regular physical laws. In steels at temperatures around ambient, hydrogen diffuses much more slowly than is calculated from these physical laws.

Liquation (Liquation cracking)
The partial fusion of minute regions rich in impurity elements (particularly S) in the HAZ close to the fusion boundary. Liquation can lead to cracking as a result of high temperature contraction strains, but the cracks are very small, although they can initiate hydrogen cracking.

Liquid metal penetration
Penetration of a solid metal, usually along its grain boundaries, by a molten metal with which it is in contact. Although cracking may not always result, such penetration is usually weakening and/or embrittling.

Local brittle zone (LBZ)
A small region leading to local brittle fracture, usually found in a fracture toughness test but possible in any type of fracture, in a generally ductile fracture surface. Such zones can seriously reduce the resistance of a material to fracture initiation. Although in a toughness test the brittle fracture may be arrested, LBZs are liable to lead to complete failure in a structure

with long-range stresses. Such zones may be found in reheated HAZ material and occasionally in weld metal.

Magnetic particle inspection (MPI)
Similar to **dye penetrant** inspection, except that the penetrating fluid contains fine particles which are ferro-magnetic. An electromagnet is used to cause these particles to adhere temporarily to crack surfaces and other breaks in the structure at the surface. Either a.c. or d.c. may be used for the electromagnet; the latter has an ability to detect faults which are just below the surface.

MAG welding (see also **MIG welding**)
Gas-shielded metal arc welding using a consumable wire electrode where the shielding gas is provided by a shroud of active or non-inert gas or mixture of gases.
 The shielding gas is 'active' (hence MAG, metal active gas) because it contains oxygen or CO_2, often with an inert gas such as Ar and/or He. The term *CO_2 welding* is used when CO_2 alone is used as the shielding gas. A solid or a cored wire may be used, although with the latter, the term *cored wire welding* is more common.

Manual metal arc welding
Metal arc welding with straight covered electrodes of a suitable length and applied by the operator without automatic or semi-automatic means of replacement. No protection in the form of gases from a separate source is applied to the arc or molten pool during welding.

Maraging
A heat treatment given to some complex nickel steels (so-called **maraging steels**) of low carbon content which are martensitic at room temperature, even with slow cooling. The treatment strengthens the steel by an ageing process involving the precipitation of intermetallic compounds.

Martensite
A body-centred tetragonal (i.e. distorted BCC) constituent, supersaturated in C, formed when steels are rapidly cooled from austenite. It is the hardest constituent possible in a rapidly cooled steel and has poor toughness and a poor resistance to hydrogen cracking unless of low carbon content.

Metallographic (examination)
To do with the (microscopic) structure of metals. Metallographic examination involves preparing a polished section of a metal and studying it by microscope, usually after it has been etched to reveal its structure.

Metal powder
An admixture of iron and other metal powders (alloying and/or deoxidising) used either as an ingredient of MMA electrode coverings, or as an addition made during submerged-arc welding (see **iron powder**).

M_F temperature (also M_S, M_{90})
The temperature at which the transformation from austenite to martensite finishes (M_F), begins (M_S), is 90% complete (M_{90}) and so on.

Microalloying
Alloying with usually less than about 0.2% of alloying elements such as Nb, V, Ti and/or B. Microalloyed steels are usually strong and tough, with good weldability for their strength.

Microvoid (coalescence)
Microscopically small voids produced when a material is stressed sufficiently to fracture or to separate (decohere) small inclusions, or brittle or hard particles from the matrix.

MIG welding (see also MAG welding)
Gas-shielded metal arc welding using a consumable wire electrode where the shielding is provided by a shroud of inert gas.

The term was originally used (and sometimes still is used) for both **MIG** and **MAG welding**.

Millscale
Oxides produced on or near the surface of steel while it is being hot processed and which can be firmly embedded in the surface.

Misalignment, angular
Misalignment when the pieces welded together are out of line before and/or on completion of welding so that a distinct angle is seen at the weld in a transverse section of the joint.

Misalignment, linear
Misalignment when the pieces welded together are out of line before and/or on completion of welding so that a distinct step is present at the joint on one or both surfaces.

Multi-arc welding
Welding with two or more arcs, usually producing one weld pool. It has the advantage of high productivity without the disadvantages of using the same total current in one arc.

NACE
National Association of [American] Corrosion Engineers, who produce recommendations and standards, including those on the selection and use of materials for sour service (NACE standard MR0175).

Necking
After reaching the maximum load during a tensile test of a sufficiently ductile metal, deformation is concentrated into a short length of the test specimen. Metal in this length contracts in the transverse direction and extends in the longitudinal direction more rapidly than the metal outside this region, thus producing a neck.

Nelson diagram
A diagram showing the maximum temperatures and partial pressures of hydrogen to which several types of steel may be exposed without them suffering hydrogen attack.

Neutral (flux)
A flux, usually for submerged arc welding, which has no effect on the major constituents (Mn, Si and major alloying elements other than C) of the weld metal.

Nital
A metallographic etch consisting of ethanol (ethyl alcohol) with between 1% and about 20% nitric acid (HNO_3). Unless otherwise stated it is the standard etch for ferritic steels, the lower concentrations being used for microscopic examination and the stronger etches for macro work.

Non-destructive examination (NDE)
Examination for defects without cutting or otherwise damaging the object being examined. Sometimes known as NDT – non-destructive testing.

Non-metallic inclusions (see Inclusions)

Normalising
Heating a ferritic steel to a temperature within the austenitising range and cooling in still air.

Notch toughness see Toughness

Nucleation
The process where a phase change is started, in an otherwise uniform structure, at an irregularity such as a grain boundary or an inclusion.

Nucleation may involve either a phase crystallising from a liquid or a new phase crystallising from the solid; in either case, the most efficient nucleus is one having the same crystal structure and closely similar lattice dimensions to the new phase, even though the nucleus and the new phase are not mutually soluble.

Overlap
An imperfection at a toe or root of a weld caused by metal flowing on to the surface of the parent metal without fusing to it.

Overmatching (weld metal)
The condition when the yield (or tensile) strength of a weld metal exceeds that of the parent metal. If requiring overmatching, it is important to be clear whether overmatching is required on yield stress or tensile strength, because the ratios of the two are likely to be different in the two types of material.

Pearlite
A transformation product resulting from slow cooling from austenite, comprising a lamellar microstructure of ferrite and iron carbide.

Peening
Hammering or blasting the surface of a metal with rounded shot in order to induce compressive residual stresses in the surface layers and reduce the risk of fatigue or other types of cracking.

Penetration (finger penetration)
The depth to which the weld pool extends. In MAG with spray transfer and submerged arc welding, the arc initially penetrates deeply (finger penetration), although more diffuse heating causes the upper part of the pool to widen.

Peritectic
Where an alloy is solidifying as one phase and another phase becomes more stable than the first, a peritectic transformation is said to take place. Above the peritectic temperature, the higher temperature phase solidifies; below it, the lower temperature phase solidifies, the peritectic being the boundary between the two types of solidification. In a binary system the peritectic has an invariant temperature and composition; in the Fe : C system it is at 0.18% C and 1492 °C between liquid, delta ferrite and austenite.

Planar defect
A defect, such as lack of fusion or a crack, which lies essentially in one plane and which provides a damaging stress concentration.

Plasma arc welding
Arc welding in which the heat for welding is produced with a constricted arc . . .; plasma being generated by the hot ionised gases issuing from the orifice and supplemented by an auxiliary source of shielding gas.

Polarity
The polarity of a welding electrode with regard to the metal being welded.

Porosity
Holes or pores in a weld metal formed by gas evolved during solidification (*gas porosity*, although in BS499 the term 'porosity' is used only for gas porosity) or by contraction which is not adequately fed (*shrinkage porosity*). The distinction between the two types is not clear cut because some gas evolution is needed to initiate shrinkage porosity. In many alloy systems gas porosity consists of spherical pores and shrinkage porosity of inter-dendritic cavities.

Post-heat
The application of heat to a weld, immediately after welding and before cooling out, so as to maintain the minimum interpass temperature, or to raise it to some higher value (well below any PWHT temperature), in order to increase the rate of diffusion of hydrogen out of the weldment.

Post-weld heat treatment (PWHT)
Heat treatment given after welding in order to reduce residual stresses (*stress relief heat treatment*) and/or to temper hard regions, usually in the HAZ. The temperatures for ferritic steels are usually between about 550 and 750 °C, i.e. below the A_{c1} and in the upper part of the tempering temperature range.

Preheating temperature (preheat)
The temperature immediately prior to the commencement of welding resulting from the heating of the parent metal in the region of the weld.

Normally, a minimum preheat temperature is required; when it is necessary to achieve particular levels of toughness and/or strength in the HAZ or weld metal a maximum value may also be required. Preheat may be local or general.

Glossary

Prior austenite grain size
The grain size of austenite before it finally transforms on cooling. The austenite grain size increases rapidly as the melting temperature is approached, but is not then reduced on cooling before transformation. A coarse prior austenite grain size can be refined by heating the steel briefly to just above the A_{r3}. Various expressions are used to quantify prior austenite grain size, one of which (the ASTM system) uses a single number, 0 being coarse and 10 fine.

Quenching
Cooling a metal rapidly either to retain a high temperature microstructure or, as in the case of ferritic steels, to produce a microstructure (typically martensite) which can only be produced by rapid cooling. Rapid cooling may be effected by immersing the metal into water or oil (cold or hot), by spraying or, in the case of air hardening steels, by a blast of air.

Radiography
Examination of an object by means of X- or gamma-rays and recording the resultant radiograph on photographic film. Radiography is only capable of detecting bulky defects; a planar defect is normally detected only if its plane is roughly parallel to the incident beam.

Refined HAZ
Part of the HAZ of a multipass weld which has been reheated to a temperature sufficient to refine its prior austenite grain size.

Reheat cracking (stress relief cracking)
Cracking that forms after welding in a restricted number of steels during PWHT by a process of creep, usually during heating to the PWHT temperature. A crack found only after PWHT is not necessarily a reheat crack.

Reheating
Heating of a weld run and/or its HAZ by a succeeding run. The term is normally used when a significant change, particularly in microstructure, results from reheating.

Residual elements
Elements, which usually originate from alloyed steel scrap and non-ferrous metals in the steel charge, of the type that can be added to steels as alloying elements. They include Ni, Cr, Mo, Cu, Nb, V and sometimes Sn. Residual elements can affect the CE value of the steel and also other properties (see also **Impurity**).

Residual hydrogen
That portion of the hydrogen content of a ferritic steel which will not be released by diffusion at ambient temperature (25±5 °C), but may be extracted at higher temperatures, e.g. 650 °C. It is believed that this hydrogen is trapped in molecular form, either in the lattice, in voids (particularly in association with non-metallic inclusions) or in chemical combination with other elements, such as carbon (when methane is formed).

Residual welding stress (residual stress)
Stress remaining in a metal part or structure as a result of welding. The stresses in a weldment result because the contraction of the hot weld is hindered by the **restraint** of the joint. Residual stresses tend to be tensile at the weld (at the weld surface in a multipass weld) and are balanced by compressive stresses in the parent steel (or in the root regions of a multipass weld).

Restraint
The extent to which the parts to be welded are secured to prevent all controllable movement during and after welding. The parent steel, unless extremely thin and narrow, provides sufficient restraint to result in a weld containing **residual stresses.**

Retained austenite
Austenite in a ferritic steel which persists on cooling to ambient temperature. After PWHT, retained austenite may transform to martensite of relatively high carbon content and a second heat treatment (*double tempering*) may be required to temper it.

Retained phase
Although most of the volume of a ferritic weld metal transforms above ~400 °C, a small volume transforms at a much lower temperature, if at all. Because retained austenite results if there is no transformation of this volume, and because all the phases have a similar shape and distribution to retained austenite, the term 'retained phase' is used.

Rimming steel
An ingot steel which has not been killed. During solidification of the ingot, the steel solidifies for some distance from the outside before dissolved gases evolve, so that the outside – the rim – has a different appearance from the centre in a polished and etched section, even after drawing to wire. This type of steel has long been used for making welding electrodes and is still employed when low contents of nitrogen and residual elements are needed.

Glossary

Root concavity
A shallow groove that may occur in the root of a butt weld.

Root run (pass)
The first run deposited in the root of a multi-run weld.

Run-off plate (block)
A piece of metal so placed as to enable the full section of weld metal to be maintained up to the end of a joint.

Its use prevents shrinkage and cracks in the crater region (which should all be on the run-off plate) from impairing the integrity of the weld itself.

Rutile electrode
A covered electrode in which the covering contains a high proportion of titanium dioxide.

A rutile electrode normally gives high weld hydrogen contents.

Sealed cored wires
Cored wires which are sealed, e.g. if they are manufactured by filling a tube with powdered flux or metal before drawing down to cored wire.

Segregation
Differences in composition between different regions in a piece of metal or, occasionally, between two pieces of metal cast from the same ladle. Segregation may be on a large scale, e.g. between the surface regions and the centre of an ingot, or on a microscopically small scale (microsegregation), e.g. between the centre of the arm of a dendrite and an interdendritic region.

Self-shielded welding
A type of welding with continuous wire feed using a cored wire which allows welding to be carried out without shielding gases or externally applied fluxes.

SEM
Scanning electron microscope; useful for fractographic examination up to high magnifications (several thousand diameters) as well as for conventional metallography. Many SEMs have a micro-analytical facility.

Short transverse
The through-thickness direction, i.e. perpendicular to a plate surface. The term is used in connection with lamellar tearing when tensile test pieces

are taken with their axes perpendicular to the surface of a plate. The reduction of area (RA) is the parameter measured in such a test and the term short transverse reduction of area is usually abbreviated to STRA.

Sigma (σ) phase
A very brittle compound between Fe and Cr which can be formed when stainless steels are exposed to high temperatures. This compound can form in the weld metal when ferritic steels are welded with austenitic stainless steel fillers and given long PWHTs.

Slag
A fused non-metallic residue produced by the welding process. With MMA and submerged arc welding, the slag cover is continuous and usually can be detached easily, possibly aided by wire brushing. With fluxless processes (e.g. TIG and MAG) isolated islands of slag can be present on the weld surface.

Slice bend test
A test for assessing the resistance of steel plate to lamellar tearing. A thin slice is cut through the thickness of the plate, both of the original surfaces are fillet welded to a steel bend test specimen and the composite specimen subjected to bending so that the test piece slice is in tension in the short transverse direction. A plate with a high susceptibility to tearing will show cracks along planes of inclusions at low bend angles.

Solidification cracking
A crack produced at a late stage in the solidification of a weld run due to rupture of liquid films which have not solidified (see **Hot cracking**).

Sour service
Service where a material is in contact with a liquid containing appreciable amounts of H_2S, particularly when the solution is acid.

Spray transfer
Metal transfer which takes place as a rapidly projected stream of droplets of diameter no larger than the consumable electrode from which they are transferred.

Step ageing
A heat treatment given to samples of steel preparatory to impact testing in order to determine their susceptibility to temper embrittlement. The treatment consists of subjecting the testpiece to progressively longer times at successively lower temperatures.

Glossary

Stickout
The length of wire in a continuous wire welding process between the contact tip and the arc. A longer stickout results in more resistive heating of the welding wire in this region, which may be sufficient to drive off a part of any hydrogenous material embedded in the wire surface.

STP
Standard temperature and pressure; the standard conditions (0 °C and 1 bar) to which measured volumes of gas are calculated so that they may be compared.

Strain-age cracking
A term which is not clearly defined, but possibly refers to hydrogen cracking of steel which has been embrittled by strain-ageing.

Strain-ageing
A hardening and embrittling phenomenon in steels at slightly elevated temperatures caused by the interaction of dislocations and atoms of N and C interstitially dissolved in the lattice.

Stress corrosion cracking (SCC)
Cracking of metals produced by the combined action of corrosion and tensile stress (residual and/or applied). In ferritic steels, one form of stress corrosion cracking (**hydrogen stress cracking**) is the result of embrittlement by hydrogen introduced chemically from the environment and residual stresses resulting from welding. Restrictions on the hardness of steel and weldments are made to avoid the problem.

Stress relief cracking see **Reheat cracking**

Stress relief heat treatment see **Post-weld heat treatment**

Structural cracking
In the context of **reheat cracking** the term is used to describe cracking in normal (butt and fillet) welds in contrast to underclad cracking.

Sub-critical HAZ
In a steel, that part of the HAZ that has been heated to temperatures below the A_{c1}. This region is outside the visible HAZ.

Submerged-arc welding
Metal arc welding in which a bare wire electrode or strip is used, the arc(s) being enveloped in a granular flux, some of which fuses to form a removable cover of slag on the weld.

Sulphide stress cracking (SSC)*
Brittle fracture by cracking under the combined action of tensile stress and corrosion in the presence of water and H_2S, the H_2S allowing the ingress of hydrogen into the steel; essentially another term for **hydrogen stress cracking**.

Surface pocking (gas flat)
Shallow depressions on the surface of a submerged arc weld caused by the pressure of gas from the arc cavity being unable to escape sufficiently rapidly through the molten slag.

Tab test
A test for the resistance of steel plate to lamellar tearing in which a small steel plate (the tab) is welded at an angle to the surface of the test plate and hammered off. The appearance of the fracture gives a guide to the susceptibility of the test plate.

Tack weld
A short length of weld used to hold components together so that they can be welded. Tack welds may be incorporated into the main weld or removed at suitable stages during welding. Care should be taken to avoid cracking of tack welds, particularly if they are incorporated in the final weld.

Tearing
A mode of ductile fracture involving brittle regions (e.g. inclusions or hard carbides) ahead of the main crack breaking or separating from the matrix. The main crack then extends to incorporate these micro-cracks (sometimes known as microvoids, hence **microvoid coalescence**). Normal ductile fracture and lamellar tearing are examples of this type of tearing; in hot tearing (liquation cracking) the initial voids are locally melted regions.

TEM
Transmission electron microscope; can be used for the examination of very thin films of metal or of metal surfaces by taking replicas (with or without shadowing) at higher magnifications than are practicable with an SEM. Most TEMs have a facility for selected area diffraction to determine crystal structures.

Temper (tempering)
The process of heating a steel, usually in the quenched condition, to a temperature below that at which transformation to austenite starts. Tem-

pering (which occurs during PWHT), usually (but not always) causes softening. During welding, successive weld beads give some tempering of the underlying runs.

Temper bead
A run of weld metal deposited on the surface of a weld with the aim of tempering the HAZ on the parent steel of the final run of weld metal. It is important to ensure that the toe of the temper bead is at a small fixed distance from the toe of the main weld, otherwise the temper bead will produce its own (hard) HAZ on the parent steel. The temper bead may, after deposition, be ground off.

Temper embrittlement
A process of embrittlement of steels caused by impurities such as P, Sn, As and Sb segregating to prior austenite boundaries at temperatures within the range 350–600 °C and giving a liability to intergranular fracture on stressing below the transition temperature. The presence of C at the boundaries inhibits the segregation of the temper-embrittling impurities.

Thermal contraction
When a weld solidifies, it undergoes a volume contraction. Further contraction occurs while it cools out, except that the transformation from austenite to lower temperature transformation products gives an expansion, counterbalancing some of the previous contraction. Thermal contraction gives rise to distortion and **residual stresses**, the balance between the two depending on the level of **restraint**.

Thermal cutting
The parting or shaping of materials by the application of heat with or without a stream of cutting oxygen.

Through-thickness direction see **Short transverse**

TIG welding
Inert-gas arc welding using a pure or activated tungsten electrode where the shielding is provided by a shroud of inert gas.

Toe
The boundary between a weld face and the parent metal or between runs. Note the term 'toe' should always be qualified according to whether it applies to the complete weld or to individual runs.

Total hydrogen
The hydrogen content of a metal, i.e. **diffusible** plus **residual hydrogen**.

Toughness (Notch toughness, see also Fracture toughness)
The ability of a material to resist deformation and fracture, particularly in the presence of a notch. The most common toughness test is the Charpy impact test which, because of its high speed of straining, is normally unable to detect the presence of hydrogen embrittlement in ferritic steels.

Transformation
The change of crystal structure which occurs in some materials in which different crystal structures are stable over different temperature (and pressure) ranges. In ferritic steels, the most important transformation is from the high temperature form, austenite, to lower temperature transformation products, such as ferrite and martensite.

Transgranular
A direction of cracking or fracture predominantly across the grains in a metal, as opposed to intergranular fracture (transgranular is sometimes, confusingly, known as *intragranular*).

Transition temperature (range)
A temperature or range of temperatures above which fracture is completely ductile and below which it is wholly brittle. The actual temperature (range), at which both or either type of fracture can occur, depends on the rate of stressing and the section thickness being stressed, as well as on the steel and its condition.

Transvarestraint test
A test for solidification cracking of which two types exist. In the test developed at TWI, a weld is made across a plate which is bent about the weld axis by a preset amount as the liquid weld pool is still on the plate; this test may be used with most welding processes that can be used in the flat position. The test developed in Japan can also be used for any process, but the weld bead is first deposited and then is remelted by a TIG arc, during which operation the specimen is similarly bent. Both tests were developed (in order to produce centre-line cracks) from a test known as the Varestraint test in which a weld is made along the centre of a plate, which is bent downwards across the welding direction during welding, to produce predominantly transverse cracking.

Two-layer technique
A welding technique designed to produce a refined HAZ on parent steel

in order to achieve acceptable toughness in repair and other welds which cannot be given a PWHT. The first two layers are deposited under closely controlled conditions, the second using a higher heat input so that it refines (but does not re-coarsen) the coarse HAZ regions on the parent steel produced by the first layer.

Ultrasonic examination
Non-destructive examination by means of a beam of ultrasonic sound transmitted through the metal. The technique is suitable for detecting planar defects and flaws located below the surface.

Underclad cracking
Cracking in the HAZ of a ferritic steel under a weld cladding. Even if found after PWHT, it may be hydrogen cracking rather than reheat cracking.

Undercut
An irregular groove at the toe of a run in the parent metal, or in previously deposited weld metal, due to welding.

Undermatching (weld metal)
The condition when the yield (or tensile) strength of a weld metal is below that of the parent metal.

Underwater welding
Welding under water, either in a hyperbaric chamber or with the arc in the water (see **wet welding**).

Vibratory stress relief
A method of stress relief in which vibrations, induced in the part to be treated, reduce 'high-spot' stresses.

Visible HAZ see **Heat-affected zone**

Visual examination
Examination for defects by eye, possibly aided by a low power lens.

Weldability
The ease with which a metal can be welded to give a joint free from unacceptable flaws, particularly cracks, and having the properties required.

Weld hydrogen concentration
A laboratory measurement made on a test weld to determine the amount of hydrogen absorbed and retained by weld metal deposited using carefully controlled welding, cooling and sampling conditions.

Weld hydrogen level
Arbitrary levels of weld hydrogen concentration into which a consumable in a particular condition can be placed in order to devise welding procedures intended to avoid hydrogen cracking.

Welding procedure
A specific course of action followed in welding, including a list of materials used and, where necessary, tools to be used.

Welding procedure, safe
A welding procedure that has been demonstrated to give a very low risk of cracking.

Weld metal
The metal in a fusion weld that has been melted. It may consist of melted parent metal or a mixture of melted parent metal and filler metal.

Wet welding
Welding in water. This process gives very fast cooling of the weld and, in the case of fusion welds, a very high level of hydrogen.

Trade and other names used in the text

AX – a designation of high strength weldable Q & T steel.
Corten – United States Steel Corporation.
HY – US Navy abbreviation denoting high yield strength (in ksi).
Maraging – see Glossary.
OX – Svenskt Stal.
T1 – United States Steel Corporation.

Index

Δ G parameter for reheat cracking, 40
acicular ferrite
   fine grain size and toughness, 25, 199
   formation, role of shielding gas or flux, 27
   influence of, Al (from parent steel), 26, 27
      alloying elements, 25
      oxygen, 26–28
   in HAZ, nucleation by vanadium, 27
   in weld metal, 23–28, 199
   nucleation, by Ti-coated inclusions, 25, 26
   role of welding process, 27
   steels, 7
acid fluxes, weld metal oxygen content, 19
advice on weldability, sources, 3
alloying ingredients
   inadequate mixing leading to weld metal hard spots, 22
   influence on weld metal strength, 18, 191, 192
   mixing of, 21, 22
alloy steels, 4
alumina
   inclusions and lamellar tearing, 104, 106
   in fluxes, influence on acicular ferrite, 27
   relative basicity, slags and fluxes, 43
aluminium as nitride-former, strain ageing, 41
aluminium, influence on,
   acicular ferrite nucleation, 26, 27
   solidification cracking, 70
   strain ageing, 41, 199, 212
aluminium in self-shielded wires, influence on acicular ferrite, 27
aluminium nitride, influence on strain ageing, 199
aluminium-treated steel, 5
   lamellar tearing 106, 115
ammonia, liquid, stress corrosion cracking, 209
annealing of steels, 8
anode/cathode reactions, 206
aqueous corrosion of weldments, 206, 207
arc blow, influence of nickel in steel, 182
arc efficiency, different welding processes, 10, 143
arc energy calculation, multiarc welding, 9
arc melting, *see also* electric steelmaking, 5
arc strike, effect on structure, 181
arcing, stray, 181
argon in pores from gas shield, 186
armour steels, 7
austenite retained in HAZ, subsequent transformation to martensite, 150
austenitic stainless steel consumables, to avoid hydrogen cracking, 152–155
   preheat for, 152
austenitic stainless steel weld metal,
   expansion different from ferritic steel, 153
   martensite at fusion boundary, 153
   stress relief, 153
   tongues of melted parent and cracks, 21
austenitic stainless steels, 4, 7, 51
austenitising heat treatments, impracticability for most welds, 14
autogenous welding, weld composition, 18
AX steels, 6

back gouging, influence on,
   solidification cracking, 76
   strain ageing, 200
bainite in weld metal, 25, 27, 199
bainitic steels, 7
balanced steel, *see also* semi-killed steel, 5
   welding and lamellar tearing, 114, 118
basic electrodes, hydrogen levels, 135
   resistance to moisture pick up, 135
   storing, re-drying and baking, 135, 147
basicity, relative of different compounds in fluxes and slags, 43
basicity index, slags and fluxes, 42
bay region of HAZ, liquation cracking, 87, 94
bead-on-plate test to assess liquation cracking, 91

271

biological fouling, possible hardness relaxation for stress corrosion cracking, 211, 212
blistering from
  hydrogen attack, 165, 166, 220
  sour service, 166
bores of heavy forgings, heat treatment before welding, 165
boron, as nitride-former and strain ageing, 41
  in consumables for weld root to minimise low toughness by strain ageing, 201
  influence on,
    hardenability, 141
    reheat cracking, 170, 172
    solidification cracking, 66, 70
    weld metal toughness and hardness, 18, 201
bright mild steel, 5
brittle fracture, factors controlling, 194
  initiation by hydrogen cracks,162
buttering
  as preparation to avoid lamellar tearing, 116-118
  of free-cutting steel, lamellar tearing, 118
  solidification cracking, 49
  of high carbon steel, solidification cracking, 85
  to repair lamellar tearing, 103, 118

calcium-treated steel, 5
calcium treatment and lamellar tearing, 104
carbon equivalent (CE), 35, 36, 141, 142
  influence on HAZ toughness, 195
  use of to estimate weld metal strength, 37, 192
carbon influence on,
  HAZ toughness, 195, 196
  liquation cracking, 87-89
  reheat cracking, 40, 172, 173
  solidification cracking, 39, 40, 66-80
  toughness of martensite in weld metal, 199
  weld metal toughness, 199
carbon: manganese steels, 4
  graphitisation, 220
  hydrogen attack, 220
  hydrogen cracking, 8, 147
carbon: molybdenum steels
  graphitisation, 220, 221
  hydrogen attack, 220
carbon-oxygen reaction, influence on porosity, 186
carbon steels, 3
  graphitisation, 220, 221
  hydrogen attack, 220
  hydrogen cracking, 147
  solidification cracking, 73
carburising steels, 7, 49

cast irons, 3, 51, 52
cast steel, 5
castings and lamellar tearing, 104, 105, 110-114, 118
castings, hot tearing of:
  similar to weld solidification cracking, 55
cathodic protection, 207
caustic cracking, 208
  effect of PWHT, 208
cavitation (creep), confirmation of reheat cracking, 178, 179
cellulosic electrodes
  hydrogen level, 135, 137
  penetration and hydrogen level, 168
Charpy upper shelf energy,
  estimation from inclusion content, 238
chevron cracking,
  see transverse 45° weld metal hydrogen cracking
chipping, damage and residual stresses, 28
chromium influence on, corrosion of weldments, 205
  creep resistance, 6
  graphitisation, 221
  hydrogen attack, 6, 165, 218
  oxidation resistance, 6, 205
  reheat cracking, 170-173
  solidification cracking, 66, 70, 73
  weld metal properties, 18
chromium:molybdenum steels, 4, 6
  reheat cracking, 172, 173
  strength for high temperature hydrogen service, 216
  temper embrittlement, 41, 42, 213, 214
chromium:molybdenum:vanadium steels, 6, 40
  reheat cracking, 40, 173
chromium:molybdenum weld metal,
  bainitic microstructure, 25, 199
  reheat cracking, 173
chromium steels, 4, 7
cladding with two layers to reduce reheat cracking, 174
clamping of high strength steels, 77, 86
clean steel, lamellar tearing resistance, 103, 118
CO_2 welding, oxygen content and toughness, 2
  weldability with, 2
coarse austenite grain size by heating, 1
coarse grained HAZ, influence on transformation microstructure, toughness and cracking, 15
coarse weld metal austenite grains, influence on toughness, 23
cobalt, influence on solidification cracking, 70, 73
cold rolling, 5
cold work from mechanical cutting, 28

Index

cold worked steels, 7
combined thickness, weld cooling, HAZ hardness and hydrogen cracking, 11, 12, 142, 143
combined water, source of weld hydrogen, 135
compensating plates and lamellar tearing, 110
composition differences from different,
  dilution in multipass weld runs, 20, 21
  shielding in multipass weld runs, 20, 21
composition of weld metal:
  steps between weld metal and HAZ, 21
  uniformity, 19–23
see also weld metal composition
compositional parameters, see Formulae for . . .
compositional steel types, 3, 4
compound preparations to reduce amount of weld metal needed, 84
concast plate (thin), hydrogen cracking at centre, 116
  lamellar tearing, 116, 164
constructional steels, see also structural steels, 5
consumable guide welding, weldability, 3
see also electroslag welding
consumables, welding, selection for subsequent heat treatment and hot work, 30, 34, 37–38
continuous cooling transformation (CCT) diagram and HAZ microstructure, 15
continuously cast steels, see also concast steels, 5
  lamellar tearing, 109, 115, 119
  need for Al addition, 109
  strain ageing, 17, 194
contraction strains, 29, 102
contraction of weld on solidifying, 54
  feeding, 54, 55
control of hydrogen level to avoid hydrogen cracking, 146, 147
control techniques for reheat cracking, 174–178
control tests for lamellar tearing, 119–128
cooling of weld bead, factors affecting, 142, 143
cooling rate, factors controlling, 1, 11, 142, 143
  influence on weld metal properties, 1, 191, 195
cooling time, influence on hydrogen diffusion 11, 12
copper pick up in submerged arc and MMA welding, 92
  influence on,
    corrosion of weldments, 205
    reheat cracking, 172, 173
    solidification cracking, 70, 73
    sources giving liquid metal penetration in HAZ and weld metal, 91, 92, 184
cored wires, hydrogen level, 135
  pick up of moisture by unprotected reels, 147
corner joints,
  joint angle location to minimise lamellar tearing, 114
  lamellar tearing, 110, 114, 118
correlations between Charpy and fracture toughness tests, 193,194
corrosion fatigue use of cathodic protection, 210
corrosion,
  in liquid metals, 207
  of weldments, 205–209, 211
  preferential, 206, 207, 211
  products, distinction from slag inclusions, 183
Corten steel, see also weathering steels, 5
cosmetic beads (low heat input), danger of hydrogen cracking, 154
cracked ammonia heat treatment atmosphere,
  hydrogen content, 168
cracking from sour service, 166, 168
cracks, tendency to over-react when discovered, 187
  types, 45
  use of fracture mechanics to assess whether acceptable, 187
Cranfield test for lamellar tearing susceptibility, 128
crater,
  contraction shrinkage and cracking, 54
  examination of for solidification cracking, 86
  pipes, 185
creep, failure Type IV, 205
  parameter, Larson-Miller, 42
  resistance of weldments, 204, 205
    influence of, graphitisation, 222
    hydrogen attack, 220
  resistant steels, 6
  service, transition joints, 205
critical hardness for hydrogen cracking, 141
  influence of,
    hydrogen level, 141
    post-weld cooling, 141
    preheat/interpass temperature, 141
    restraint, 141
critical HAZ hardness, for hydrogen cracking, 141
see also critical hardness
cross rolling of plate and lamellar tearing, 106
cruciform joints and lamellar tearing, 114, 115
CSF (crack susceptibility factor) for solidification cracking, 40, 73

current type and polarity,
  influence on depth/width ratio, 83
cut surfaces, left unwelded and welding
  onto, 28
cutting, 28
  thermal, *see* thermal cutting
de-aerator corrosion, 207
decarburisation resulting from hydrogen
  attack, 220
decarburised plate surface, influence on
  reheat cracking, 174
decohered inclusions, ultrasonic indications, 116, 119
decohesion of inclusions,
  cold work and PWHT, 102, 119
  first stage of lamellar tearing, 102, 105, 106
  influence of thermal expansion coeffi-
    cient, 105, 106
deep drawing steels, 7
deeply penetrating welds and lamellar
  tearing 107, 109
delay before inspecting for hydrogen
  cracking, 159, 189
depth/width ratio, influence on
  correction for fillet weld shape, 58, 59, 75
  solidification cracking, 56–59, 70, 74–86
descriptive steel types, 5–7
destructive control tests for lamellar tearing, 119–128
detection of,
  45° transverse weld metal hydrogen
    cracking, 159
  hydrogen attack, 220
  hydrogen cracking, 159
    at multipass weld roots, 159
    time delay to let cracks develop, 159,
      189
    with non-ferritic weld metals, 159
  lamellar tearing, 128
  liquation cracking, 93
  reheat cracking, 178, 179
    under cladding, 178
  solidification cracking, 93
diffusible hydrogen, *see* weld hydrogen
  content
diffusion of hydrogen in steel and welds,
  see hydrogen diffusion
dilution, and weld penetration, 19, 81
  influence of,
    metal powder additions, 19
    preheat/interpass temperature, 19
    welding parameters, current type and
      polarity, 19
  influence on,
    solidification cracking, 69, 72, 80, 82, 86
    weld composition, 69, 72
  in weld surfacing, 19
  of parent steel into weld with different
    welding processes, 19
  techniques to minimise, 78, 81–83

dislocation network in weld metal,
  influence of,
    austenising (normalising), 30, 191
    heat treatment, 30, 191
  influence on strength, 30, 191
dissimilar metal, joining steel to, 50–53
dissimilar metal joints,
  diffusion across joint line, 51
  property considerations, 50, 51
  transition pieces for, 50, 51
  with austenitic stainless steels, 51
  with clad steels, 53
  with copper alloys, 52
  with light alloys, 52, 53
  with nickel alloys, 52
  with non-metals, 53
  with weldable cast irons, 51, 52
  with wrought iron, 52
dissimilar steel joints,
  austenitic stainless steel fillers, 48
  nickel alloy fillers, 48
  PWHT of, 48, 49
  selection of welding procedures, 47, 48
  strength and toughness consider-
    ations, 47
  ultra low hydrogen consumables, 48
  with case hardened steels, 49
  with coated steels, 49, 50
  with free-cutting steels, 49
  with plated steels, 49, 50
dissimilar steels, joining, 47–50
ductility dip, 93, 193
  cracking, 93, 231, 232
    impurities and strain required for, 93,
      232
    welds made under fluctuating stress,
      93, 231, 232
  problems during hot working, 93
ductility of weld metal, 193
  influence of hydrogen, 193
  when notched, 193
ductility reduction, at high temperatures, 193
  near melting temperature, 193
  see also ductility dip, liquation cracking
dye penetrant (DP) inspection, range of
  use, 188

egg-box construction and lamellar tearing,
  110
electric steelmaking, 4, 5
electrode coverings, functions, 19
electrogas welding, *see* electroslag welding,
electron beam welding,
  normalising for good toughness, 2
  solidification cracking, 2, 73, 74
  weldability with, 2
  weld metal composition, 18
electroslag melting, 5
electroslag or electrogas welding for T-
  joints, lamellar tearing, 114

Index

electroslag or electrogas welds,
  normalising for good toughness, 3
  weldability with, 3
  welding speed limitation, solidification cracking, 3
embrittlement by hydrogen,
  influence of,
    HAZ hardness, 140
    inherent toughness, 140
    strain rate, 133, 145
  mechanism, 133
  susceptibility of steel, 140-144
embrittlement, types likely in welds, 199, 203, 212-218
engineering steels, 3, 5, 9
  hydrogen cracking, 9
  liquation cracking, 9
  solidification cracking, 9, 86
epitaxial growth of weld metal from HAZ, 23
etching to reveal solidification structure, 94
eutectic temperature,
  influence on solidification cracking, 55, 66-68
eutectic, liquation of, 86, 87
excess metal, 180
excess penetration, 180
  influence on ultrasonic inspection, 180
exothermic gas as heat treatment atmosphere, hydrogen content, 168
extrusions, 5
  lamellar tearing, 102, 104, 115

fatigue, 210
  cracks, initiation by hydrogen cracks, 162
  of weldments in high strength steels, 210
  resistance, weld profile blending, 210
    influence of metallurgical factors, 210
ferrite, different forms in weld metal, 14, 15
ferro-hardeners, alloying additions, 19
fillet weld size, influence on lamellar tearing, 116
fine grained heat-affected zone, 15, 16
fisheyes, 162-164
  in all weld tensile tests, influence of hydrogen removal heat treatment, 164
  influence of PWHT, 164
  in transverse tensile tests, influence of undermatching weld metal, 164
  size of flaw, 164
flare angle of HAZ, solidification cracks in, 94
  see also bay region
fluxes welding, 19, 42, 43
  functions, 19
  influence on weld composition, 19, 69-72
forging, 9
  lamellar tearing, 104, 114, 118

formulae for,
  creep, 42
  embrittlement, 41, 42
  hardenability, 35-37
  hardness and strength, 37, 38, 204
  HAZ hardness (martensite and bainite), 37
  heat treatment, 42
  hydrogen cracking, 35-37
  lamellar tearing, 41
  reheat cracking, 40, 214, 215
  slags and fluxes, 42
  solidification cracking, 39, 40, 98-106
  strain ageing, 41
  temper embrittlement, 41, 42, 213, 214
  underclad reheat cracking, 40
  weld metal mechanical properties, 37-39
  weld metal microstructures, 39
foundry repairs, normalising after, 38
fracture mechanics to assess safety of cracks, 187
fracture toughness testing of HAZs, 196
free aluminium, influence on strain ageing, 41
  free-cutting steels, 6, 9, 49
    lamellar tearing, 118
    liquation cracking, 6, 9, 49, 88
    solidification cracking, 6, 9, 49
  free carbon, influence on strain ageing, 212
  free nitrogen, influence on strain ageing, 41, 199, 212
freezing range, influence on solidification cracking, 54, 66-69
friction welding under water, 233
fusion boundary parallel to plate surface, 102

gas flats, see surface pocking
gas porosity, see porosity
general preheat compared with local preheat, 143
gouging, 28
grain refinement, influence on,
  HAZ toughness, 196-198
  solidification cracking, 80
  solidification structure of weld metal, 80
grain size, influence on,
  weld metal strength, 191
  weld metal temper embrittlement, 41
graphitisation, 220-222
  description, 221, 222
  'eyebrow' type, 222
  incubation period, 221
  influence of, composition and microstructure, 221
    prior plastic deformation, 221
  influence on, creep strength, 222
    ductility and toughness, 222
    strength, 222

in HAZ and weld metal, 222
possible effect on temper embrittlement, 215, 221
temperature range, 221
grinding damage, residual stresses, 28
grinding of gouged preparation when repairing lamellar tearing, 118
grinding of weld toes to combat reheat cracking, 178
grooving and buttering, preparation,
to repair lamellar tearing, 118
to avoid lamellar tearing, 118

H_2S service, *see* sour service.
Hadfield's manganese steel, 7
half bead technique influence on HAZ toughness, 198
hard spots in weld metal, from partially dissolved,
alloy metals, 23
deoxidants, 23, 184
influence on stress corrosion cracking, 23, 184
hardened surfaces from thermal cutting and gouging, 28
hardness measurement in multipass welds, 140
HAZs, 204
factors controlling, 140
hardness of weldments, 204
for corrosive service, 204
for sour (H_2S) service, 208, 209, 211, 212
HAZ, 14-18
description, 14-18
grain size, influence on reheat cracking, 173
near fusion boundary, 15
hardenability, influence of alloy content, 141
hardening behaviour, division of by microstructure, 36, 37
hardness, influence on reheat cracking, 174
microstructure, 15-18
acicular ferrite, 25-28
refinement, 17, 173-178, 229, 230
reheating and intercritical reheating, 17-18
temperature reached near fusion boundary, 15
softening of cold worked steels, 7
transformation of on cooling, 15
HAZ hydrogen cracking, in previously deposited weld metal, 161, 162
cracks within visible HAZ, 160
HAZ refinement with automatic welding, to reduce risk of reheat cracking, 176, 178
weld preparation, 176, 177
see also two layer refinement and half bead technique

HAZ toughness, 194-198
of older steels, strain ageing, 194
recommendations, 195-197
strain ageing outside visible HAZ, 194
heat-affected zone, *see* HAZ
heat flow paths, 11, 12
heat input,
determination, 9, 10
difference from arc energy, 10, 143
influence on,
HAZ toughness, 195-197
reheat cracking, 174
weld cooling, HAZ hardness and hydrogen cracking, 143, 144
heat resistant steels, 6, 7
heat treatment of steels, 7, 8
for removal of hydrogen, 34
for tempering, 32-34
heat treatment parameter, Larson-Miller, 42
heating rate to PWHT temperature,
influence on reheat cracking, 174
HIC in sour service, 208, 209
influence of, hardness, 208, 209
inclusions, 208, 209
see also hydrogen stress corrosion cracking
high alloy steels, 4-7
high carbon steels, 7
high hardenability steels, 6, 7
high heat input welding processes,
influence on lamellar tearing, 116
high melting point inclusions,
influence on lamellar tearing, 106
high pressure hydrogen storage, 165, 215-220
influence of steel strength, 165, 216, 217
high strength low alloy (HSLA) steels, 6
high strength steels, 6, 8, 9, 79
clamping, 79
high strength structural steels, reheat cracking, 173
high strength weld metal, toughness, 199
homogenising of steels, 7, 8
hot cracking, *see* solidification, liquation, or ductility dip cracking
hot rolling, 5
hot tapping, 13, 14, 231
influence of fluid on cooling, 13,14
hot tearing in welds, *see* liquation cracking
hot wire additions to reduce dilution, 82
hot working of partly welded items, normalising after, 38
humid ambient conditions, pick up of moisture by consumables, 147
hydrogen attack, 165, 218-220
blistering, 166, 220
description, 165, 220
incubation time, 220
influence of Cr in steel, 165, 218
of C:½Mo steel, 220
hydrogen, behaviour in steel and weld metal,136-140

influence on, penetration, in two sided welding 168
in wet welding, 168, 234
welding behaviour, 135, 168
entry into steel at different temperatures and pressures,137, 215, 216
measurement of in welding, 134, 135
molecular, entry into steel, 165, 166
reduction by diffusion, 13, 139, 140, 147-158
influence of interpass temperature,13, 150
solubility, in austenite and ferrite, 137, 216
in molten steel (weld metal), 137, 216
sources in welding, 134-136
condensed moisture, 136
grease and oil, 136, 147
paint and weld-through primers, 49, 136, 147
parent steel, 136, 147
problems during friction and TIG welding, 136
influence of section thickness, 136
rust, 136
the atmosphere, absolute humidity, 136, 147
influence of pressure in hyperbaric welding, 136, 147, 232
steel in hydrogen service, 136, 218
steel in sour (H_2S) service, 136
welding consumables, 135, 136
hydrogen blistering, influence of inclusions, 166
hydrogen build up in multipass welding, 150
hydrogen cracking,
conditions for, 134
during welding, 133-162
from sour service, 166, 167, 208, 209, 211, 212
in centreline segregates, 164
initiated by liquation cracks, 86, 141
slow growth, 133
techniques for avoiding, 146-158
hydrogen cracks, morphology, 160, 161
hydrogen damage, see hydrogen attack
hydrogen diffusion, 133, 138-140
variability near ambient temperature, 140
to high stress concentrations, 133
hydrogen diffusion rate, influence of microstructure, 140
second phase particles (sulphides), 140
temperature, 138-140
hydrogen embrittlement, 140-146, 203, 204, 215-218
by hydrogen left after welding, 140-146
in heavy sections, 203
in hydrogen service at high temperature, 215-218
hydrogen flaking in heavy forging billets, 164
hydrogen in heat treatment, 168
hydrogen in parent steel, lower near free surface after heat treatment, 136, 165
hydrogen in welding, influence on lamellar tearing, 115, 116, 168
hydrogen levels, 135
hydrogen service, precautions for repair, 165, 218, 230
precautions on shutdown, 165, 218
risk of steel suffering, creep damage, 218, 220
temper embrittlement, 218
strengths for Cr:Mo steel, 216, 217
types of vessel involved, 218
hydrogen stress corrosion cracking in sour service, 166, 209, 211, 212
influence of hardness, 209, 211, 212
possible hardness relaxation, 212
PWHT with alloy and high strength steels, 209, 211
see also HIC, 165, 166, 208, 209, 211, 212
hydrogen sulphide, role in hydrogen stress corrosion, cracking,165, 166, 208, 211
hydrogen traps, 140
hydrogen-containing compounds as source of weld hydrogen, 135
hydrogen-induced cracking, see also HIC,166, 167
hyperbaric chamber, atmosphere and humidity,147, 232
influence on weld cooling, 14
hyperbaric welding for repair, 232, 233
processes and consumables, 232, 233

I-beams, fillet welded, lamellar tearing, 109
identification of,
hydrogen cracks, 160-162
liquation cracks, 93-100
reheat cracks, 179
solidification cracks, 93-100
weld metal hydrogen cracks, 161, 162
lamellar tears, 130
IIW carbon equivalent, 35, 141
impact tests, insensitivity to hydrogen, 203
impurities influence on,
temper embrittlement, 41, 42, 212, 213
weld metal strength, 18
weld metal toughness and cracking, 18
impurities, segregation to grain boundaries and temper embrittlement, 212, 213
inclusion distribution variability, 107, 121
influence of position of plate in ingot, 107
influence on lamellar tearing, 105, 107
inclusions, as notches to initiate brittle

fracture 196
  at centreline in concast plate, 105
  attenuation of ultrasonic beam, 128
  decohesion from matrix 102, 105, 106, 128, 189
  influence on,
    ductile fracture, 38, 196
    HAZ ferrite nucleation, 141
    HAZ toughness, 195, 196
    HAZ transformation and hardenability, 141
    hydrogen cracking susceptibility, 141
    lamellar tearing, 102, 104–109
    in weld metal, influence on ductility, 38, 104
  of metallic particles, 183, 184
  of welding slag, 183
  see also non-metallic and metallic inclusions 183, 184
inclusion shape control, for sour (H$_2$S) service, 5, 166
  for lamellar tearing, 5, 8, 104, 106, 115
incomplete penetration, 181
independent nucleation of grains in high heat input weld metal, 23
induction surface hardening, 7
ingot cast steels, lamellar tearing, 107
inspectability, influence of,
  coarse grained metal, 189
  compressive residual stresses, 189
  non-ferritic weld metal, 188, 189
  PWHT, 159, 189, 190
  steel of high inclusion content, 189
inspection delay for hydrogen cracking, 159, 189
inspection for defects, 46, 187–190
inter-run time to allow hydrogen reduction by diffusion, 150
intercritically heated HAZ, 16–18, 198
intergranular cracking and fracture,
  from temper embrittlement, 213
  from hydrogen cracks, 161
  in weld metal after PWHT, 202, 203
intermediate PWHT,
  influence on,
    inclusion decohesion, 116
    lamellar tearing, 116
    reheat cracking, 178
intermittent solidification cracking, 55, 96
interpass temperature,
  influence on hydrogen diffusion, 13, 150
  measurement, 13
  need for controls, 13, 143
  see also preheat/interpass temperature
iron oxide, relative basicity in slags and fluxes, 43
iron powder, incompletely dissolved, soft spots, 23
  see also metal powder

isothermal transformation method, bainite formation, 151
  to avoid hydrogen cracking, 51, 152

J-factor for temper embrittlement, 41, 173, 213
joint design, influence on lamellar tearing, 111–114
joint integrity, 46, 191–209
joint types, influence on lamellar tearing, 109–114, 118, 128

lack of fusion, 181
lamellar tearing, 8, 102–132
  detection of, 128
  ductile fracture mechanism, 103, 130
  extrusions and plate, 8, 102, 104, 115
  identification, 130–132
  influence of steel strength, 8, 115
  long range stresses, 102, 103
  outside visible HAZ, 102, 120
  repair, 118
  tests, for control, 119–128
    large scale, 128
  two stages of, 102
lamellar tears, visible at weld toes, 103
Larson-Miller parameter for heat treatment and creep, 42
laser surface hardening, 7
laser welding, weldability with, 2
  autogenous, weld metal composition, 18
LBZ: see local brittle zones
lead, influence on solidification cracking, 70
  on or near solidification crack surfaces, 98, 100
lifting lugs and lamellar tearing, 109
lime, relative basicity in slags and fluxes, 43
line pipe steels, solidification cracking, 73
liquation cracking, 86–91, 93
  assessment of, 91
  free-cutting steels, 49, 88
  influence of,
    compositional factors, 87–89
    heat input, 89
    nickel alloy fillers, 89
    stainless steel fillers, 89
    weld metal solidification temperature, 89
  medium carbon steels, 86
  oxy-acetylene welds, 86
  premature failure in bend tests, 86, 88
  segregate bands, 87, 96
  submerged arc welds, 87
liquation cracks, detection, 93
  healing by low melting weld metal, 89
  identification, 93–100

Index

influence on other cracking, 86–88, 141
　size of, 88
liquation crack tips, bluntness, 88
liquid ammonia and stress corrosion
　cracking, 209
liquid metal penetration by,
　copper during welding, 91, 92, 184
　zinc during welding, 50, 184
local brittle zones, influence of PWHT,
　198
　influence on HAZ toughness, 18, 198
local normalising, by TIG remelting of
　cladding, 178
　to refine HAZ and reduce reheat cracking risk, 178
local PWHT, 34, 229
long range stresses, *see* stresses, long range
long welding cables, voltage drops, 143, 232
long welds, solidification cracking, 76, 77, 79, 84
low alloy steels, 4, 9
low carbon content,
　influence on solidification cracking, 40, 66–69
low hydrogen consumable use,
　influence on lamellar tearing, 115
low hydrogen consumables to reduce
　preheat and distortion, 84
low inclusion steel to resist lamellar tearing, 107, 118
low oxygen (Al treated) steels,
　Type II MnS inclusions, 69, 106
low strength weld metal,
　influence on,
　　lamellar tearing, 116
　　solidification cracking, 84

machinable steels, *see also* free-cutting steels, 6
MAG welding, shielding gas weld oxygen
　content and toughness, 2
　solidification cracking, 78, 80, 84
　weldability with, 2
magnesia, relative basicity in slags and
　fluxes, 43
magnetic particle inspection (MPI), range
　of use, 188
manganese:nickel:molybdenum steels,
　reheat cracking, 173
manganese oxide, relative basicity in slags
　and fluxes, 43
manganese steels,
　see Hadfield's manganese and C:Mn steels
manganese sulphide,
　films influence on liquation cracking, 89, 193
　inclusions and lamellar tearing, 104, 106
manganese sulphides on or near solidification/liquation cracks, 98–100
manganese,
　influence on,
　　liquation cracking, 87–89
　　reheat cracking, 173
　　solidification cracking, 40, 66–74, 80
　　temper embrittlement, 41, 42, 212–214
　　weld metal properties and cracking, 18
maraging steels, 4, 7
martensite hardness, influence of carbon
　content, 37, 141
martensite in weld metal, 25, 26, 199
martensitic steels, 7
martensitic weld metal, advantage of low
　carbon, 199
maximum HAZ hardness, 140
maximum residual stress, influence of
　weld metal yield stress, 32
melted parent steel, tongues of, hard
　microstructures and cracking, 21
metal composition factor (MCF) for reheat
　cracking, 172
metal powder, influence on dilution, 82
metallic inclusions,
　from undissolved alloying elements and
　　deoxidants, 21, 22, 184
　in TIG welds, 183, 184
metallic spatter, 184
metallurgical aspects of inspection, 189, 190
methane in pores from hydrogen, 186
micro-alloyed steels, 4
micro-alloyed weldable steels,
　hydrogen cracking, 8
　reheat cracking with B and Mo, 8
　toughness and restriction of heat input, 8
microstructural control by control of cooling to avoid hydrogen cracking, 147
microstructural steel types, 7
microstructure, influence on HAZ toughness, 15–18, 196–198
MIG welding, *see* MAG welding
mild steel, 3, 4
　hydrogen cracking, 8
minerals, for fluxes and electrode coverings, 19, 42, 43
　basic, sulphur reduction by, 19
　relative basicity, 19, 43
misalignment, angular or linear,
　stress concentration, 181
mixed inclusions, influence on lamellar
　tearing, 106
MMA electrodes, storing and re-drying, 135, 147
MMA welding, free-cutting steels, 2
　solidification cracking, 73–76, 81–85
　weldability with, 2
Mn/S ratio, influence on liquation cracking, 88

Mn/Si ratio, influence on liquation cracking, 87
moisture, source of weld hydrogen, 135, 136
molecular hydrogen, *see* Hydrogen, molecular
molybdenum steels, 6
see also carbon:molybdenum steels
molybdenum influence on,
 reheat cracking, 172, 173
 solidification cracking, 66, 70, 73
 weld metal properties and cracking, 18
molybdenum:boron steels, reheat cracking, 173
MPI, *see* Magnetic particle inspection
multipass welds, change of residual stress during welding, 30–32
multiple arc submerged arc welding, solidification cracking, 78, 81–84, 86

Nelson diagram for hydrogen attack, 165, 218–220
nickel alloy consumables,
 solidification cracking, 153
 to avoid hydrogen cracking, 152–155
nickel:chromium:molybdenum steels, 4
 temper embrittlement, 203, 214
nickel:chromium:molybdenum: vanadium steels,
 temper embrittlement, 42
nickel:cobalt (9Ni:4Co) steel, 4
nickel influence on, solidification cracking, 66, 70, 71, 73
 weld metal properties and cracking, 18
nickel steels, 4
 temper embrittlement, 42, 203, 214
niobium carbide on or near solidification or liquation cracks, 100
niobium influence on,
 HAZ toughness, 195
 solidification cracking, 66, 69
nitride-formers and strain ageing, 41
nitriding steels, 7
nitrogen influence on,
 HAZ toughness, 195
 porosity, 185
 strain ageing, 44, 190–201, 212
 weld metal toughness, 185, 199, 200
non-destructive control tests for lamellar tearing, 119
non-destructive examination (NDE), 187–190
non-ferritic consumables influence on,
 hydrogen cracking, 152–155
 NDE, 154, 188, 189
non-metallic inclusions, *see also* inclusions, 183
normalised rolled steels, Ti addition to avoid strain ageing, 8, 17, 201
normalising heat treatment,
 of steels, 8
 of welds, 34, 35
 after foundry repair, 38
 to refine HAZ grain size and reheat cracking, 178
notch ductility of weld metal, 193, 194
 see also fracture toughness
notches (at weld toes), reheat cracking, 174
nucleation of ferrite at grain boundaries and within grains, 25, 26

open hearth steelmaking, 4
outside of pipes, relaxed hardness requirement for stress corrosion, 212
overlap, 181
overmatching weld metal, residual stress, 191, 192
 hydrogen cracking, 192
OX steels, 6
oxidation of weldments, 205
 influence of chromium, 205
oxidation potential, slags and fluxes, 43
oxide, influence on hydrogen entering steel, 215
oxygen in weld metal influence on,
 acicular ferrite nucleation, 26, 27
 solidification cracking, 40, 69, 70, 73

painted steels, influence on hydrogen cracking 49, 136
parameters, *see* formulae for
parent steel strength, influence on restraint and solidification cracking, 79, 84
partial melting of alloying ingredients leading to hard spots in welds, 21, 22, 184
P_{cm} formula for hardenability, 36, 141
P_E parameter for temper embrittlement, 41, 173
peening weld toes, influence on,
 lamellar tearing, 116
 reheat cracking, 178
penetration into side of preparation, influence on solidification cracking, 81
penetration, weld, influence of hydrogen, 168, 181, 234
 relation to dilution, 19, 85
penetrators, risk of lamellar tearing, 110
peritectic reaction, influence on solidification cracking, 66–69
phosphorus segregation, influence on intergranular fracture after PWHT, 202, 203
phosphorus influence on,
 corrosion of weldments, 205
 liquation cracking, 87
 solidification cracking, 39, 40, 69–74
plasma welding, autogenous weld metal composition, 18
 weldability with, 2
plate, 5

Index

lamellar tearing, 102, 106, 115
plate surface, decarburised, influence on reheat cracking, 174
plate thickness, influence on lamellar tearing, 107, 108
pores, analysis of gas in, 186
porosity, 185, 186
   due to, lack of shielding, 185
   trapped shielding gas, 186
   in TIG welds due to lack of deoxidants, 185
positional welding, difficulty of microstructural control, 147
post-weld heat treatment, see PWHT
postheating, influence on lamellar tearing, 115, 116
potassium oxide,
   relative basicity in slags and fluxes, 43
preheat temperature, measurement, 13, 143
   need for controls on, 13, 47, 48
preheat/interpass temperature,
   influence on, hardness, 13, 143
      HAZ toughness, 196
      hydrogen cracking, 13, 143, 146-154
      hydrogen diffusion, 13, 147
      microstructure and properties, 6, 13
      weldment properties, 13
   selection to avoid exceeding M_f, 150
   selection to avoid tempering steel, 150
preheating temperature difference in welding, 115
   penetrators, influence on lamellar tearing, 115
   set-through nozzles, influence on lamellar tearing, 115
preheating, influence on, dilution, 79, 80
   lamellar tearing, 115
preparation, see weld preparation
pressure vessel steels, reheat cracking with V and Mo, 9
pressure, influence on welding, 232-234
prior austenite grain size, coarsened by heating, 1
   influence on reheat cracking, 173
protective coatings, welding onto, 49, 50, 36, 211
   against corrosion, 211
P_{SR} formula for reheat cracking, 40
PWHT, 32-34, 46, 198, 201-203
   in alloy steels, 33
   influence on,
      HAZ toughness, 198, 202
      in-service hydrogen cracking, 33, 167, 208, 209, 211, 212, 218
      inspectability, 159, 189, 190
      residual stress, 33, 198, 201, 202
      strain ageing, 199, 200, 212
      stress corrosion cracking, 33, 207-209
      weld metal toughness, 202
   intergranular fracture after, 202, 203
   of thick C and C:Mn steels, 32
   procedure if temperatures different for different steels, 48, 49, 202
   repeated, 202
   secondary hardening, 33, 198, 202
   temperatures, 33, 198, 202
      influence on reheat cracking, 174
      relation to A_{c1} temperature, 33
   tempering and softening, 33, 48
   uniform heating, 34

QT (Q&T) steels, see quenched and tempered steels
quench cracks, role of hydrogen, 168
quenched and tempered steels, 6
   selection of PWHT temperature, 33
quenched and tempered weldable steels, hydrogen cracking in weld metal, 9
quenching and tempering of,
   steels, 8
   welds, 34, 35

radiography, range of use, 188
rail steels, 7
rare earth metal (rEM) treatment, lamellar tearing, 104
RE-treated steel, 5
reduction, HAZ hardness below critical, factors 142, 143
reduction of residual stress by low strength weld metal, 115, 145
reduction of stress concentration by grinding or peening weld toes,145, 174, 210
reduction of weld metal strength by austenitising heat treatments, 34, 35, 37
refined weld metal microstructure, 28
refinement of HAZ grain size, see HAZ refinement
reheat cracking, 40, 41, 46, 170-179
   during heating to PWHT temperature, 170
   failure by creep cavitation mechanism, 172
   in coarse grained regions, 170
   influence of,
      composition and impurities, 172, 173
      relative strengths of HAZ and weld metal 170, 171
      strengths of grain boundaries and interiors, 170
   in HAZs and weld metals, 170
   in heavy sections, 170
   in segregate bands, 172
   in structural welds, 170
   possible confusion with hydrogen and liquation cracking, 179
   reduction of grain boundary ductility by impurities, 170
   under cladding, 170, 171

removal of hydrogen from steel after high temperature hydrogen service, 216, 218
repair by grinding out and blending, 226
repair under water, 232–234
repair welding, 224–235
  and joining in space, possibility and processes, 234, 235
  expected life, 224, 228, 233
  influence of steel composition and strength, 226, 227
    matching strength, very high strength steels, 226, 227
  methods, 226–228
  need, 224, 225
  of brittle fracture failure, 225
  of brittle material, 226
  of defects, 224, 225
  of fatigue failure, 224, 225
  of lamellar tearing, 103, 108
  of overload failure, 224
  of steel components containing fluids, 14, 231
  of steel containing hydrogen, 136, 215, 218, 230
  of steel not previously welded, 226
  of steel of doubtful toughness, 226
  of temper embrittled steel, containing hydrogen with creep damage, 215
  of worn or corroded surfaces, 228
  processes, 227, 228
  technique, limitations, 47, 227, 228
  without preheat, 228
  without PWHT, 177, 228, 229
residual hydrogen, see weld hydrogen content, residual residual stress values, for fracture mechanics analysis, 32
for tolerable defect size calculation, 32
small scale variability, 32
residual stress, direction of and lamellar tearing, 102, 115
residual stresses, 29–32
  build up on cooling 29–31
  from mechanical cutting 28
  influence of transformation strains and temperatures 30
  influence on, hydrogen cracking, 144,145
  yield strength, 30
  leading to deformation, 30
  restraint, 30, 79, 84
  influence on,
    lamellar tearing, 102, 105, 118
    solidification cracking, 79, 84, 86
retained austenite in HAZ, subsequent transformation to martensite, 150, 202
retained phases in weld metal, 24
rimming steels, 5
  for welding wires, low level of impurities, 18

strain aging in sub-critical HAZ, 17
rod, 5
roller quenched steels, 8
rolling, influence of degree of on lamellar tearing, 107
root concavity, 180
root face, influence on welding current for second side root pass, 85
root gaps, opening during welding, influence on solidification cracking, 57, 58, 70, 76, 77, 84, 86
root region,
  compressive residual stresses, 30–32
  difficulty of NDE, 31, 189
RQT steels, 6
run-off blocks length to avoid,
  crater cracking on main weld, 84
  solidification cracking, 84
rutile electrodes, hydrogen level, 135
  relative basicity in slags and fluxes, 43

secondary hardening during PWHT, 33, 198, 208
sections, hot rolled, 5
segregates in concast steel, hydrogen cracking, 142
segregation, beyond tip of solidification crack, 94
  difference between ladle and product analysis, 141
  on microscopic scale, 19
self shielded wire weld metal, refinement of grain size for toughness, 2, 28
self shielded wire welding, weldability with, 2
solidification cracking, 80
semi-killed steels, 5
  lamellar tearing, 106, 115
  strain ageing in sub critical HAZ, 17
serrated load/extension behaviour from strain ageing, 200, 201
service problems, 47, 210–222
set-on nozzles, lamellar tearing, 110
set-through nozzles, lamellar tearing, 110
shape, faults of, 180–183
  influence on weld and fatigue cracking, 181
sheet, cold rolled, 5
sheet steels, 5, 7
short transverse direction and ductility, 102
short transverse tensile test, assessment of, 121
  for lamellar tearing susceptibility, 107, 120, 121
  measurement of STRA, 121
  specimen, normalising when testing thin plate, 120
  specimen numbers, 120
  testpiece, extension pieces, 120

Index

hydrogen removal from, 120, 121
shrinkage porosity, 185
silica, relative basicity in slags and fluxes, 43
silicate/sulphide inclusions,
   influence on lamellar tearing, 106
silicon influence on,
   HAZ hardenability and hydrogen cracking, 36
   reheat cracking, 172, 173
   solidification cracking, 69, 70, 72, 80
   temper embrittlement, 41, 42, 212–214
silicon-killed steels, 5
   lamellar tearing, 106,115
   strain ageing in sub critical HAZ, 17
slag, welding, 42, 43
slice bend test for lamellar tearing, 121–123
   assessment, 123
   welding, 123
sodium oxide, relative basicity in slags and fluxes, 43
soft spots in weld metal from incompletely dissolved iron powder, 23
soft weld metal, for buttering, 103, 115, 116, 118
solid wires, hydrogen level, 136
solidification crack surfaces, oxidised, 93, 97
solidification cracking, 54–86, 93–100
   below weld surface, 59, 61–63, 93
   double fillet welds, 79
   identification, 93–100
   influence of,
      backing strips, 57, 84
      buttering high carbon parent steel, 85
      clamping, 76, 85, 86
      composition, 56, 66–74, 80, 85, 86
      depth/width ratio, 56–58, 70, 74–76, 83, 85, 86
      lack of root fusion, 57, 76
      low melting point eutectic, 55, 67, 68
      preheat, 79, 80
      restraint, 79, 84, 86
      root defects, 57, 76
      root gaps, 57, 70, 76, 77, 84, 86
      shape of weld surface, 58, 59, 75, 86
      steel strength and thickness, 79
      tacking, 76, 84, 86
      welding current, 74, 79, 81, 83–86
      welding current type and polarity, 82, 83
      welding process, 69, 74, 80, 81, 84, 85
      welding sequence and practice, 79, 81, 82, 84
      welding speed, 58, 74, 78, 85
      weld pool shape, 76, 77, 84–86
      weld preparation, 81, 83, 84, 86
      weld size, 77, 78
      wire composition, 80, 85
   intermittent, 55, 93
   nucleation by gases, 55
   open to surface, 58
   remedial measures, 85
   repair, 85
      by melting out, 85
   resulting from deviation of welding procedures, 84
   root pass of second side, 79
   root runs of butt welds, 74, 80
   very shallow welds, 76
solidification cracks,
   detection, 93
   rounded tips, 60
   subsequent propagation, 60
solidification pattern, influence on solidification cracking, 55–58, 83, 74
sour service, 165, 166, 208, 209, 211, 212
   maximum hardness for, 204, 208, 211, 212
space, possibility of repair welding and joining, 234, 235
spikes on solidification crack surfaces, 98
stainless steel surfacing, underclad cracking, 154, 170, 171, 174, 178
stainless steel, see austenitic stainless steel
start porosity, 185
steel cleanliness, influence on weld hydrogen pick up, 147
steel,
   compositions suitable for wet welding, 234
   grades with low levels of inclusions, 5, 103, 118
   made from scrap, influence of residual elements, on carbon equivalent, 142
   manufacturing, methods, 4, 5
   strength, influence on solidification cracking, 79, 84
   type influence on,
      lamellar tearing, 115, 118, 119,
      reheat cracking, 173
      weldability, 3–9
steels, different types, see under type name
steels of different thickness, welding and solidification cracking, 79
steels with low M_F temperatures, procedures to avoid hydrogen cracking, 150
step ageing test for temper embrittlement, 213
stickout, influence on dilution, 82
stiffeners, stiffened diaphragms and lamellar tearing, 110–112
storage of high pressure hydrogen, see high pressure hydrogen
STRA, see also short transverse ductility, 115
STRA values, influence of sulphur content, 107, 108
   scatter, 120
   significance, 121
strain ageing, 17, 189–201, 212
   cracking, 45
   in service, 212
   influence of PWHT, 199

stray arcing, 181, 182
strength level of steel, 3
  estimation from hardness, 204, 227
stress concentrations, hydrogen cracking, 144
  root gaps, inclusion stringers, undercut, 144
  weld toe intrusions, 144, 145
stress corrosion cracking, initiation at hard spots, 23, 184
see also HIC and hydrogen stress corrosion cracking, 166, 167, 208, 209, 211, 212
stress, influence on hydrogen cracking, 144, 145
stress relief cracking, see reheat cracking
stresses, long range, influence on lamellar tearing, 102, 105, 107
strip, cold rolled, 5
structural steels, 5
sub-critical HAZ, strain ageing, 17
submerged arc fluxes, hydrogen level, 135
  storing and re-drying, 135, 147
submerged arc weld metal, formulae for strength, 37, 38
  influence of heat input on toughness, 2
submerged arc welding, solidification cracking, 40, 69–72, 74–84
  weldability with, 2
sulphides, influence on hydrogen stress corrosion cracking, 166, 208, 211, 212
  mixed, influence on solidification cracking, 68
sulphur in steel, influence on,
  lamellar tearing, 115, 118, 119
  liquation cracking, 87–89
  STRA levels for Al-treated steels, 103–109, 115
sulphur in weld metal, reduction by basic fluxes and basic electrodes, 49, 73
  influence on, solidification cracking, 40, 69–72, 74–84
surface hardened steels, 7
surface pocking, 180, 182, 183

T1 steel, 6
$T_{800-500}$ (T8/5, etc); see cooling time
tab test for lamellar tearing susceptibility, 123, 124
  assessment, 124
tack welds, sudden failure and solidification, cracking, 77, 84
taper sectioning of bead-on-plate test to assess, liquation cracking, 91
tearing, from mechanical cutting, 28
tearing, lamellar, see lamellar tearing
T-butt and T-fillet joints,
  lamellar tearing, 109–114, 118
temper bead, technique for wet welding repair, 234
  influence on, HAZ hardness, 150, 151, 196
  HAZ toughness, 196, 198
temper embrittlement, 41, 42, 203, 212–215
  during PWHT, 203
  influence of, free carbon, 214, 215
    impurities, 41, 42, 213
  in service, 212–215
  of weld metals, 41, 203, 213
  possibility if steel graphitised, 215, 221
  reversibility by PWHT, 213
  steel types, 41, 42, 203, 213, 214
  step ageing test, 213
temperature control method, hard HAZs, 150, 151
  to avoid hydrogen cracking, 147–151
  to estimate preheat/interpass temperature, 147
temperature of hydrogen embrittlement and cracking, 145, 146
tests, large scale, for lamellar tearing, 127, 128
thermal cutting and gouging, carbon pickup, 28
  grinding off damage, 28
  hardened surfaces, 28
  influence on residual stresses, 28
  preheating to avoid hydrogen cracking, 28
thermal facetting on surfaces of solidification and liquation cracks, 96
thermal striations, see thermal facetting
TIG welding, autogenous weld metal composition, 18
solidification cracking, 39, 40, 68–74, 80–86
  weldability with, 2
time delay before inspection for hydrogen cracking, 159, 189
  effect of ambient temperature, 159
titanium as nitride-former and strain ageing, 41
titanium carbo-sulphides on or near solidification or liquation cracks, 100
titanium on inclusions, influence on acicular ferrite nucleation, 25, 26
titanium, addition to nucleate acicular ferrite, 27
  excessive, influence on toughness, 27
  influence on,
    HAZ refinement and toughness, 195, 197
    solidification cracking, 70
TMCP steels, 5
tool steels, 7
  hydrogen cracking, 9
total aluminium and strain ageing, 41
total hydrogen, see weld hydrogen content, total
toughness, loss of in service, 212–218
toughness of welds, estimation, 38

Index

toughness, resistance to brittle or ductile (tearing) fracture, 194
toughness testing of welds containing hydrogen, 194, 196
toughness tests, correlations between, 193, 194
 see also Charpy tests, impact tests, fracture, toughness tests
toughness, HAZ, influence of carbon and microstructure, 1
tramline corrosion, 206
transformation from austenite, expansion, influence on residual stress, 30
transformation temperature, influence on residual stress, 30
transition joints for creep service, 205
transvarestraint test for solidification cracking, 40, 69
transverse 45° weld metal hydrogen cracking, 157, 158
travel speed, influence on HAZ width, 195, 197
tungsten inclusions in TIG welds, 183, 184
tungsten, influence on solidification cracking, 73
two layer HAZ refinement,
 for as-welded repair, 198, 229, 230
 influence of,
  angle of attack, 177, 229
  preheat temperature, 178, 196, 230
  weld bead overlap, 177, 195, 230
 influence on,
  HAZ toughness, 198, 229
  reheat cracking, 177, 178
 with automatic and semi-automatic welding, 178, 230
 with MMA welding, 177, 231
two-stage heating to PWHT temperature, influence, on reheat cracking, 178
Type II MnS inclusions and lamellar tearing, 106

UCS (units of crack susceptibility) for solidification cracking, 40, 69-72, 74, 80, 86
UCS values, critical, solidification cracking, 70, 74, 80
ultrasonic examination of steel and lamellar tearing, 119, 121, 189
ultrasonic inspection, range of use, 188, 189
underclad cracking: hydrogen or reheat cracking, 154
undercut, 144, 181
undermatching weld metal, 191, 192
 with 9% nickel steel, 192
underwater welding for repair, 232-234
 influence on weld cooling, 14, 233
uneven heating to PWHT temperature, influence, on residual stresses, 174

unexpected solidification cracking, suggested causes, 85

vanadium influence on,
 acicular ferrrite nucleation, 26
 HAZ toughness, 195
 reheat cracking, 40, 170, 172, 173
 solidification cracking, 70, 73
very low hydrogen levels, maintaining on site, 147
VIM-VAR steels, 5
visual examination, range of use, 188

water traps, avoidance in corrosive situations, 211
weathering (weather resistant) steels, 5
 corrosion in, marine atmospheres, 205, 206
 moist air, 205, 206
 painting, 205, 206
 weldability, 205, 206
weld cavities, 205, 206
weld cooling cycle,
 influence of,
  external factors, 13, 14
  heat input, 9
  steel thickness, 10-12
 influence on HAZ and weld metal microstructure, 9
weld cooling rate or time, see cooling rate or time
weld crater, see crater
weld drop through opening root gap, 76, 77
weld hydrogen content,
 diffusible, residual and total, 134, 135, 165
 related to deposited or fused metal, 134, 135
weld hydrogen contents, 134-136
weld metal (austenite) grain size, growth, 23
weld metal composition, 18-23
 autogenous welds, 18
 control of, 71, 72, 80
 influence of flux and shielding gas, 19, 39, 72
weld metal grains, independent nucleation during solidification, 23
weld metal hardness, influence on weld metal hydrogen cracking, 143, 144, 155, 156
weld metal, hydrogen cracking, 8, 9, 143, 144, 155-158, 161, 162
 in alloyed weld metal, 156, 157
 in C:Mn weld metal in thick section, 157, 158
 influence of,
  distance for hydrogen diffusion, 158
  heat input, 158
  hydrogen diffusion, 155-158
 very low carbon steels, 156, 157

with high hydrogen levels, 156
with HSLA steels, 156, 157
weld metal inclusion content,
 estimation from composition, 38, 39
weld metal microstructure, 23–28
 influence of reheating, 28
weld metal strength, at different temperatures, 192
 influence, of PWHT, 193
  strain ageing, 192
 influence on, solidification cracking, 85,
  toughness, 25, 199
weld metals of low solidification temperature and liquation cracks, 89
weld preparation, influence on,
 HAZ refinement and toughness, 198
 lamellar tearing, 114, 118
weld root, *see also* root (region)
 gouging of to remove strain aged material, 200, 201
 selection of consumables, 200, 201
 strain ageing, 200, 201
weld size, influence on lamellar tearing, 107
weld strain, influence on solidification cracking, 78,79
weld-through primers, 49, 136
weld toe grinding, influence on reheat cracking, 174, 178
weldability, sources of advice on, 3
 influence of welding process, 2, 3
weldable steels, 5
welder qualification test failures, liquation cracks, 86, 88
welding consumables: *see* consumables, (welding)
welding defects: *see* welding faults
welding diagrams for heat input and preheat to avoid hydrogen cracking, 147, 148
welding faults, 45, 46, 180–186
welding of steel containing hydrogen, 168, 218, 230
welding fluxes, *see* fluxes, welding
welding onto pipelines containing fluids, 13, 14
 see also hot tapping, welding parameters, influence on lamellar tearing, 116
welding process, influence on,
 HAZ toughness, 2, 3

hydrogen cracking, 2, 3
lamellar tearing, 2, 3, 116
liquation cracking, 2, 3
reheat cracking, 2, 3
solidification cracking, 2, 3
weldability, 2, 3
weld metal toughness, 2, 3
welding sequence, influence on solidification cracking, 79, 84
welding speed, influence on,
 depth/width ratio and penetration, 78
 solidification cracking, 56, 74, 78, 85
welding technique, influence on lamellar tearing, 116–118
wet welding for repair, 228, 233, 234
 consumables and processes, 233, 234
 electrodes, waterproofing, 234
 influence of current type and polarity, 234
 influence on,
  dilution and penetration, 234
  weld cooling, 14, 233
 hydrogen cracking, 233
 suitable depths, 234
 unsuitability of austenitic stainless steel electrodes, 234
window test for lamellar tearing, 128
wire drawing, 5
woody fracture surface, lamellar tearing, 125, 129, 130
wrought irons, 3, 52

X parameter for temper embrittlement, 42, 173, 213

Y parameter for temper embrittlement, 42
yield strength of HAZs, estimation from hardness, 38
yield strength of weld metals, estimation from hardness, 38

Z parameter for temper embrittlement, 42
Z-direction, *see* short transverse direction.
Z-grades, *see* low inclusion level.
zinc penetration, welding zinc coated steel, 50, 184
zinc-coated steels welding of,
 influence on welding fume, 49, 50
 risk of liquid metal penetration, 50, 184
zirconia, relative basicity in slags and fluxes, 43